T0324864

TRIGONOMETRIC FUNCTIONS AND COMPLEX NUMBERS

WORLD CENTURY MATHEMATICAL OLYMPIAD SERIES

Series Editor: Shan Zun *(Nanjing Normal University, China)*

Vol. 1   Trigonometric Functions and Complex Numbers
by Desheng Yang (Shanghai Xiangming High School, China)
translated by Chunhui Shen (Shanghai Xiangming High School, China)

World Century Mathematical Olympiad Series – Vol. 1

TRIGONOMETRIC FUNCTIONS AND COMPLEX NUMBERS

Desheng Yang

translated by **Chunhui Shen**

Shanghai Xiangming High School, China

Published by

World Century Publishing Corporation
27 Warren Street, Suite 401-402, Hackensack, NJ 07601, USA

Library of Congress Cataloging-in-Publication Data
Names: Yang, Desheng.
Title: Trigonometric functions and complex numbers / Desheng Yang,
    Shanghai Xiangming High School, China.
Description: New Jersey : World Century, 2016. | Series: World century mathematical
    olympiad series ; vol. 1 | Includes bibliographical references and index.
Identifiers: LCCN 2016026528 | ISBN 9781938134869 (alk. paper)
Subjects: LCSH: Trigonometry. | Numbers, Complex. | Functions of complex variables.
Classification: LCC QA531 .Y345 2016 | DDC 516.24/6--dc23
LC record available at https://lccn.loc.gov/2016026528

British Library Cataloguing-in-Publication Data
A catalogue record for this book is available from the British Library.

Originally published in Chinese by Shanghai Scientific & Technological Education
Publishing House, 2010

Published by arrangement with Shanghai Century Publishing Group.

Typeset by Stallion Press
Email: enquiries@stallionpress.com

Printed in Singapore

Preface

The year 2016 is officially the 30th anniversary of China joining the IMO (International Mathematical Olympiad). Looking back on these 30 years, the Chinese team has made remarkable achievements: a total of 142 gold medals, 32 silver medals, 5 bronze medals, and 19 times to be the first in total score. These achievements could not be separated from all the participants in the field of Chinese Mathematical Olympiad.

This series is a summary of their work, and the authors are the representative of all the participants in the Chinese Olympic mathematics. For example, Lu Hongwen, a tutor of postgraduate candidates, not only has made remarkable achievements in the field of algebraic number theory, but also very concerned about the mathematics contest. Chen Ji, who is recognized as one of the leading experts in the field of inequality. Other members are the leading figures in the current Chinese Mathematical Olympiad, such as Xiong Bin and Feng Zhigang who repeatedly served as the leader or deputy leader of Chinese IMO national team, and served as members of the Chinese Mathematical Olympiad committee. They have made a great contribution to the Chinese Mathematical Olympiad and trained a lot of talents.

Based on the authors' experience and original work, there are many new ideas, new problems, new solutions, and new methods in this series.

Would these cause the series too hard for readers?

Of course, this series has a little difficulty. But as the authors fully understand the background of these IMO problems, they have the ability to provide the contents with breathtaking lightness.

However, reading is a hard work, in particular reading a math book. Never give up whatever difficulty you meet. It requires a lot of courage and strength. In this way, you not only read a classic book, but also enjoy the fun of reading.

The authors of the book, of course, have to work hard to write well. But it is not perfect, even the classics have occasional omissions. In a sense, this provides the space for readers to think, imagine, and gallop.

If you can think of some new problems and solutions and find the deficiency or improved results in the series, congratulations! That is what the ancients called "Reading to find the trick". We welcome all of you to put forward suggestions and criticisms to this series.

The authors and the titles of this series are as follows:

Analytic Geometry (Huang Libing and Lu Hongwen)
Function Iteration and Function Equation (Wang Weiye and Xiong Bin)
Algebraic Inequalities (Chen Ji and Ji Chaocheng)
Circular (Tian Tingyan)
Elementary Theory of Numbers (Feng Zhigang)
Set and Correspondence (Shan Zun)
Sequence of Number and Mathematical Induction (Shan Zun)
Combinatorial Problem (Liu Peijie and Zhang Yongqin)
Vector and Solid Geometry (Tang Lihua)
Trigonometric Functions and Complex Numbers (Yang Desheng)

This volume of *Trigonometric Functions and Complex Numbers* is composed of two parts: trigonometric functions and complex numbers. All parts are based on the basic knowledge from the shallower to the deeper, which covers all levels of mathematical contest content. Area method, triangle substitution and triangle inequality in *trigonometric functions*, complex numbers and trigonometric and complex numbers and equations in *complex numbers* which reach the difficulty of IMO.

Yang Desheng, the author of this book, born in August 1957, was a Bachelor of Science of Hubei University, one of the Guangdong Provincial People's Government Special-grade Secondary School teachers and one of the Shanghai Municipal People's Government Special-grade Secondary School teachers. Now teaching in Shanghai Xiangming Middle School. He is also the deputy secretary general of the Chinese National Research Institute of Elementary Mathematics.

For a long time he was engaged in mathematical contest counseling, and is rated as an excellent coach of Chinese Mathematical Olympiad. Under his guidance, hundreds of students won the National High School Math League first, second and third prize. He has published over 60 papers and 63 titles on Mathematical Olympiad.

Shen Chunhui, the translator of the book, born in 1987, was Bachelor of Science and Master of Education of East China Normal University. Now teaching in Shanghai Xiangming Middle School. Also, engaged in math contest counseling, TI graphics calculator contest counseling, etc. He has repeatedly been rated as excellent guidance teachers of contests. He published a number of articles in *Journal of Mathematics Education, Mathematics Teaching*, etc.

Contents

PART I
Trigonometric Function

Chapter 1

Trigonometric Identity

1.1. Trigonometric Ratios for Any Angle

Any angle and its measures

1. Positive angle, negative angle and zero angle

If a ray rotating around the endpoint counterclockwise formed as the positive angle, its measurement is positive, and if a ray rotating around the endpoint clockwise formed as the positive angle, its measurement is negative (see Fig. 1.1):

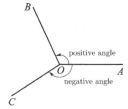

Fig. 1.1.

2. Coterminal angle

In the process of angle forming, we can find there are numerous angles whose initial and terminal sides coincide with a certain angle α. Their range is $360°$ integer times that of α in size. We put the set of the angles (including α itself) with the same edge as α:

$$\{\beta|\beta = k \cdot 360° + \alpha, \ k \in Z\}.$$

Namely, any angle with the same terminal side as angle α can be expressed as a total of angle α and integer number of perigons.

3. Quadrant angle

If the terminal ray of an angle in standard position lies in the first quadrant, then the angle is said to be a first-quadrant angle. Similarly, second-, third-, and fourth-quadrant angles are defined as follows:

First-quadrant angles: $\{\alpha | k \cdot 360° < \alpha < k \cdot 360° + 90°, \ k \in Z\}$;

Second-quadrant angles:

$\{\alpha | k \cdot 360° + 90° < \alpha < k \cdot 360° + 180°, k \in Z\}$;

Third-quadrant angles:

$\{\alpha | k \cdot 360° + 180° < \alpha < k \cdot 360° + 270°, k \in Z\}$;

Fourth-quadrant angles:

$\{\alpha | k \cdot 360° + 270° < \alpha < k \cdot 360° + 360°, k \in Z\}$.

4. Quadrantal angle

If the terminal ray of an angle in standard position lies along an axis, the angle is called a quadrantal angle.

- If the terminal ray of the angle lies along positive y-axis, then $\theta = 360°k + 90°, \ k \in Z$.
- If the terminal ray of the angle lies along negative y-axis, then $\theta = 360°k - 90°, \ k \in Z$.
- If the terminal ray of the angle lies along y-axis, then $\{\theta | \theta = 180°k + 90°, \ k \in Z\}$.
- If the terminal ray of the angle lies along positive x-axis, then $\theta = 2 \times 180°k + 0°, \ k \in Z$.
- If the terminal ray of the angle lies along negative x-axis, then $\theta = 2 \times 180°k + 180°, \ k \in Z$.
- If the terminal ray of the angle lies along x-axis, then $\{\theta | \theta = 180°k, \ k \in Z\}$.

Exercise

1. Find angles between $0°$ with $360°$, that are coterminal with each given angle, and judging in which quadrantal angle.
   (1) $-120°$; (2) $640°$; (3) $-950°12'$.

Solution: (1) Since $-120° = -360° + 240°$, it follows that $240°$ is coterminal with $-120°$, which is the third-quadrant angle.
(2) Since $640° = 360° + 280°$, it follows that $280°$ is coterminal with $640°$, which is the fourth-quadrant angle.
(3) Since $-950°12' = -3 \times 360° + 129°48'$, it follows that $129°48'$ is coterminal with $-950°12'$, which is the third-quadrant angle.

2. Give a set S for the measure of all angles that are coterminal with each given angle, find angles between $-360°$ and $720°$ in S.
(1) $60°$; (2) $-21°$; (3) $363°14'$.

Solution: (1) $S = \{\beta | \beta = 60° + k \cdot 360°, \ k \in Z\}$.
The angles between $-360°$ and $720°$ in S are as follows:

$$-1 \times 360° + 60° = -300°;$$

$$0 \times 360° + 60° = 60°;$$

$$1 \times 360° + 60° = 420°.$$

(2) $S = \{\beta | \beta = -21° + k \cdot 360°, \ k \in Z\}$.
The angles between $-360°$ and $720°$ in S are as follows:

$$0 \times 360° - 21° = -21°;$$

$$1 \times 360° - 21° = 339°;$$

$$2 \times 360° - 21° = 699°.$$

(3) $S = \{\beta | \beta = 363°14' + k \cdot 360°, \ k \in Z\}$.
The angles between $-360°$ and $720°$ in S are as follows:

$$-2 \times 360° + 363°14' = -356°46';$$

$$-1 \times 360° + 363°14' = 3°14';$$

$$0 \times 360° + 363°14' = 363°14'.$$

3. Find the set for the angle, which displays the terminal side of the angle in the set of angles in the shadow area (exclusive of boundaries). (See Fig. 1.2.)

Solution:.

(A) $\{\alpha | 60° + k \cdot 360° < \alpha < 255° + k \cdot 360°, \ k \in Z\}$.
(B) $\{\alpha | - 120° + k \cdot 360° < \alpha < 45° + k \cdot 360°, \ k \in Z\}$.

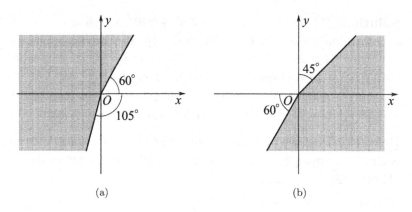

(a)   (b)

Fig. 1.2.

4. If α is a second-quadrant angle, how about $\frac{\alpha}{2}$? How about 2α?

Solution: If α is a second-quadrant angle, then
$k \cdot 360° + 90° < \alpha < k \cdot 360° + 180°, \ k \in Z$.

Hence $k \cdot 180° + 45° < \frac{\alpha}{2} < k \cdot 180° + 90°$, because $k \in Z$. Therefore $k = 2n$ or $k = 2n + 1$.

When $k = 2n$, $n \cdot 360° + 45° < \frac{\alpha}{2} < n \cdot 360° + 90°$, therefore $\frac{\alpha}{2}$ is a first-quadrant angle.

When $k = 2n + 1$, $n \cdot 360° + 225° < \frac{\alpha}{2} < n \cdot 360° + 270°$, therefore $\frac{\alpha}{2}$ is a third-quadrant angle.

Hence when α is a second-quadrant angle, therefore $\frac{\alpha}{2}$ is a first-quadrant angle or third-quadrant angle. Similarly, $180° + k \cdot 720° < 2a < 360° + k \cdot 720°, k \in Z$. Hence, when a is the second-quadrant angle, $2a$ is a third-quadrant angle or the fourth-quadrant angle.

Radian

(1) Radian: When an arc of a circle has the same length as the radius of the circle, as shown in Fig. 1.3, the measure of the central angle $\angle AOB$ is by definition 1 radian.

(2) Radian measure: The radian to measure angle.

(3) Equation for the radian measure of the central angle α: $|\alpha| = \frac{l}{r}$, where α is the subtended angle in radians, l the arc length, and r the radius.

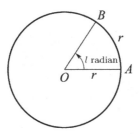

Fig. 1.3.

(4) Conversion formulas between degree measure and radian measure: One revolution measured in radians is 2π and measured in degrees is $360°$. We therefore have the conversion formulas:

$$360° = 2\pi \text{ radians}$$

$$180° = \pi \text{ radians}$$

$$1° = \frac{\pi}{180} \text{ radians} \approx 0.017 \text{ radians}$$

$$1 \text{ radian} = \left(\frac{180}{\pi}\right)° \approx 57°17'44.8'' \approx 57.3°.$$

Because the radian number is the ratio of the arc length to radius, so it is a real number. Therefore, in expressing the range of angle in radian measure, usually omit the word "radian". Due to different needs, the angle is not only measured in degree, but also in radian. For example, if we want to learn the trigonometric function, we need to draw its image, using the degree measure, which will cause inconvenience. Since the introduction of radian, the set of angles corresponds with the set of real numbers as shown in Fig. 1.4:

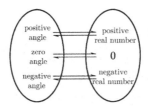

Fig. 1.4.

Degree	0°	15°	30°	45°	60°	75°	90°	105°	120°	135°	150°	180°	210°
Radian	0	$\dfrac{\pi}{12}$	$\dfrac{\pi}{6}$	$\dfrac{\pi}{4}$	$\dfrac{\pi}{3}$	$\dfrac{5\pi}{12}$	$\dfrac{\pi}{2}$	$\dfrac{7\pi}{12}$	$\dfrac{2\pi}{3}$	$\dfrac{3\pi}{4}$	$\dfrac{5\pi}{6}$	π	$\dfrac{7\pi}{6}$

Degree	225°	240°	270°	300°	315°	330°	360°
Radian	$\dfrac{5\pi}{4}$	$\dfrac{4\pi}{3}$	$\dfrac{3\pi}{2}$	$\dfrac{5\pi}{3}$	$\dfrac{7\pi}{4}$	$\dfrac{11\pi}{6}$	2π

(5) Formulas for the arc length and area of a sector of a circle: The following formulas are arc length and area of a sector of a circle, where α $(0 < \alpha < 2\pi)$ is the central angle of a sector, r the radius, l is the length, and S the area (see Fig. 1.5):

$$(1)\ l = \alpha r;$$

$$(2)\ S = \frac{1}{2}\alpha r^2;$$

$$(3)\ S = \frac{1}{2}lr.$$

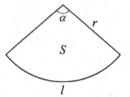

Fig. 1.5.

Exercise

5. The rope is twisted around the rim of 40-cm radius. The lower end B of the rope is hanging an object W (Fig. 1.6). If the wheel rotates counterclockwise by six laps every min, how many seconds does it take to lift objects upward 100 cm from position W?

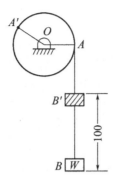

Fig. 1.6.

Solution: When the wheel is rotated in a clockwise direction, the arc length AA' from A to A' is equal to the distance from B to B'. When $BB' = 100$ cm, then $AA' = 100$ cm time arc, and the center angle $\angle AOA'$ corresponding to arc AA' is given by $\frac{100}{40}$ radians.

Since the wheel rotates six times per minute at a uniform speed, the radian $\frac{6 \times 2\pi}{60}$ per second is given by $\frac{\pi}{5}$. When T seconds have passed, the radian that passed is given by $\frac{\pi}{5}t$. (The bigger the T is, the bigger the radian $\frac{\pi}{5}t$ is, which may exceed an arbitrary angle of 2π). Therefore, it takes 4 s to lift the object W position upward 100 cm.

6. As shown in Fig. 1.7, r is a radius of circular, x the length of arc ACB, y the area of arch. Find the equation of the function.

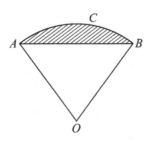

Fig. 1.7.

Solution: Area of sector $OAB = \dfrac{1}{2} \cdot x \cdot 1 = \dfrac{x}{2}.$

$\because \alpha = \dfrac{l}{r}$.

Suppose $\angle AOB = x$ (rad).

Hence $S_{\triangle OAB} = \dfrac{1}{2} \cdot 1 \cdot 1 \cdot \sin x = \dfrac{1}{2} \sin x$.

Therefore, $y = \dfrac{x}{2} - \dfrac{\sin x}{2} \quad (0 < x < \pi)$.

Trigonometric ratios for any angle

As shown in Fig. 1.8, put the right-angled triangle in rectangular coordinate system. Supposing $P(x, y)$, then $OQ = x$, $QP = y$, $|OP| = r = \sqrt{x^2 + y^2} \ (r > 0)$, where α have six trigonometric ratios:

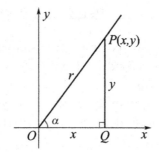

Fig. 1.8.

$$\sin \alpha = \frac{y}{r}, \quad \cos \alpha = \frac{x}{r},$$

$$\tan \alpha = \frac{y}{x}, \quad \alpha \neq k\pi + \frac{\pi}{2}, \quad k \in Z,$$

$$\cot \alpha = \frac{x}{y}, \quad \alpha \neq k\pi, \quad k \in Z,$$

$$\sec \alpha = \frac{r}{x}, \quad \alpha \neq k\pi + \frac{\pi}{2}, \quad k \in Z,$$

$$\csc \alpha = \frac{r}{y}, \quad \alpha \neq k\pi, \quad k \in Z.$$

Sign of six trigonometric ratios in each quadrant

Quadrant	Coordinate of P		$\sin \alpha$	$\cos \alpha$	$\tan \alpha$	$\cot \alpha$	$\sec \alpha$	$\csc \alpha$
	x	y						
I	+	+	+	+	+	+	+	+
II	−	+	+	−	−	−	−	+
III	−	−	−	−	+	+	−	−
IV	+	−	−	+	−	−	+	−

The unit circle line to represent the above trigonometric functions

1. Directed line segment

The directed line segment: as we know, the coordinate axis prescribes the direction of the straight line. A line segment parallel to the coordinate axis can also be prescribed in two opposite directions. As shown in Fig. 1.9, for the line MN on the axis, we can specify two opposite directions of this line from point M to point N or from point N to point M. They each have a direction, respectively named by line MN and line NM, and the line PQ parallel to axis y, that can also be prescribed with two opposite directions. If such a line segment direction is consistent with the direction of coordinate axis, it is positive; otherwise it is negative.

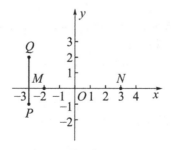

Fig. 1.9.

2. Trigonometric function line

As shown in Fig. 1.10, set arbitrary angle vertex as O, with starting line not coinciding with non-negative x-axis axle and terminal side intersecting at $P(x, y)$. Draw a line perpendicular to the x-axis through P whose pedal is M; draw a tangent to a unit circle through A (1, 0). The tangent line is surely to be parallel to the axis y. Suppose that it intersects at point T with the terminal line of angle α (when α for the first, four quadrant angle) or its reverse extension line (when α for second, three quadrant angle).

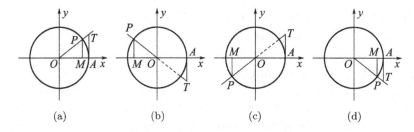

(a)　　　　　(b)　　　　　(c)　　　　　(d)

Fig. 1.10.

Then $OM = x$, $MP = y$.

According to the definition of sine and cosine:

$$\sin \alpha = \frac{y}{r} = y = MP,$$

$$\cos \alpha = \frac{x}{r} = x = OM.$$

The two directed lines related to the unit circle, namely MP and OM, are called **sine function and cosine function** line of angle α.

Similarly, we see AT as a directed line segment. According to the definition of tangent function and knowledge of similar triangle, there exists

$$\tan \alpha = \frac{y}{x} = \frac{MP}{OM} = AT.$$

If the directed line AT is a tangent, when AT is above the axle x, it is positive. If the directed line AT is under axle x, it is negative. When the angle α of the terminal side is on x axle, the sinusoidal line and tangent line will, respectively, change into a point; when the angle of the terminal side is on the y-axis, cosine line will change into a single point while tangent line will not exist.

We call the three strips MP, OM, AT related to the unit circle as trigonometric function line of angle α.

Exercise

7. If the particle $P(-3m, 4m)$, $m \neq 0$, is on the terminal ray of α, find six trigonometric functions of α.

Solution:

(1) When $m > 0$,

$$\because x = -3m, \quad y = 4m.$$

$$\therefore r = \sqrt{(-3m)^2 + (4m)^2} = 5|m| = 5m;$$

$$\sin\alpha = \frac{4m}{5m} = \frac{4}{5}; \quad \cos\alpha = \frac{-3m}{5m} = -\frac{3}{5};$$

$$\tan\alpha = \frac{4m}{-3m} = -\frac{4}{3}; \quad \cot\alpha = \frac{-3m}{4m} = -\frac{3}{4};$$

$$\sec\alpha = \frac{5m}{-3m} = -\frac{5}{3}; \quad \csc\alpha = \frac{5m}{4m} = \frac{5}{4}.$$

(2) when $m < 0$, $r = 5|m| = -5m$;

$$\therefore \sin\alpha = -\frac{4}{5}; \quad \cos\alpha = \frac{3}{5}; \quad \tan\alpha = -\frac{4}{3};$$

$$\cot\alpha = -\frac{3}{4}; \quad \sec\alpha = \frac{5}{3}; \quad \csc\alpha = -\frac{5}{4}.$$

Comment: In this question, coordinates of point P contain m, and $r > 0$, so m should be classified and discussed we have to prevent missing $m < 0$. Further discussion: when $m > 0$, α is the second-quadrant angle; when $m < 0$, α is the fourth-quadrant angle. There are two situations of sign of trigonometric ratios.

8.   (1) Supposing α is acute angle, prove $\sin\alpha < \alpha < \tan\alpha$.

       (2) Supposing α is acute angle, prove $\sin\alpha + \cos\alpha > 1$.

Proof: (1) As shown in Fig. 1.11, imagine on the unit circle a particle P that starts at $A(1,0)$. Moreover $PM \perp OA$, AT is parallel to y-axis.

Using the trigonometric function line: $\sin\alpha = MP$, $\overgroup{AP} = \alpha$, $AT = \tan\alpha$, and $MP < \overgroup{AP} < AT$, so $\sin\alpha < \alpha < \tan\alpha$.

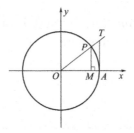

Fig. 1.11.

(2) As shown in Fig. 1.11, we have the following.
Using the trigonometric function line: $OM = \cos\alpha$, $MP = \sin\alpha$.
In $\triangle POM$, $OM + MP > OP$, so $\sin\alpha + \cos\alpha > 1$.   □

9. Supposing $\sin\alpha > 0$, $\cos\alpha < 0$, and $\sin\frac{\alpha}{3} > \cos\frac{\alpha}{3}$, find the domain of $\frac{\alpha}{3}$.

Solution:

$$2k\pi + \frac{\pi}{2} < \alpha < 2k\pi + \pi$$

$$\frac{2k\pi}{3} + \frac{\pi}{6} < \frac{\alpha}{3} < \frac{2k\pi}{3} + \frac{\pi}{3}, \quad k \in Z.$$

Now

$$\sin\frac{\alpha}{3} > \cos\frac{\alpha}{3},$$

$$2k\pi + \frac{\pi}{4} < \frac{\alpha}{3} < 2k\pi + \frac{5\pi}{4}, \quad k \in Z;$$

$$\frac{\alpha}{3} \in \left(2k\pi + \frac{\pi}{4}, 2k\pi + \frac{\pi}{3}\right) \cup \left(2k\pi + \frac{5\pi}{6}, 2k\pi + \pi\right), \quad k \in Z.$$

10. Let $0 < b < 1$, and $0 < a < \frac{\pi}{4}$, compare the size: $x = (\sin a)^{\log_b \sin a}$, $y = (\cos a)^{\log_b \cos a}$, $z = (\sin a)^{\log_b \cos a}$.

Solution:

$$\because 0 < b < 1,$$

$$\therefore f(x) = \log_b x \text{ is decreasing function,}$$

$$\because 0 < \alpha < \frac{\pi}{4},$$

$$\therefore 0 < \sin \alpha < \cos \alpha < 1,$$

$$\therefore \log_b \sin \alpha > \log_b \cos \alpha > 0,$$

$$\therefore (\sin \alpha)^{\log_b \sin \alpha} < (\sin \alpha)^{\log_b \cos \alpha},$$

$$\therefore x < z;$$

$$\because (\sin \alpha)^{\log_b \cos \alpha} < (\cos \alpha)^{\log_b \cos \alpha},$$

$$\therefore z < y, \quad x < z < y.$$

11. Supposing θ is the repeated root of equation $x^2 + 4x \cos \theta + \cot \theta = 0$, find θ.

Solution: Equation $x^2 + 4x \cos \theta + \cot \theta = 0$ has repeated roots;

$$\therefore \Delta = 16 \cos^2 \theta - 4 \cot \theta = 0,$$

$$\therefore 4 \cot \theta (2 \sin 2\theta - 1) = 0,$$

$$\because \theta \text{ is the acute angle,}$$

$$\therefore \sin 2\theta = \frac{1}{2},$$

$$\therefore 2\theta = \frac{\pi}{6} \quad \text{or} \quad 2\theta = \frac{5\pi}{6},$$

$$\therefore \theta = \frac{\pi}{12} \quad \text{or} \quad \frac{5\pi}{12}.$$

1.2. The Induction Formulas and the Relationship Between Trigonometric Ratios of the Same Angle

The induction formulas

1. Induction formula One

$$\sin(2k\pi + \alpha) = \sin\alpha, \quad \cos(2k\pi + \alpha) = \cos\alpha,$$
$$\tan(2k\pi + \alpha) = \tan\alpha, \quad \cot(2k\pi + \alpha) = \cot\alpha$$
$$(k \in Z, \alpha \in [0, 2\pi).$$

2. Induction formula two

$$\sin(-\alpha) = -\sin\alpha, \quad \cos(-\alpha) = \cos\alpha,$$
$$\tan(-\alpha) = -\tan\alpha, \quad \cot(-\alpha) = -\cot\alpha.$$

3. Induction formula three

$$\sin(\pi + \alpha) = -\sin\alpha, \quad \cos(\pi + \alpha) = -\cos\alpha,$$
$$\tan(\pi + \alpha) = \tan\alpha, \quad \cot(\pi + \alpha) = \cot\alpha.$$

4. Induction formula four

$$\sin(\pi - \alpha) = \sin\alpha, \quad \cos(\pi - \alpha) = -\cos\alpha,$$
$$\tan(\pi - \alpha) = -\tan\alpha, \quad \cot(\pi - \alpha) = -\cot\alpha.$$

5. Induction formula five

$$\sin(2\pi - \alpha) = -\sin\alpha, \quad \cos(2\pi - \alpha) = \cos\alpha,$$
$$\tan(2\pi - \alpha) = -\tan\alpha, \quad \cot(2\pi - \alpha) = -\cot\alpha.$$

6. Induction formula six

$$\sin\left(\frac{\pi}{2}+\alpha\right) = \cos\alpha, \quad \cos\left(\frac{\pi}{2}+\alpha\right) = -\sin\alpha,$$

$$\tan\left(\frac{\pi}{2}+\alpha\right) = -\cot\alpha, \quad \cot\left(\frac{\pi}{2}+\alpha\right) = -\tan\alpha.$$

7. Induction formula seven

$$\sin\left(\frac{\pi}{2}-\alpha\right) = \cos\alpha, \quad \cos\left(\frac{\pi}{2}-\alpha\right) = \sin\alpha,$$

$$\tan\left(\frac{\pi}{2}-\alpha\right) = \cot\alpha, \quad \cot\left(\frac{\pi}{2}-\alpha\right) = \tan\alpha.$$

8. Induction formula eight

$$\sin\left(\frac{3\pi}{2}-\alpha\right) = -\cos\alpha, \quad \cos\left(\frac{3\pi}{2}-\alpha\right) = -\sin\alpha,$$

$$\tan\left(\frac{3\pi}{2}-\alpha\right) = \cot\alpha, \quad \cot\left(\frac{3\pi}{2}-\alpha\right) = \tan\alpha.$$

9. Induction formula nine

$$\sin\left(\frac{3\pi}{2}+\alpha\right) = -\cos\alpha, \quad \cos\left(\frac{3\pi}{2}+\alpha\right) = \sin\alpha,$$

$$\tan\left(\frac{3\pi}{2}+\alpha\right) = -\cot\alpha, \quad \cot\left(\frac{3\pi}{2}+\alpha\right) = -\tan\alpha.$$

The induction formula $k \cdot \frac{\pi}{2} + \alpha$ $(k \in Z)$ can be summed up as: "when the odd element changes, the even element does not change. The sign depends on the quadrant." When k is odd, we get complementary function. When k is even, we have trigonometric function of the same name. If we see α as an acute angle, by the quadrant of the primary function of symbol.

Exercise

1. Suppose

$$f(x) = \begin{cases} \sin \pi x & (x < 0), \\ f(x-1) + 1 & (x \geq 0), \end{cases}$$

$$g(x) = \begin{cases} \cos \pi x & \left(x < \dfrac{1}{2}\right), \\ g(x-1) + 1 & \left(x \geq \dfrac{1}{2}\right). \end{cases}$$

Find

$$g\left(\frac{1}{4}\right) + f\left(\frac{1}{3}\right) + g\left(\frac{5}{6}\right) + f\left(\frac{3}{4}\right).$$

Solution:

$$f\left(\frac{1}{3}\right) = f\left(-\frac{2}{3}\right) + 1 = \sin\left(-\frac{2}{3}\pi\right) + 1 = -\frac{\sqrt{3}}{2} + 1,$$

$$f\left(\frac{3}{4}\right) = f\left(-\frac{1}{4}\right) + 1 = \sin\left(-\frac{1}{4}\pi\right) + 1 = -\frac{\sqrt{2}}{2} + 1,$$

$$g\left(\frac{1}{4}\right) = \cos\frac{\pi}{4} = \frac{\sqrt{2}}{2},$$

$$g\left(\frac{5}{6}\right) = g\left(-\frac{1}{6}\right) + 1 = \cos\left(-\frac{\pi}{6}\right) + 1 = \frac{\sqrt{3}}{2} + 1,$$

so the solution is 3.

2. Simplify:

$$\frac{\cos(2\pi - \alpha)\cot(\pi + \alpha)\tan(-\alpha - \pi)}{\sin(\pi - \alpha)\cot(3\pi - \alpha)}.$$

Solution:

$$\because \; \tan(-\alpha - \pi) = \tan[-(\pi + \alpha)] = -\tan(\pi + \alpha) = -\tan\alpha,$$

$$\cot(3\pi - \alpha) = \cot[2\pi + (\pi - \alpha)] = \cot(\pi - \alpha) = -\cot\alpha$$

$$\therefore \; = \frac{\cos\alpha \cdot \cot\alpha \cdot (-\tan\alpha)}{\sin\alpha \cdot (-\cot\alpha)} = \frac{\cos\alpha}{\sin\alpha} \cdot \frac{\sin\alpha}{\cos\alpha} = 1.$$

3. Suppose $k \in Z$, prove:

(1) $\cos(k\pi + \alpha) = (-1)^k \cos\alpha$;

(2) $\sin(k\pi + \alpha) = (-1)^k \sin\alpha$.

Proof: When k is even number, $k = 2n \; (n \in Z)$;

$$\cos(k\pi + \alpha) = \cos(\alpha + 2n\pi) = \cos\alpha = (-1)^{2n}\cos\alpha = (-1)^k\cos\alpha,$$

$$\sin(k\pi + \alpha) = \sin(\alpha + 2n\pi) = \sin\alpha = (-1)^{2n}\sin\alpha = (-1)^k\sin\alpha.$$

When k is odd number, $k = 2n + 1 \; (n \in Z)$,

$$\cos(k\pi + \alpha) = \cos[\alpha + 2(n + 1)\pi] = \cos(\alpha + \pi)$$

$$= -\cos\alpha = (-1)^{2n+1}\cos\alpha = (-1)^k\cos\alpha;$$

$$\sin(k\pi + \alpha) = \sin[\alpha + 2(n + 1)\pi] = \sin(\alpha + \pi)$$

$$= -\sin\alpha = (-1)^{2n+1}\sin\alpha = (-1)^k\sin\alpha.$$

$$\therefore \; \forall k \in Z \Rightarrow \cos(k\pi + \alpha) = (-1)^k\cos\alpha;$$

$$\sin(k\pi + \alpha) = (-1)^k\sin\alpha. \qquad \square$$

4. (1) Suppose

$$f(x) = \frac{\sin(n\pi - x)\cos(n\pi + x)}{\cos[(n+1)\pi - x]} \cdot \tan(x - n\pi)$$

$$\cdot \cot(n\pi - x)(n \in Z),$$

Evaluation $f\left(\frac{7\pi}{6}\right)$.

(2) Suppose α is a third-quadrant angle, and $\sin\left(\alpha - \frac{7\pi}{2}\right) = -\frac{1}{5}$, evaluate

$$f(\alpha) = \frac{\sin(\pi - \alpha)\cos(2\pi - \alpha)\tan\left(-\alpha + \frac{3\pi}{2}\right)}{\cot(-\alpha - 3\pi)\sin\left(-\frac{\pi}{2} - \alpha\right)}.$$

Solution:

(1) By simplification and evaluation, we have the following.

$$\because \tan(x - n\pi) \cdot \cot(n\pi - x) = -\tan(n\pi - x)\cot(n\pi - x) = -1,$$

$$\therefore f(x) = -\frac{\sin(n\pi - x)\cos(n\pi + x)}{\cos[(n+1)\pi - x]}.$$

When $n = 2k$ $(k \in Z)$, $f(x) = -\dfrac{\sin(-x)\cos x}{\cos(\pi - x)} = -\sin x.$

When $n = 2k+1$ $(k \in Z)$, $f(x) = -\dfrac{\sin(\pi - x)\cos(\pi + x)}{\cos(-x)} = \sin x.$

$$\therefore f(x) = (-1)^{n-1}\sin x.$$

$$\therefore f\left(\frac{7\pi}{6}\right) = (-1)^{n-1}\sin\frac{7\pi}{6} = (-1)^{n-1}\sin\left(x + \frac{\pi}{6}\right)$$

$$= (-1)^n \sin\frac{\pi}{6} = \frac{(-1)^n}{2}$$

$$= \begin{cases} \dfrac{1}{2}, & \text{when } n \text{ is even number;} \\ -\dfrac{1}{2} & \text{When n is odd number.} \end{cases}$$

(2)

$$f(\alpha) = \frac{\sin(\pi - \alpha)\cos(2\pi - \alpha)\tan\left(-\alpha + \frac{3\pi}{2}\right)}{\cot(-\alpha - 3\pi)\sin\left(-\frac{\pi}{2} - \alpha\right)}$$

$$= \frac{\sin\alpha\cos\alpha\cot\alpha}{-\cot\alpha(-\cos\alpha)} = \sin\alpha.$$

$$\because \sin\left(\alpha - \frac{7\pi}{2}\right) = \sin\left(\alpha + \frac{\pi}{2}\right) = \cos\alpha,$$

$$\therefore \cos\alpha = -\frac{1}{5}.$$

$\because \alpha$ is the third-quadrant angle.

$$\therefore \sin\alpha < 0, \sin\alpha = -\sqrt{1 - \cos^2\alpha} = -\frac{2\sqrt{6}}{5}.$$

$$\therefore f(\alpha) = -\frac{2\sqrt{6}}{5}.$$

Comments: If we directly substitute the value for evaluation, it will obviously be very complex. It is a common practice to simply and then to get the answers. Emphasis should be put on the classified discussion. This makes full use of principle "the big horn is transformed into the small angle" and the formula "when the odd element changes, the even element does not change. The sign depends on the quadrant." If we should transform the angle into $\frac{k\pi}{2} \pm \alpha(k \in Z)$, when k is even, we get the value of trigonometric function of the same name. When k is odd, we get its different name of α (sine to cosine, cosine to sine, tangent to cotangent and cotangent to tangent). Then the previously added α serves as the sign for the original trigonometric function value. Pay attention to the definition of odd and even when solving such problems.

5. According to the condition, find the angle x.

(1) Suppose $\sin x = \dfrac{1}{2}, x \in [0, 2\pi)$.

(2) Suppose $\tan x = \dfrac{\sqrt{3}}{3}, x \in [-\pi, \pi)$.

Solution: (1) $\because$ $\sin x = \dfrac{1}{2}$, $x \in [0, 2\pi)$, $\therefore$ $x \in [0, \pi)$;

$\because$ $\sin \dfrac{\pi}{6} = \dfrac{1}{2}$, $\dfrac{\pi}{6} \in \left[0, \dfrac{\pi}{2}\right)$, $\therefore$ $x = \dfrac{\pi}{6}$;

$\because$ $\sin\left(\pi - \dfrac{\pi}{6}\right) = \sin\dfrac{\pi}{6} = \dfrac{1}{2}$, $\pi - \dfrac{\pi}{6} \in \left[\dfrac{\pi}{2}, \pi\right)$,

$\therefore$ $x = \dfrac{5\pi}{6}$.

So $x = \dfrac{\pi}{6}$ or $x = \dfrac{5\pi}{6}$.

(2) $\because$ $\tan x = \dfrac{\sqrt{3}}{3}$, $x \in [-\pi, \pi)$,

$\therefore$ $x \in \left[-\pi, -\dfrac{\pi}{2}\right) \cup \left(0, \dfrac{\pi}{2}\right)$.

$\because$ $\tan \dfrac{\pi}{6} = \dfrac{\sqrt{3}}{3}$, $\dfrac{\pi}{6} \in \left(0, \dfrac{\pi}{2}\right)$, $\therefore$ $x = \dfrac{\pi}{6}$;

$\because$ $\tan \dfrac{7\pi}{6} = \tan\left(\pi + \dfrac{\pi}{6}\right) = \tan\dfrac{\pi}{6} = \dfrac{\sqrt{3}}{3}$, $\dfrac{7\pi}{6} \notin [-\pi, \pi)$,

$\because$ $\tan\left(-2\pi + \dfrac{7\pi}{6}\right) = \tan\dfrac{7\pi}{6} = \dfrac{\sqrt{3}}{3}$,

$\because$ $-2\pi + \dfrac{7\pi}{6} = -\dfrac{5\pi}{6} \in \left[-\pi, -\dfrac{\pi}{2}\right)$, $\therefore$ $x = -\dfrac{5\pi}{6}$.

So $x = \dfrac{\pi}{6}$ or $x = -\dfrac{5\pi}{6}$

The Relationship between Trigonometric Ratios of the Same Angle

1. **Reciprocal Relationships:** $\sin\alpha \cdot \csc\alpha = 1$, $\cos\alpha \cdot \sec\alpha = 1$, $\tan\alpha \cdot \cot\alpha = 1$;
2. **Quotient Relationships:** $\tan\alpha = \dfrac{\sin\alpha}{\cos\alpha}$, $\cot\alpha = \dfrac{\cos\alpha}{\sin\alpha}$;

3. **Pythagorean Relationships**: $\sin^2 \alpha + \cos^2 \alpha = 1$, $1 + \tan^2 \alpha = \sec^2 \alpha$, $1 + \cot^2 \alpha = \csc^2 \alpha$.

Exercise

6. Let $\cos \alpha = \frac{4}{5}$, and α is the fourth-quadrant angle, find the other trigonometric functions of α.

   Solution:

   $$\because \ \sin^2 \alpha + \cos^2 \alpha = 1$$

   $$\therefore \ \sin \alpha = \pm\sqrt{1 - \cos^2 \alpha}$$

   $\because \ \alpha$ is the fourth-quadrant angle,

   $\therefore \ \sin \alpha < 0, \text{so}$

   $$\sin \alpha = -\sqrt{1 - \cos^2 \alpha} = -\sqrt{1 - \left(\frac{4}{5}\right)^2} = -\frac{3}{5},$$

   $$\tan \alpha = \frac{\sin \alpha}{\cos \alpha} = \frac{-\frac{3}{5}}{\frac{4}{5}} = -\frac{3}{4},$$

   $$\cot \alpha = \frac{1}{\tan \alpha} = -\frac{4}{3},$$

   $$\sec \alpha = \frac{1}{\cos \alpha} = \frac{5}{4},$$

   $$\csc \alpha = \frac{1}{\sin \alpha} = -\frac{5}{3}.$$

7. Let $\tan \alpha = \frac{5}{12}$, find $\sin \alpha, \cos \alpha$ and $\cot \alpha$.

   Solution:

   $$\cot \alpha = \frac{1}{\tan \alpha} = \frac{12}{5}.$$

   $$\because \ 1 + \tan^2 \alpha = \sec^2 \alpha = \frac{1}{\cos^2 \alpha},$$

$$\therefore \cos^2 \alpha = \frac{1}{1 + \tan^2 \alpha} = \frac{1}{1 + \left(\frac{5}{12}\right)^2} = \left(\frac{12}{13}\right)^2,$$

$$\because \tan \alpha = \frac{5}{12} > 0,$$

$\therefore \alpha$ is the first-quadrant angle or the third-quadrant angle.

(1) When α is the first-quadrant angle, then $\sin \alpha > 0$ and $\cos \alpha > 0$,

$$\therefore \cos \alpha = \frac{12}{13}, \quad \sin \alpha = \sqrt{1 - \cos^2 \alpha} = \frac{5}{13}.$$

(2) when α is the third-quadrant angle, then $\sin \alpha < 0$ and $\cos \alpha < 0$,

$$\therefore \cos \alpha = -\frac{12}{13}, \quad \sin \alpha = -\sqrt{1 - \cos^2 \alpha} = -\frac{5}{13}.$$

8. If the terminal ray of α in standard position does not lie on the axis of coordinates, and $\sin \alpha = m$, find $\tan \alpha, \csc \alpha$.

Analysis: According to one trigonometric ratio of α, find the other trigonometric ratios: if we know its quadrant, other trigonometric ratios are unique; if its quadrant is uncertain, find other trigonometric ratios according to all conditions of its terminal sides.

Generally, if $\sin \alpha$ is known, find $\cos \alpha$ firstly; if $\cos \alpha$ is known, find $\sin \alpha$ firstly; if $\tan \alpha$ is known, find $\sec \alpha$ firstly; if $\cot \alpha$ is known, find $\csc \alpha$ firstly. Use the Pythagorean relationships firstly, and then use reciprocal relationships and quotient relationships to find other trigonometric ratios. Pay attention to the sign ahead root.

Solution:

$$\begin{cases} \sin^2 \alpha + \cos^2 \alpha = 1 \\ \sin \alpha = m \end{cases} \Rightarrow \cos^2 \alpha = 1 - \sin^2 \alpha = 1 - m^2.$$

(1) When α is a first-quadrant angle or a fourth-quadrant angle, then $\cos \alpha = \sqrt{1 - m^2}$.

$$\therefore \tan \alpha = \frac{\sin \alpha}{\cos \alpha} = \frac{m\sqrt{1 - m^2}}{1 - m^2}, \quad \csc \alpha = \frac{1}{\sin \alpha} = \frac{1}{m}.$$

(2) When α is a second-quadrant angle or a third-quadrant angle, then $\cos \alpha = -\sqrt{1 - m^2}$.

$$\therefore \ \tan \alpha = \frac{\sin \alpha}{\cos \alpha} = -\frac{m\sqrt{1 - m^2}}{1 - m^2}, \quad \csc \alpha = \frac{1}{\sin \alpha} = \frac{1}{m}.$$

9. Let $\tan \alpha = 2$.

(1) find $\dfrac{\sin \alpha + 3 \cos \alpha}{3 \sin \alpha - 4 \cos \alpha}$;

(2) find $\dfrac{\sin^2 \alpha + 8 \sin \alpha \cos \alpha - 6 \cos^2 \alpha}{3 \sin^2 \alpha - 4 \cos^2 \alpha}$;

(3) find $\sin^2 \alpha - 3 \sin \alpha \cos \alpha + 4 \cos^2 \alpha - 2$.

Solution:

(1) $\dfrac{\sin \alpha + 3 \cos \alpha}{3 \sin \alpha - 4 \cos \alpha} = \dfrac{\tan \alpha + 3}{3 \tan \alpha - 4} = \dfrac{5}{2}$.

(2) $\dfrac{\sin^2 \alpha + 8 \sin \alpha \cos \alpha - 6 \cos^2 \alpha}{3 \sin^2 \alpha - 4 \cos^2 \alpha} = \dfrac{\tan^2 \alpha + 8 \tan \alpha - 6}{3 \tan^2 \alpha - 4} = \dfrac{7}{4}$.

(3) $\sin^2 \alpha - 3 \sin \alpha \cos \alpha + 4 \cos^2 \alpha - 2$

$$= \frac{\sin^2 \alpha - 3 \sin \alpha \cos \alpha + 4 \cos^2 \alpha}{\sin^2 \alpha + \cos^2 \alpha} - 2$$

$$= \frac{\tan^2 \alpha - 3 \tan \alpha + 4}{\tan^2 \alpha + 1} - 2 = -\frac{8}{5}.$$

10. Simplify $\sin^3 \alpha (1 + \cot \alpha) + \cos^3 \alpha (1 + \tan \alpha)$.

Solution:

$$\sin^3 \alpha (1 + \cot \alpha) + \cos^3 \alpha (1 + \tan \alpha)$$

$$= \sin^3 \alpha \left(1 + \frac{\cos \alpha}{\sin \alpha} \right) + \cos^3 \alpha \left(1 + \frac{\sin \alpha}{\cos \alpha} \right)$$

$$= \sin^3 \alpha \frac{\sin \alpha + \cos \alpha}{\sin \alpha} + \cos^3 \alpha \frac{\cos \alpha + \sin \alpha}{\cos \varepsilon}$$

$$= \sin^2 \alpha (\sin \alpha + \cos \alpha) + \cos^2 \alpha (\sin \alpha + \cos \alpha)$$

$$= (\cos \alpha + \sin \alpha)(\sin^2 \alpha + \cos^2 \alpha)$$

$$= \sin \alpha + \cos \alpha.$$

11. If $-270° < x < -180°$, simplify

$$\sqrt{\left(1 + \tan \frac{x}{2}\right)^2 + \left(1 - \tan \frac{x}{2}\right)^2}.$$

Solution:

$$= \sqrt{1 + 2\tan \frac{x}{2} + \tan^2 \frac{x}{2} + 1 - 2\tan \frac{x}{2} + \tan^2 \frac{x}{2}}$$

$$= \sqrt{2\left(1 + \tan^2 \frac{x}{2}\right)}$$

$$= \sqrt{2 \sec^2 \frac{x}{2}}$$

$$= \sqrt{2} \left| \sec \frac{x}{2} \right|.$$

When $-270° < x < -180°$, $-135° < \frac{x}{2} < -90°$, $\sec \frac{x}{2} < 0$,

$$= -\sqrt{2} \sec \frac{x}{2}$$

12. (1) If $\sin(3\pi - \theta) - \cos(5\pi + \theta) = \frac{1}{5}, \theta \in (0, \pi)$, find
$3\sin^2(k\pi - \theta) - 2\sin[(k+1)\pi + \theta]\cos[(k-1)\pi - \theta] - 2\cos^2(k\pi + \theta)$
$(k \in Z)$.
(2) If $f(\cos x) = \cos 17x$, find $[f(\cos x)]^2 + [f(\sin x)]^2$.

Solution 1: $\sin \theta + \cos \theta = \dfrac{1}{5}$ (1)

$$(\sin \theta + \cos \theta)^2 = \frac{1}{25}$$

$$2 \sin \theta \cos \theta = -\frac{24}{25} < 0.$$

$$\therefore \frac{\pi}{2} < \theta < \pi, \ \therefore \ \sin \theta - \cos \theta > 0.$$

$$\therefore (\sin \theta - \cos \theta)^2 = 1 - 2 \sin \theta \cos \theta = \frac{49}{25},$$

then,

$$\sin \theta - \cos \theta = \frac{7}{5}. \tag{2}$$

According to Eqs. (1) and (2), $\sin \theta = \frac{4}{5}, \cos \theta = -\frac{3}{5}$:

$$3 \sin^2(k\pi - \theta) - 2 \sin[(k+1)\pi + \theta]$$
$$\times \cos[(k-1)\pi - \theta] - 2 \cos^2(k\pi + \theta)$$
$$= 3[(-1)^k(-\sin \theta)]^2 - 2(-1)^{k+1} \sin \theta \cdot (-1)^{k-1}$$
$$\times \cos(-\theta) - 2[(-1)^k \cos \theta]^2$$
$$= 3 \sin^2 \theta - 2 \sin \theta \cos \theta - 2 \cos^2 \theta = \frac{54}{25}.$$

Solution 2: (1) When $0 < \theta < \frac{\pi}{2}$, $\sin \theta + \cos \theta > 1$,

$$\because \ \sin \theta + \cos \theta = \frac{1}{5} < 1, \ \therefore \ \frac{\pi}{2} < \theta < \pi.$$

$$\therefore \ 5 \sin \theta = 1 - 5 \cos \theta,$$

$$\therefore \ 25 \cos^2 \theta - 5 \cos \theta - 12 = 0,$$

$$\therefore \ \cos \theta = -\frac{3}{5}, \cos \theta = \frac{4}{5} \text{ (rejection), so } \sin \theta = \frac{4}{5}.$$

(2) $\because f(\sin x) = f \left[\cos \left(\frac{\pi}{2} - x \right) \right] = \cos 17 \left(\frac{\pi}{2} - x \right) = \sin 17x,$

$$\therefore [f(\cos x)]^2 + [f(\sin x)]^2 = \cos^2 17x + \sin^2 17x = 1.$$

Comment: The key of question (1) is to find $\sin \theta$ and $\cos \theta$ firstly. It is necessary to reduce the range of the angle θ, and then find $\sin\theta$ and $\cos\theta$ through the idea of equation. Solution 1 constructs equation about $\sin\theta$ and $\cos\theta$. Generally, for three types of $\sin\theta + \cos\theta$, $\sin\theta - \cos\theta$ and $\sin\theta\cos\theta$, if value of one type is known, we can find values of other two types by quadratic sum and the relationship between trigonometric ratios of the same angle. Just pay attention to the choice of positive and negative when the evaluation after squaring. Solution 2 constructs the equation about $\cos\theta$, and it is difficult to simplify this type in this question. Mastering basic induction formula is the key to solve the problem, but also to learn trigonometric functions.

13. Proving identity: $\frac{\cos\alpha}{1+\sin\alpha} - \frac{\sin\alpha}{1+\cos\alpha} = \frac{2(\cos\alpha-\sin\alpha)}{1+\sin\alpha+\cos\alpha}$.

Analysis: In this question, there are three ideas as follows:

(1) From left to right, reduction of fractions to a common denominator of left-type and numerator factorization to produce factor $2(\cos a - \sin a)$, and then decomposition to factor $1 + \sin a + \cos a$, after the denominator appropriates deformation.

(2) From left to right, because there are denominator factor $1 + \sin a + \cos a$ in the right type, numerator and denominator are multiplied by $1 + \sin\alpha + \cos\alpha$.

(3) Turn left and right sides to the same denomination, so $1 + \sin a + \cos a$ is the simplest form. Solve the question by $\frac{\cos a}{1+\sin a} = \frac{1-\sin a}{\cos a}$, $\frac{\sin a}{1+\cos a} = \frac{1-\cos a}{\sin a}$ and geometric theorem.

Proving 1:

$$\text{left} = \frac{\cos\alpha(1 + \cos\alpha) - \sin\alpha(1 + \sin\alpha)}{(1 + \sin\alpha)(1 + \cos\alpha)}$$

$$= \frac{2(\cos\alpha - \sin\alpha)(1 + \sin\alpha + \cos\alpha)}{2 + 2(\sin\alpha + \cos\alpha) + 2\sin\alpha \cdot \cos\alpha}$$

$$= \frac{2(\cos\alpha - \sin\alpha)(1 + \sin\alpha + \cos\alpha)}{1 + \sin^2\alpha + \cos^2\alpha + 2\sin\alpha + 2\cos\alpha + 2\sin\alpha\cos\alpha}$$

$$= \frac{2(\cos\alpha - \sin\alpha)(1 + \sin\alpha + \cos\alpha)}{(1 + \sin\alpha + \cos\alpha)^2} = \text{right.}$$

Proving 2:

$$\text{left} = \frac{1+\sin\alpha+\cos\alpha}{1+\sin\alpha+\cos\alpha}\left(\frac{\cos\alpha}{1+\sin\alpha} - \frac{\sin\alpha}{1+\cos\alpha}\right)$$

$$= \frac{1}{1+\sin\alpha+\cos\alpha}\left[\frac{(1+\sin\alpha+\cos\alpha)\cos\alpha}{1+\sin\alpha}\right.$$

$$\left. -\frac{(1+\sin\alpha+\cos\alpha)\sin\alpha}{1+\cos\alpha}\right]$$

$$= \frac{1}{1+\sin\alpha+\cos\alpha}\left(\cos\alpha + \frac{\cos^2\alpha}{1+\sin\alpha} - \sin\alpha - \frac{\sin^2\alpha}{1+\cos\alpha}\right)$$

$$= \frac{1}{1+\sin\alpha+\cos\alpha}\left(\cos\alpha + \frac{1-\sin^2\alpha}{1+\sin\alpha} - \sin\alpha - \frac{1-\cos^2\alpha}{1+\cos\alpha}\right)$$

$$= \frac{1}{1+\sin\alpha+\cos\alpha}\left(\cos\alpha + 1 - \sin\alpha - \sin\alpha - 1 + \cos\alpha\right)$$

$$= \frac{2(\cos\alpha-\sin\alpha)}{1+\sin\alpha+\cos\alpha} = \text{right.}$$

Proving 3:

$$\because \quad \frac{\cos\alpha}{1+\sin\alpha} = \frac{1-\sin\alpha}{\cos\alpha} = \frac{\cos\alpha+1-\sin\alpha}{1+\sin\alpha+\cos\alpha}$$

$$\frac{\sin\alpha}{1+\cos\alpha} = \frac{1-\cos\alpha}{\sin\alpha} = \frac{\sin\alpha+1-\cos\alpha}{1+\cos\alpha+\sin\alpha}$$

$$\therefore \quad \frac{\cos\alpha}{1+\sin\alpha} - \frac{\sin\alpha}{1+\cos\alpha} = \frac{\cos\alpha+1-\sin\alpha}{1+\sin\alpha+\cos\alpha}$$

$$-\frac{\sin\alpha+1-\cos\alpha}{1+\cos\alpha+\sin\alpha} = \frac{2(\cos\alpha-\sin\alpha)}{1+\sin\alpha+\cos\alpha}.$$

Comment: To prove the trigonometric identity, it is essential to pay attention to the basic relationship of the same angle of trigonometric ratio and the application of the induced formula. In the process of deformation, it is proved generally from complex to simple. The above methods are proved by the right type as the target, so the deformation process as much as possible will be constructed as the denominator $1 + \sin a + \cos a$ and the numerator $2(\cos a - \sin a)$. If the proving formula contains both cut and string, turn to string generally. When both sides are complex, it can be considered that two sides of which are equal to another formula. Proving 3 flexibility use scaling properties, so proving is enough simple.

14. Proving: $\cos \alpha(2 \sec \alpha + \tan \alpha)(\sec \alpha - 2 \tan \alpha) = 2 \cos \alpha - 3 \tan \alpha$.

Prove:

$$\cos \alpha(2 \sec \alpha + \tan \alpha)(\sec \alpha - 2 \tan \alpha)$$

$$= \cos \alpha \left(\frac{2}{\cos \alpha} + \frac{\sin \alpha}{\cos \alpha} \right) \left(\frac{1}{\cos \alpha} - \frac{2 \sin \alpha}{\cos \alpha} \right)$$

$$= \frac{1}{\cos \alpha}(2 + \sin \alpha)(1 - 2 \sin \alpha)$$

$$= \frac{1}{\cos \alpha}(2 - 2 \sin^2 \alpha - 3 \sin \alpha)$$

$$= \frac{1}{\cos \alpha}(2 \cos^2 \alpha - 3 \sin \alpha)$$

$$= 2 \cos \alpha - 3 \tan \alpha = \text{right.}$$

Exercises One

1. Let $\tan \theta = 2$, find $\sin^2 \theta + \sin \theta \cos \theta - 2 \cos^2 \theta$.

2. Let $\sin \theta, \cos \theta$ be two roots of equation $2x^2 - (\sqrt{3} + 1)x + m = 0$, find $\frac{\sin \theta}{1 - \cot \theta} + \frac{\cos \theta}{1 - \tan \theta}$.

3. Let $3 \sin^2 \alpha + 2 \sin^2 \beta = 2 \sin \alpha$, find the range of values of $\sin^2 \alpha + \sin^2 \beta$.

4. Let $\{x \mid \cos^2 x + \sin x + m = 0\} \neq \phi$, find the range of values of m.

5. Let $\cos x + \cos y = 1$, find the range of values of $\sin x - \sin y$.

6. If $\sqrt{\frac{1+\sin\alpha}{1-\sin\alpha}} - \sqrt{\frac{1-\sin\alpha}{1+\sin\alpha}} = 2\tan\alpha$ is identity, find the range of values of α.

7. In $\triangle ABC$, $\sin A(\sin B + \cos B) - \sin C = 0$, $\sin B + \cos 2C = 0$, find A, B, C.

8. Let $-\frac{\pi}{2} < x < 0$, $\sin x + \cos x = \frac{1}{5}$. Find $\dfrac{3\sin^2\frac{x}{2} - 2\sin\frac{x}{2}\cos\frac{x}{2} + \cos^2\frac{x}{2}}{\tan x + \cot x}$.

9. Let $f(t) = \sqrt{\frac{1-t}{1+t}}$, $g(x) = \cos x \cdot f(\sin x) + \sin x \cdot f(\cos x)$, $x \in (\pi, \frac{17\pi}{12})$.

   (1) Simply $g(x)$ to the expression of $A\sin(\omega x + \varphi) + B$ ($A > 0, \omega > 0, \varphi \in [0, 2\pi)$).

   (2) Find the range of $g(x)$.

Chapter 2

Trigonometric Identity

2.1. Cosines, Sines and Tangents of Addition and Subtraction of Two Angles

Cosines and sines of addition and subtraction of two angles

$$\sin(\alpha + \beta) = \sin \alpha \cos \beta + \cos \alpha \sin \beta,$$

$$\sin(\alpha - \beta) = \sin \alpha \cos \beta - \cos \alpha \sin \beta,$$

$$\cos(\alpha + \beta) = \cos \alpha \cos \beta - \sin \alpha \sin \beta,$$

$$\cos(\alpha - \beta) = \cos \alpha \cos \beta + \sin \alpha \sin \beta.$$

Exercise

1. Find $\cos \left(\alpha + \frac{5\pi}{12}\right) \cos \left(\alpha + \frac{\pi}{6}\right) + \cos \left(\frac{\pi}{12} - \alpha\right) \cos \left(\frac{\pi}{3} - \alpha\right)$.

Analysis: Uniform angle $\frac{\pi}{12} - \alpha$ and $\alpha + \frac{\pi}{6}$ by induction formula, and then use the sines of addition of two angles. Uniform angle $\frac{5\pi}{12} + \alpha$ and $\alpha + \frac{\pi}{6}$ by induction formula, and then use the cosines of addition of two angles.

Solution 1:

$$\cos \left(\alpha + \frac{5\pi}{12}\right) \cos \left(\alpha + \frac{\pi}{6}\right) + \cos \left(\frac{\pi}{12} - \alpha\right) \cos \left(\frac{\pi}{3} - \alpha\right)$$

$$= \sin \left[\frac{\pi}{2} - \left(\alpha + \frac{5\pi}{12}\right)\right] \cos \left(\alpha + \frac{\pi}{6}\right)$$

$$+ \cos \left(\frac{\pi}{12} - \alpha\right) \sin \left[\frac{\pi}{2} - \left(\frac{\pi}{3} - \alpha\right)\right]$$

$$= \sin \left(\frac{\pi}{12} - \alpha\right) \cos \left(\alpha + \frac{\pi}{6}\right) + \cos \left(\frac{\pi}{12} - \alpha\right) \sin \left(\alpha + \frac{\pi}{6}\right)$$

$$= \sin\left[\left(\frac{\pi}{12} - \alpha\right) + \left(\alpha + \frac{\pi}{6}\right)\right]$$

$$= \sin\frac{\pi}{4} = \frac{\sqrt{2}}{2}.$$

Solution 2:

$$\cos\left(\alpha + \frac{5\pi}{12}\right)\cos\left(\alpha + \frac{\pi}{6}\right) + \cos\left(\frac{\pi}{12} - \alpha\right)\cos\left(\frac{\pi}{3} - \alpha\right)$$

$$= \cos\left(\alpha + \frac{5\pi}{12}\right)\cos\left(\alpha + \frac{\pi}{6}\right)$$

$$+ \sin\left[\frac{\pi}{2} - \left(\frac{\pi}{12} - \alpha\right)\right]\sin\left[\frac{\pi}{2} - \left(\frac{\pi}{3} - \alpha\right)\right]$$

$$= \cos\left(\alpha + \frac{5\pi}{12}\right)\cos\left(\alpha + \frac{\pi}{6}\right) + \sin\left(\alpha + \frac{5\pi}{12}\right)\sin\left(\alpha + \frac{\pi}{6}\right)$$

$$= \cos\left[\left(\alpha + \frac{5\pi}{12}\right) - \left(\alpha + \frac{\pi}{6}\right)\right]$$

$$= \cos\frac{\pi}{4} = \frac{\sqrt{2}}{2}.$$

2. Suppose $\cos\phi = -\frac{3}{5}, \phi \in \left(\frac{\pi}{2}, \pi\right)$, find $\sin\left(\frac{\pi}{6} + \phi\right)$.

Analysis: $\sin\left(\phi + \frac{\pi}{6}\right) = \sin\phi\cos\frac{\pi}{6} + \cos\phi\sin\frac{\pi}{6}$ and $\cos\phi = -\frac{3}{5}$, so only we need to find $\sin\phi$.

Solution:

$$\cos\phi = -\frac{3}{5}, \phi \in \left(\frac{\pi}{2}, \pi\right),$$

$$\sin\left(\phi + \frac{\pi}{6}\right) = \sin\phi\cos\frac{\pi}{6} + \cos\phi\sin\frac{\pi}{6}$$

$$= \frac{4}{5} \times \frac{\sqrt{3}}{2} + \left(-\frac{3}{5}\right) \times \frac{1}{2}$$

$$= \frac{4\sqrt{3} - 3}{10}.$$

3. Prove $\cos(\alpha + \beta)\cos(\alpha - \beta) = \cos^2\alpha - \sin^2\beta$.

Proving:

$$\text{Left} = (\cos\alpha\cos\beta - \sin\alpha\sin\beta)(\cos\alpha\cos\beta + \sin\alpha\sin\beta)$$
$$= \cos^2\alpha\cos^2\beta - \sin^2\alpha\sin^2\beta$$
$$= \cos^2\alpha(1 - \sin^2\beta) - (1 - \cos^2\alpha)\sin^2\beta$$
$$= \cos^2\alpha - \cos^2\alpha\sin^2\beta - \sin^2\beta + \cos^2\alpha\sin^2\beta$$
$$= \cos^2\alpha - \sin^2\beta = \text{Right}.$$

Comment: The proof of identity is generally followed the principle from complex to simple.

4. Suppose $\cos(\alpha - \frac{\beta}{2}) = -\frac{1}{9}, \sin(\frac{\alpha}{2} - \beta) = \frac{2}{3}$, and $\frac{\pi}{2} < \alpha < \pi, 0 < \beta < \pi$, find $\cos\frac{\alpha+\beta}{2}$.

Solution:

$$\cos\left[\left(\alpha - \frac{\beta}{2}\right) - \left(\frac{\alpha}{2} - \beta\right)\right]$$

$$= \cos\left(\alpha - \frac{\beta}{2}\right)\cos\left(\frac{\alpha}{2} - \beta\right) - \sin\left(\alpha - \frac{\beta}{2}\right)\sin\left(\frac{\alpha}{2} - \beta\right).$$

$$\because \frac{\pi}{2} < \alpha < \pi, 0 < \beta < \pi, \text{ so}$$

$$0 < \alpha - \frac{\beta}{2} < \pi, -\frac{3\pi}{4} < \frac{\alpha}{2} - \beta < \frac{\pi}{2}, \sin\left(\frac{\alpha}{2} - \beta\right) = \frac{2}{3} > 0,$$

$$\therefore 0 < \frac{\alpha}{2} - \beta < \pi, \therefore 0 < \frac{\alpha}{2} - \beta < \frac{\pi}{2}.$$

$$\therefore \sin\left(\alpha - \frac{\beta}{2}\right) = \sqrt{1 - \cos^2\left(\alpha - \frac{\beta}{2}\right)} = \frac{4\sqrt{5}}{9}$$

$$\cos\left(\frac{\alpha}{2} - \beta\right) = \cdots = \frac{\sqrt{5}}{3}, \therefore \cos\frac{\alpha+\beta}{2} = \frac{7\sqrt{5}}{27}.$$

5. (1) Suppose $\frac{\pi}{2} < \beta < \alpha < \frac{3\pi}{4}, \cos(\alpha - \beta) = \frac{12}{13}, \sin(\alpha + \beta) = -\frac{3}{5}$, find $\sin 2\alpha$.

(2) Let $\cos(x + \frac{\pi}{4}) = \frac{3}{5} \cdot \frac{17\pi}{12} < x < \frac{7\pi}{4}$, find $\frac{\sin x + \cos x}{1 - \tan x}$.

Solution:

(1) Since $\dfrac{\pi}{2} < \beta < \alpha < \dfrac{3\pi}{4}, \cos(\alpha - \beta) = \dfrac{12}{13} > 0,$

$\therefore \ 0 < \alpha - \beta < \dfrac{\pi}{4}, \sin(\alpha - \beta) = \dfrac{5}{13},$

$\therefore \ \pi < \alpha + \beta < \dfrac{3\pi}{2}, \sin(\alpha + \beta) = -\dfrac{3}{5}, \ \therefore \ \cos(\alpha + \beta) = -\dfrac{4}{5},$

$\therefore \ \sin 2\alpha = \sin[(\alpha + \beta) + (\alpha - \beta)] = \cdots = -\dfrac{56}{65}.$

(2) Since $\cos\left(x + \dfrac{\pi}{4}\right) = \dfrac{3}{5} \cdot \dfrac{17\pi}{12} < x < \dfrac{7\pi}{4},$

$\therefore \ \cos x = \cos\left[\left(\dfrac{\pi}{4} + x\right) - \dfrac{\pi}{4}\right] = -\dfrac{\sqrt{2}}{10}, \sin x = -\dfrac{7\sqrt{2}}{10}, \tan x = 7,$

$\therefore \ \dfrac{\sin x + \cos x}{1 - \tan x} = \dfrac{-\frac{7\sqrt{2}}{10} + (-\frac{\sqrt{2}}{10})}{1 - 7} = \dfrac{2\sqrt{2}}{15}.$

Comment: Pay attention to overall substitution of the trigonometric function values after angle transformation.

6. Suppose $\sin x + \sin y = \frac{\sqrt{2}}{2}$, find the range of values of $\cos x + \cos y$.

Solution:

$$\sin x + \sin y = \frac{\sqrt{2}}{2}; \tag{1}$$

$$\text{let } m = \cos x + \cos y; \tag{2}$$

$(1)^2 + (2)^2, \ 2 + 2(\cos x \cos y + \sin x \sin y) = \dfrac{1}{2} + m^2,$

$\cos(x - y) = \dfrac{1}{2}m^2 - \dfrac{3}{4}, \ \because \ |\cos(x - y)| \le 1,$

$\therefore \ \left|\dfrac{1}{2}m^2 - \dfrac{3}{4}\right| \le 1 \Rightarrow -\dfrac{\sqrt{14}}{2} \le m \le \dfrac{\sqrt{14}}{2}.$

The comprehensive application of triangle trigonometry

In $\triangle ABC$:

$$A + B + C = \pi, A + B = \pi - C, \frac{A + B}{2} = \frac{\pi}{2} - \frac{C}{2},$$

$$\sin(A + B) = \sin C, \cos(A + B) = -\cos C, \ldots$$

$$\sin \frac{A + B}{2} = \cos \frac{C}{2}, \cos \frac{A + B}{2} = \sin \frac{C}{2}.$$

Exercise

7. In $\triangle ABC$, let $\cos A = \frac{4}{5}, \cos B = \frac{12}{13}$, find $\sin C$ and $\cos C$.

Solution:

$\because$ A, B, C are internal angles of $\triangle ABC$,

$\therefore$ $A, B, C \in (0, \pi)$, $\therefore$ $\sin A, \sin B, \sin C > 0$,

$\therefore$ $\sin A = \sqrt{1 - \cos^2 A} = \frac{3}{5}, \sin B = \sqrt{1 - \cos^2 B} = \frac{5}{13}.$

$\because$ $A + B + C = 180°$,

$\therefore$ $\sin C = \sin[180° - (A + B)] = \sin(A + B) = \cdots = \frac{56}{65}.$

$$\cos C = \cos[180° - (A + B)] = -\cos(A + B) = \cdots = -\frac{33}{65}.$$

Tangents of addition and subtraction of two angles

$$\tan(\alpha + \beta) = \frac{\tan \alpha + \tan \beta}{1 - \tan \alpha \tan \beta},$$

$$\tan(\alpha - \beta) = \frac{\tan \alpha - \tan \beta}{1 + \tan \alpha \tan \beta}.$$

Exercise

8. (1) Using tangents of addition of two angles, find $\frac{1+\tan 75°}{1-\tan 75°}$.

(2) Calculate: $\tan 15° + \tan 30° + \tan 15° \tan 30°$.

(3) $\tan 20° + \tan 40° + \sqrt{3} \tan 20° \tan 40°$.

Solution:

(1) $\because \tan 45° = 1$;

$$\frac{1 + \tan 75°}{1 - \tan 75°} = \frac{\tan 45° + \tan 75°}{1 - \tan 45° \tan 75°}$$

$$= \tan(45° + 75°)$$

$$= \tan 120° = -\sqrt{3}.$$

(2) $\tan 15° + \tan 30° + \tan 15° \tan 30°$

$$= \tan(15° + 30°)(1 - \tan 15° \tan 30°)$$

$$+ \tan 15° \tan 30° = 1.$$

(3) $\tan 20° + \tan 40° + \sqrt{3} \tan 20° \tan 40°$

$$= \tan(20° + 40°)(1 - \tan 20° \tan 40°)$$

$$+ \sqrt{3} \tan 20° \tan 40° = \sqrt{3}.$$

Comment: Use the deformation of formula:
$\tan \alpha + \tan \beta = \tan(\alpha + \beta)(1 - \tan \alpha \tan \beta)$ to make the problem simple.

9. Suppose $\sin(30° + \alpha) = \frac{3}{5}, 60° < \alpha < 150°$, find $\tan(75° + \alpha)$.

Solution:

$$\tan(75° + \alpha) = \tan[45° + (30° + \alpha)] = \frac{1 + \tan(30° + \alpha)}{1 - \tan(30° + \alpha)};$$

$$\because \sin(30° + \alpha) = \frac{3}{5}, 60° < \alpha < 150°, \ 90° < 30° + \alpha < 180°,$$

$$\therefore \tan(30° + \alpha) = -\frac{3}{4}, \tan(75° + \alpha) = \frac{1}{7}.$$

10. In isosceles right $\triangle ABC$ (see Fig. 2.1), $\angle C = 90°$, point D, E are two trisection points of BC, find $\tan \alpha, \tan \beta, \tan \gamma$.

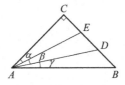

Fig. 2.1.

Analysis: Find $\tan \alpha, \tan(\alpha + \beta)$ firstly, and then use the transformation

$$\beta = (\alpha + \beta) - \alpha, \gamma = \frac{\pi}{4} - (\alpha + \beta).$$

Solution:

$$\because \ \tan \alpha = \frac{1}{3}, \tan(\alpha + \beta) = \frac{2}{3},$$

$$\therefore \ \tan \beta = \tan[(\alpha + \beta) - \alpha]$$

$$= \frac{\tan(\alpha + \beta) - \tan \alpha}{1 - \tan(\alpha + \beta) \tan \alpha}$$

$$= \frac{\frac{2}{3} - \frac{1}{3}}{1 - \frac{2}{3} \times \frac{1}{3}} = \frac{3}{11};$$

$$\tan \gamma = \tan \left[\frac{\pi}{4} - (\alpha + \beta) \right]$$

$$= \frac{\tan \frac{\pi}{4} - \tan(\alpha + \beta)}{1 + \tan \frac{\pi}{4} \tan(\alpha + \beta)}$$

$$= \frac{1 - \frac{2}{3}}{1 + 1 \times \frac{2}{3}} = \frac{1}{5},$$

so $\tan \alpha = \frac{1}{3}, \tan \beta = \frac{3}{11}, \tan \gamma = \frac{1}{5}.$

Auxiliary angle formula and its application

Convert $a \sin \alpha + b \cos \alpha \, (a \cdot b \neq 0)$ to the expression of $A \sin(\alpha + \phi)$ $(A > 0)$:

$$\because \left(\frac{a}{\sqrt{a^2 + b^2}} \right)^2 + \left(\frac{b}{\sqrt{a^2 + b^2}} \right)^2 = 1,$$

extracting common factor $\sqrt{a^2 + b^2}$.
Suppose

$$\frac{a}{\sqrt{a^2 + b^2}} = \cos \phi, \qquad \frac{b}{\sqrt{a^2 + b^2}} = \sin \phi,$$

$$a \sin \alpha + b \cos \alpha = \sqrt{a^2 + b^2}(\sin \alpha \cos \phi + \cos \alpha \sin \beta)$$

$$= \sqrt{a^2 + b^2} \sin(\alpha + \phi).$$

Namely, $a \sin \alpha + b \cos \alpha = \sqrt{a^2 + b^2} \sin(\alpha + \phi)(0 \leq \phi < 2\pi)$,

$$\cos \phi = \frac{a}{\sqrt{a^2 + b^2}}, \quad \sin \phi = \frac{b}{\sqrt{a^2 + b^2}} \quad \text{or} \quad \tan \phi = \frac{b}{a},$$

$a \sin \alpha + b \cos \alpha = \sqrt{a^2 + b^2} \sin(\alpha + \phi)$ *is named* auxiliary angle formula.

Exercise

11. If equation $2 \sin x + \sqrt{5} \cos x = \frac{1}{k}$ has solutions, find the range of values of k.

    Solution: $2 \sin x + \sqrt{5} \cos x = \frac{1}{k} \Rightarrow 3 \sin(x + \phi) = \frac{1}{k}$, and $\tan \phi = \frac{\sqrt{5}}{2}$, therefore $\sin(x + \phi) = \frac{1}{3k}$.
    If equation $2 \sin x + \sqrt{5} \cos x = \frac{1}{k}$ has solutions,
    $|\sin(x + \phi)| = \left| \frac{1}{3k} \right| \leq 1 \Rightarrow k \geq \frac{1}{3}$ or $k \leq -\frac{1}{3}$.

    Comment: Generally, $a \sin x + b \cos x = c$ has solution $\Leftrightarrow a^2 + b^2 \geq c^2$.

12. If $n \in N^*$ and $M > 0$, the elements in arithmetic progression $a_1, a_2, a_3, \ldots$, all meet $a_1^2 + a_{n+1}^2 \leq M$, find the maximum of $S = a_{n+1} + a_{n+2} + \cdots + a_{2n+1}$.

Solution: Suppose $a_1 = k\cos\theta, a_{n+1} = k\sin\theta, k \in [0,\sqrt{m}], \theta \in [0,2\pi)$,

$$S = \frac{(n+1)(a_{n+1} + a_{2n+1})}{2} = \frac{(n+1)(a_{n+1} + 2a_{n+1} - a_1)}{2}$$

$$= \frac{(n+1)}{2}(3a_{n+1} - a_1) = \frac{(n+1)(3k\sin\theta - k\cos\theta)}{2}$$

$$= \frac{(n+1)\sqrt{10}k\sin(\theta - \varphi)}{2} \quad \left(\tan\phi = \frac{1}{3}\right).$$

When $\sin(\theta - \phi) = 1, k = \sqrt{M}$, namely $a_1 = -\frac{\sqrt{10}}{10}\sqrt{M}, a_{n+1} = \frac{3\sqrt{10M}}{10}$, the maximum of S is $\frac{n+1}{2}\sqrt{10M}$.

Comment: $a_1^2 + a_{n+1}^2 \le M$ is a disc in cartesian coordinate system. Use the parameter equations of the disc $a_1 = k\cos\theta, a_{n+1} = k\sin\theta, k \in [0,\sqrt{m}], \theta \in [0,2\pi)$ and auxiliary angle formula to make the solution more concise.

13. Suppose $\sin\alpha = \sqrt{2}\cos\beta, \tan\alpha = \sqrt{3}\cot\beta, -\frac{\pi}{2} < \alpha < \frac{\pi}{2}, 0 < \beta < \pi$. Find α, β.

Solution:

$$\because \sin\alpha = \sqrt{2}\cos\beta, \tag{3}$$

$$\tan\alpha = \sqrt{3}\cot\beta. \tag{4}$$

Thus $\alpha = 0, \beta = \frac{\pi}{2}$ meet the problem set.

When $\alpha \ne 0, \beta \ne \frac{\pi}{2}, \frac{(3)^2}{(4)^2}$, can get $\cos^2\alpha = \frac{2}{3}\sin^2\beta$, (5)

$(3)^2$ can get $\sin^2\alpha = 2\cos^2\beta$; (6)

Equations $(5) + (6)$ can get $2\cos^2\beta + \frac{2}{3}\sin^2\beta = 1$,

$$\therefore \sin^2\beta = \frac{3}{4},$$

$$\because \; 0 < \beta < \pi \; \therefore \; \sin\beta = \frac{\sqrt{3}}{2}, \text{ put } \beta = \frac{\pi}{3} \text{ or } \beta = \frac{2}{3}\pi$$

into (1) can get $\alpha = \pm\dfrac{\pi}{4}$,

$$\therefore \;
\begin{cases} \alpha = \dfrac{\pi}{4} \\[2mm] \beta = \dfrac{\pi}{3} \end{cases} \text{ or } \quad
\begin{cases} \alpha = -\dfrac{\pi}{4} \\[2mm] \beta = \dfrac{2}{3}\pi \end{cases} \text{ or } \quad
\begin{cases} \alpha = 0, \\[2mm] \beta = \dfrac{\pi}{2} \end{cases}.$$

14. If α, β are acute angles, and $\cos\alpha + \cos\beta - \cos(\alpha+\beta) = \frac{2}{3}$, find α, β.

Analysis: This problem is an equation with two unknowns, which belongs to the type of indeterminate equation. To solve this problem, it should be solved by method of completing the square from triangular transformation.

Solution 1:

$$\because \; 2\cos\frac{\alpha+\beta}{2}\cos\frac{\alpha-\beta}{2} - 2\cos^2\frac{\alpha+\beta}{2} + 1 = \frac{3}{2}$$

$$\therefore \; 4\cos^2\frac{\alpha+\beta}{2} - 4\cos\frac{\alpha-\beta}{2}\cos\frac{\alpha+\beta}{2} + 1 = 0$$

$$\therefore \; \left(2\cos\frac{\alpha+\beta}{2} - \cos\frac{\alpha-\beta}{2}\right)^2 + \sin^2\frac{\alpha-\beta}{2} = 0$$

$$\therefore \; 2\cos\frac{\alpha+\beta}{2} - \cos\frac{\alpha-\beta}{2} = 0, \quad \sin\frac{\alpha-\beta}{2} = 0$$

$$\because \; \alpha, \beta \text{ are acute angles}$$

$$\therefore \; -\frac{\pi}{2} < \alpha - \beta < \frac{\pi}{2}, \quad 0 < \alpha + \beta < \pi$$

$$\therefore \;
\begin{cases} \dfrac{\alpha+\beta}{2} = \dfrac{\pi}{3} \\[2mm] \alpha = \beta \end{cases}, \quad \therefore \; \alpha = \beta = \frac{\pi}{3}.$$

Solution 2: According to the condition,

$$\sin \alpha \sin \beta + (1 - \cos \alpha) \cos \beta + \left(\cos \alpha - \frac{3}{2} \right) = 0.$$

Suppose $P(\sin \beta, \cos \beta)$; then circle $x^2 + y^2 = 1$ and line $(\sin \alpha)x + (1 - \cos \alpha)y + \cos \alpha - \frac{3}{2} = 0$ intersect at P.

$$\therefore d = \frac{\left| \cos \alpha - \frac{3}{2} \right|}{\sqrt{\sin^2 \alpha + (1 - \cos \alpha)^2}} \le 1$$

$$\text{simply} \left(\cos \alpha - \frac{1}{2} \right)^2 \le 0$$

$$\therefore \cos \alpha = \frac{1}{2}, \ \alpha = \frac{\pi}{3}. \text{ Similarly } \beta = \frac{\pi}{3}.$$

15. Suppose α, β are different roots of the equation

$a \cos x + b \sin x - c = 0 \ (a^2 + b^2 \ne 0)$, and $\alpha \ne \beta + 2k\pi \ (k \in Z)$.
Prove $\cos^2 \frac{\alpha - \beta}{2} = \frac{c^2}{a^2 + b^2}$.

Proof: Suppose $P_1(\cos \alpha, \sin \alpha), P_2(\cos \beta, \sin \beta)$; then circle $x^2 + y^2 = 1$ and line $ax + by = c$ intersect at different points P_1 and P_2.
Simultaneous

$$\begin{cases} x^2 + y^2 = 1, \\ ax + by = c, \end{cases} \quad \text{eliminate } y,$$

$$\therefore (b^2 + a^2)x^2 - 2acx + c^2 - b^2 = 0$$

$$\therefore \cos \alpha + \cos \beta = \frac{2ac}{a^2 + b^2}$$

$$\therefore 2 \cos \frac{\alpha + \beta}{2} \cos \frac{\alpha - \beta}{2} = \frac{2ac}{a^2 + b^2}. \tag{7}$$

Similarly, $(b^2 + a^2)y^2 - 2bcy + c^2 - a^2 = 0,$

$$\therefore \ \sin\alpha + \sin\beta = \frac{2bc}{a^2 + b^2}$$

$$\therefore \ 2\sin\frac{\alpha + \beta}{2}\cos\frac{\alpha - \beta}{2} = \frac{2bc}{a^2 + b^2} \qquad (8)$$

$$(7)^2 + (8)^2 \Rightarrow 4\cos^2\frac{\alpha - \beta}{2} = \frac{4c^2}{a^2 + b^2}$$

$$\therefore \ \cos^2\frac{\alpha - \beta}{2} = \frac{c^2}{a^2 + b^2}.$$

□

16. If $\frac{\sin^4\alpha}{\cos^2\beta} + \frac{\cos^4\alpha}{\sin^2\beta} = 1$, prove $\alpha + \beta = \frac{\pi}{2}$.

Proof 1:

Suppose $\frac{\sin^2\alpha}{\cos\beta} = \sin\theta, \ \frac{\cos^2\alpha}{\sin\beta} = \cos\theta \ \left(0 < \theta < \frac{\pi}{2}\right),$

$\sin\theta\cos\beta + \cos\theta\sin\beta = \sin^2\alpha + \cos^2\alpha = 1$

$\sin(\theta + \beta) = 1$

$\therefore \ \theta + \beta = \dfrac{\pi}{2}$

$\therefore \ \sin\theta = \cos\beta$

$\therefore \ \sin^2\alpha = \cos^2\beta$

$\therefore \ \sin\alpha = \cos\beta$

$\therefore \ \alpha + \beta = \dfrac{\pi}{2}.$

□

Proof 2:

$$\because \ \frac{\sin^4\alpha}{\cos^2\beta} + \cos^2\beta \geq 2\sin^2\alpha$$

$$\frac{\cos^4\alpha}{\sin^2\beta} + \sin^2\beta \geq 2\cos^2\alpha$$

By adding two formulas

$$\frac{\sin^4\alpha}{\cos^2\beta} + \frac{\cos^4\alpha}{\sin^2\beta} \geq 1.$$

So two formulas must be established at the same time when

$$\therefore \frac{\sin^4 \alpha}{\cos^2 \beta} = \cos^2 \beta, \quad \frac{\cos^4 \alpha}{\sin^2 \beta} = \sin^2 \beta$$

$$\therefore \cos \alpha = \sin \beta, \ \sin \alpha = \cos \beta$$

$$\therefore \alpha + \beta = \frac{\pi}{2}. \qquad\qquad\qquad \square$$

2.2. Double Angles, Half-Angle, Sum-to-Product Formula and Product-to-Sum Formula

Double angles formula:

$$\sin 2\alpha = 2 \sin \alpha \cos \alpha; \ \cos 2\alpha = \cos^2 \alpha - \sin^2 \alpha$$

$$= 2 \cos^2 \alpha - 1 = 1 - 2 \sin^2 \alpha;$$

$$\tan 2\alpha = \frac{2 \tan \alpha}{1 - \tan^2 \alpha}.$$

Convert double angles formula

$$\cos 2\alpha = 1 - 2 \sin^2 \alpha \quad \text{and} \quad \cos 2\alpha = 2 \cos^2 \alpha - 1 \text{ to}$$

$$\sin^2 \alpha = \frac{1 - \cos 2\alpha}{2}, \quad \cos^2 \alpha = \frac{1 + \cos 2\alpha}{2}.$$

Half-angles formula:

$$\sin \frac{\alpha}{2} = \pm\sqrt{\frac{1 - \cos \alpha}{2}}; \quad \cos \frac{\alpha}{2} = \pm\sqrt{\frac{1 + \cos \alpha}{2}};$$

$$\tan \frac{\alpha}{2} = \pm\sqrt{\frac{1 - \cos \alpha}{1 + \cos \alpha}} = \frac{\sin \alpha}{1 + \cos \alpha} = \frac{1 - \cos \alpha}{\sin \alpha}.$$

Universal replacement formula:

$$\sin \alpha = \frac{2 \tan \frac{\alpha}{2}}{1 + \tan^2 \frac{\alpha}{2}}; \quad \cos \alpha = \frac{1 - \tan^2 \frac{\alpha}{2}}{1 + \tan^2 \frac{\alpha}{2}}; \quad \tan \alpha = \frac{2 \tan \frac{\alpha}{2}}{1 - \tan^2 \frac{\alpha}{2}}.$$

Sum-to-product formula and product-to-sum formula:

(1) Product-to-sum formula

$$\sin\alpha\cos\beta = \frac{1}{2}[\sin(\alpha+\beta)+\sin(\alpha-\beta)],$$

$$\cos\alpha\sin\beta = \frac{1}{2}[\sin(\alpha+\beta)-\sin(\alpha-\beta)],$$

$$\cos\alpha\cos\beta = \frac{1}{2}[\cos(\alpha+\beta)+\cos(\alpha-\beta)],$$

$$\sin\alpha\sin\beta = -\frac{1}{2}[\cos(\alpha+\beta)-\cos(\alpha-\beta)].$$

(2) Sum-to-product formula

$$\sin x + \sin y = 2\sin\frac{x+y}{2}\cos\frac{x-y}{2},$$

$$\sin x - \sin y = 2\cos\frac{x+y}{2}\sin\frac{x-y}{2},$$

$$\cos x + \cos y = 2\cos\frac{x+y}{2}\cos\frac{x-y}{2},$$

$$\cos x - \cos y = -2\sin\frac{x+y}{2}\sin\frac{x-y}{2}.$$

Others:

(1) $(\sin\alpha\pm\cos\alpha)^2 = 1\pm\sin 2\alpha.$

(2) $\dfrac{1+\tan\alpha}{1-\tan\alpha} = \dfrac{\sin\alpha+\cos\alpha}{\cos\alpha-\sin\alpha}$

$$= \tan\left(\alpha+\frac{\pi}{4}\right)\left(\alpha\neq k\pi+\frac{\pi}{2}, k\pi+\frac{\pi}{4}, k\in Z\right).$$

(3) $\tan\alpha+\cot\alpha = \dfrac{2}{\sin 2\alpha}, \quad \alpha\neq\dfrac{k\pi}{2}, k\in Z;$

$$\tan\alpha-\cot\alpha = -2\cot 2\alpha, \quad \alpha\neq\frac{k\pi}{2}, \ k\in Z.$$

(4) $\sin(\alpha + \beta)\sin(\alpha - \beta) = \sin^2 \alpha - \sin^2 \beta = \cos^2 \beta - \cos^2 \alpha;$

$\cos(\alpha + \beta)\cos(\alpha - \beta) = \cos^2 \alpha - \sin^2 \beta.$

(5) $\alpha + \beta + \gamma = k\pi (k \in Z) \Leftrightarrow \tan \alpha + \tan \beta + \tan \gamma$

$= \tan \alpha \tan \beta \tan \gamma \left(\alpha, \beta, \gamma \neq n\pi + \dfrac{\pi}{2}, k \in Z \right).$

(6) $\alpha + \beta = k\pi + \dfrac{\pi}{4}(k \in Z) \Leftrightarrow (1 + \tan \alpha)(1 + \tan \beta)$

$= 2 \left(\alpha, \beta \neq k\pi + \dfrac{\pi}{2}, k \in Z \right).$

(7) $\sin 3\theta = 4\sin(60° - \theta)\sin \theta \sin(60° + \theta);$

$\cos 3\theta = 4\cos(60° - \theta)c\cos \theta \cos(60° + \theta);$

$\tan 3\theta = \tan(60° - \theta)\tan \theta \tan(60° + \theta).$

Exercise

1. If $\cos \left(\dfrac{\pi}{4} + x \right) = \dfrac{3}{5}, \dfrac{17\pi}{12} < x < \dfrac{7\pi}{4}$, find $\dfrac{\sin 2x + 2\sin^2 x}{1 - \tan x}$.

Solution:

$$\dfrac{\sin 2x + 2\sin^2 x}{1 - \tan x} = \dfrac{2\sin x \cos x(1 + \tan x)}{1 - \tan x} = \sin 2x \tan \left(\dfrac{\pi}{4} + x \right)$$

$$\sin 2x = \sin \left[2\left(\dfrac{\pi}{4} + x \right) - \dfrac{\pi}{2} \right] = -\cos 2\left(\dfrac{\pi}{4} + x \right)$$

$$= -\left[2\cos^2 \left(\dfrac{\pi}{4} + x \right) - 1 \right] = \dfrac{7}{25}$$

$$\tan \left(\dfrac{\pi}{4} + x \right) = \dfrac{\sin \left(\frac{\pi}{4} + x \right)}{\cos \left(\frac{\pi}{4} + x \right)} = -\dfrac{4}{3}.$$

So

$$\dfrac{\sin 2x + 2\sin^2 x}{1 - \tan x} = \dfrac{7}{25}\left(-\dfrac{4}{3} \right) = -\dfrac{28}{75}.$$

2. If $\sin \alpha + \sin \beta = m, \cos \alpha + \cos \beta = n(mn \neq 0)$, find
(1) $\cos(\alpha - \beta)$; (2) $\sin(\alpha + \beta)$; (3) $\tan(\alpha + \beta)$.

Solution:

(1) $(\sin \alpha + \sin \beta)^2 + (\cos \alpha + \cos \beta)^2$

$$= 2 + 2(\cos \alpha \cos \beta + \sin \alpha \sin \beta) = m^2 + n^2$$

$$\cos(\alpha - \beta) = \frac{m^2 + n^2}{2} - 1.$$

(2) $\dfrac{\sin \alpha + \sin \beta}{\cos \alpha + \cos \beta} = \dfrac{2 \sin \frac{\alpha+\beta}{2} \cos \frac{\alpha-\beta}{2}}{2 \cos \frac{\alpha+\beta}{2} \cos \frac{\alpha-\beta}{2}} = \tan \dfrac{\alpha + \beta}{2} = \dfrac{m}{n}.$

According to the universal replacement formula

$$\sin(\alpha + \beta) = \frac{2\frac{m}{n}}{1 + \left(\frac{m}{n}\right)^2} = \frac{2mn}{m^2 + n^2}.$$

(3) $\tan(\alpha + \beta) = \dfrac{2\frac{m}{n}}{1 - \left(\frac{m}{n}\right)^2} = \dfrac{2mn}{n^2 - m^2}.$

3. If $\sin \alpha + \sin \beta = \frac{1}{4}, \tan(\alpha + \beta) = \frac{24}{7}$, find $\cos \alpha + \cos \beta$.

Solution:

$$\because \sin \alpha + \sin \beta = \frac{1}{4}, \therefore 2 \sin \frac{\alpha + \beta}{2} \cos \frac{\alpha - \beta}{2} = \frac{1}{4}. \qquad (1)$$

$$\text{Suppose } \cos \alpha + \cos \beta = \alpha, \therefore 2 \cos \frac{\alpha + \beta}{2} \cos \frac{\alpha - \beta}{2} = \alpha. \quad (2)$$

$$\text{If } \alpha = 0, \therefore \cos \frac{\alpha + \beta}{2} = 0,$$

$$\therefore \alpha + \beta = 2k\pi + \pi, \tan(\alpha + \beta) = 0 \text{ (contradict with condition)}$$

$\because$ $(1) \div (2) \Rightarrow \tan \dfrac{\alpha + \beta}{2} = \dfrac{1}{4\alpha}$

$\because$ $\tan(\alpha + \beta) = \dfrac{24}{7}$

$\therefore$ $\dfrac{2 \tan \frac{\alpha+\beta}{2}}{1 - \tan^2 \frac{\alpha+\beta}{2}} = \dfrac{2 \times \frac{1}{4\alpha}}{1 - \frac{1}{16\alpha^2}} = \dfrac{8\alpha}{16\alpha^2 - 1} = \dfrac{24}{7}$

$\therefore$ $\alpha = \dfrac{1}{3}$ or $-\dfrac{3}{16}$

$\therefore$ $\cos \alpha + \cos \beta = \dfrac{1}{3}$ or $-\dfrac{3}{16}$.

4. In cartesian coordinate, $A(\cos \alpha, \sin \alpha)$, $B(\cos \beta, \sin \beta)$, $C\left(\frac{4\sqrt{3}}{3}, 2\sqrt{2}\right)$; $\left(\frac{2}{3}\sqrt{3}, \sqrt{2}\right)$ is center of gravity of $\triangle ABC$, find $\cos(\alpha + \beta)$ and $\tan \alpha + \tan \beta$.

Solution: According to the condition

$$\begin{cases} \dfrac{2}{3}\sqrt{3} = \dfrac{\cos \alpha + \cos \beta + \dfrac{4}{3}\sqrt{3}}{3} \\[4mm] \sqrt{2} = \dfrac{\sin \alpha + \sin \beta + 2\sqrt{2}}{3} \end{cases},$$

$\therefore$ $\begin{cases} \sin \alpha + \sin \beta = \sqrt{2}, & \text{(3)} \\[2mm] \cos \alpha + \cos \beta = \dfrac{2}{3}\sqrt{3} & \text{(4)} \end{cases}$

$(3)^2 + (4)^2 = 2 + 2\cos(\alpha - \beta) = 2 + \dfrac{4}{3}$, so $\cos(\alpha - \beta) = \dfrac{2}{3}$

$(3) \div (4) = \tan \dfrac{\alpha + \beta}{2} = \dfrac{\sqrt{6}}{2}$

$\therefore$ $\sin(\alpha + \beta) = \dfrac{2 \times \frac{\sqrt{6}}{2}}{1 + \frac{6}{4}} = \dfrac{2}{5}\sqrt{6}$, $\cos(\alpha + \beta) = \dfrac{1 - \frac{6}{4}}{1 + \frac{6}{4}} = -\dfrac{1}{5}$

$$\therefore \ \tan\alpha + \tan\beta = \frac{\sin(\alpha+\beta)}{\cos\alpha\cos\beta} = \frac{2\sin(\alpha+\beta)}{\cos(\alpha+\beta) + \cos(\alpha-\beta)}$$

$$= \frac{2 \times \frac{2}{5}\sqrt{6}}{-\frac{1}{5} + \frac{2}{3}} = \frac{12}{7}\sqrt{6}.$$

5. If $\sin A + \sin B = \sin C$, $\cos A + \cos B = \cos C$, find $\sin^2 A + \sin^2 B + \sin^2 C$.

Solution:

$$\because \ \sin A + \sin B = \sin C, \tag{5}$$

$$\cos A + \cos B = \cos C \tag{6}$$

$$\because \ (5)^2 + (6)^2 = 2\cos(A-B) = -1$$

$$\therefore \ \cos(A-B) = -\frac{1}{2}$$

$$\because \ (6)^2 - (5)^2 = \cos(A+B) = \cos 2C$$

$$\therefore \ \sin^2 A + \sin^2 B + \sin^2 C$$

$$= \frac{1-\cos 2A}{2} + \frac{1-\cos 2B}{2} + \frac{1-\cos 2C}{2}$$

$$= \frac{3}{2} - \frac{1}{2}(\cos 2A + \cos 2B + \cos 2C)$$

$$= \frac{3}{2} - \cos(A+B)\cos(A-B) - \frac{1}{2}\cos 2C$$

$$= \frac{3}{2} - \left(-\frac{1}{2}\right)\cos 2C - \frac{1}{2}\cos 2C = \frac{3}{2}.$$

6. If $\sin x + \sin y + \sin z = \cos x + \cos y + \cos z = 0$, find $S = \tan(x+y+z) + \tan x \tan y \tan z$.

Solution: According to the condition

$$\begin{cases} \sin x + \sin y = -\sin z, & (7) \\ \cos x + \cos y = -\cos z, & (8) \end{cases}$$

$$\because (7)^2 + (8)^2 = 2 + 2\cos x \cos y + 2\sin x \sin y$$

$$\therefore \cos(x - y) = -\frac{1}{2}$$

$$\therefore \cos(y - z) = -\frac{1}{2}, \quad \cos(z - x) = -\frac{1}{2}$$

Suppose

$$x = y + \frac{2}{3}\pi + 2k_1\pi, \quad k_1 \in Z,$$

$$y = z + \frac{2}{3}\pi + 2k_2\pi, \quad k_2 \in Z$$

$$x = z + \frac{4}{3}\pi + 2(k_1 + k_2)\pi,$$

$$x + y + z = 3z + 2(k_1 + 2k_2 + 1)\pi$$

$$\therefore S = \tan(x + y + z) + \tan x \tan y \tan z$$

$$= \tan 3z + \tan\left(z + \frac{4}{3}\pi\right)\tan\left(z + \frac{2}{3}\pi\right)\tan z$$

$$= \tan 3z + \tan\left(z + \frac{1}{3}\pi\right)\tan\left(z - \frac{1}{3}\pi\right)\tan z$$

$$= \tan 3z - \tan z \tan\left(z + \frac{1}{3}\pi\right)\tan\left(\frac{\pi}{3} - z\right) = 0.$$

7. Find $\cos^4 20° + \cos^4 40° + \cos^4 80°$.

Solution:

$$\because \cos 2\theta = 2\cos^2\theta - 1$$

$$\therefore \cos^4\theta = \left(\frac{1 + \cos 2\theta}{2}\right)^2 = \frac{1}{4}(1 + 2\cos 2\theta + \cos^2 2\theta)$$

$$= \frac{1}{4} + \frac{1}{2}\cos 2\theta + \frac{1}{4} \cdot \frac{1+\cos 4\theta}{2}$$

$$= \frac{3}{8} + \frac{1}{2}\cos 2\theta + \frac{1}{8}\cos 4\theta$$

$\therefore \ \cos^4 20° + \cos^4 40° + \cos^4 80°.$

$$= \frac{3}{8} \times 3 + \frac{1}{2}(\cos 40° + \cos 80° + \cos 160°)$$

$$+ \frac{1}{8}(\cos 80° + \cos 160° + \cos 320°)$$

$$= \frac{9}{8} + \frac{5}{8}(\cos 40° + \cos 80° + \cos 160°)$$

$$= \frac{9}{8} + \frac{5}{8}(2\cos 60° \cos 20° - \cos 20°) = \frac{9}{8}.$$

8. Whether there are acute angles α, β which satisfy (1) $\alpha + 2\beta = \frac{2\pi}{3}$; (2) $\tan\frac{\alpha}{2}\tan\beta = 2 - \sqrt{3}$. If α, β are obtained, find them. If not, give the reason.

Solution: $\because \ \dfrac{\alpha}{2} + \beta = \dfrac{\pi}{3}$

$$\therefore \ \tan\left(\frac{\alpha}{2} + \beta\right) = \frac{\tan\frac{\alpha}{2} + \tan\beta}{1 - \tan\frac{\alpha}{2}\tan\beta} = \sqrt{3}$$

$$\because \ \tan\frac{\alpha}{2}\tan\beta = 2 - \sqrt{3}$$

$$\therefore \ \tan\frac{\alpha}{2} + \tan\beta = 3 - \sqrt{3}$$

$$\therefore \ \tan\frac{\alpha}{2} \ \text{and} \ \tan\beta \ \text{are two roots of}$$

$$x^2 - (3 - \sqrt{3})x + 2 - \sqrt{3} = 0$$

$$\therefore \ x_1 = 1, x_2 = 2 - \sqrt{3}$$

$$\because \ 0 < \frac{\alpha}{2} < \frac{\pi}{4}$$

$$\therefore \ \tan \frac{\alpha}{2} \neq 1$$

$$\therefore \ \tan \frac{\alpha}{2} = 2 - \sqrt{3}, \ \tan \beta = 1$$

$$\therefore \ \alpha = 30°, \ \beta = 45°.$$

9. Find $\sin 10° \sin 50° \sin 70°$.

Analysis: Because $\sin 10° \ \sin 50° \ \sin 70° = \cos 80° \ \cos 40° \ \cos 20°$ and $20°, 40°, 80°$ are multiple relationship, consider the double angles formula: $\sin 10° \sin 50° \sin 70°$ and $\cos 10° \ \cos 50° \ \cos 70°$ are dual form, so solve them by matching method.
In addition, $10°, 50°, 70°$ are $10°, 60° - 10°, 60° + 10°$ and triple angles formula:
$\sin 3\theta = 3 \sin \theta - 4 \sin^3 \theta = 4 \sin \theta \sin(60° - \theta) \sin(60° + \theta)$, so find the solution.

Solution 1:

$$\sin 10° \sin 50° \sin 70°$$

$$= \cos 20° \cos 40° \cos 80°$$

$$= \frac{\sin 20° \cos 20° \cos 40° \cos 80°}{\sin 20°}$$

$$= \frac{\frac{1}{8} \sin 160°}{\sin 20°} = \frac{1}{8}.$$

Solution 2: Suppose $A = \sin 10° \sin 50° \sin 70°$,
$B = \cos 10° \cos 50° \cos 70°$

$$AB = \sin 10° \cos 10° \sin 50° \cos 50° \sin 70° \cos 70°$$

$$= \frac{1}{8} \sin 20° \sin 100° \sin 140°$$

$$= \frac{1}{8} \cos 70° \cos 10° \cos 50°$$

$$= \frac{1}{8} B$$

$$\therefore A = \frac{1}{8}.$$

Solution 3: $\sin 10° \sin 50° \sin 70°$

$$= \sin 10° \sin(60° - 10°) \sin(60° + 10°)$$

$$= \frac{1}{4} \sin 30°$$

$$= \frac{1}{8}.$$

Comment: Solution 1 has generality. Solve the question $\cos \theta$ $\cos 2\theta \cos 2^2\theta \cdots \cos 2^n\theta = \frac{1}{2^{n+1}} \frac{\sin 2^{n+1}\theta}{\sin \theta}$ by this method. Solution 3 uses $\sin \theta \sin(60° - \theta) \sin(60° + \theta) = \frac{1}{4} \sin 3\theta$. Similarly, $\cos \theta \cos(60° - \theta) \cos(60° + \theta) = \frac{1}{4} \cos 3\theta$ and $\tan \theta$ $\tan(60° - \theta) \tan(60° + \theta) = \tan 3\theta$.

10. Prove $\sin 1° \sin 2° \sin 3° \cdots \sin 89° = (\frac{1}{4})^{45} \cdot 6\sqrt{10}$.

Proof:

$$\sin 1° \sin 2° \sin 3° \cdots \sin 89°$$

$$= (\sin 1° \sin 59° \sin 61°)(\sin 2° \sin 58° \sin 62°)$$

$$\times \cdots (\sin 29° \sin 31° \sin 89°) \cdot \sin 30° \sin 60°$$

$$= \left(\frac{1}{4}\right)^{30} \cdot \sqrt{3}(\sin 3° \sin 6° \sin 9° \cdots \sin 87°)$$

$$= \left(\frac{1}{4}\right)^{30} \cdot \sqrt{3}(\sin 3° \sin 57° \sin 63°)(\sin 6° \sin 54° \sin 66°)$$

$$\times \cdots (\sin 27° \sin 33° \sin 87°) \cdot \sin 30° \sin 60°$$

$$= \left(\frac{1}{4}\right)^{40} \cdot 3(\sin 9° \sin 18° \sin 27° \cdots \sin 27° \sin 81°)$$

$$\sin 9° \sin 18° \sin 27° \cdots \sin 72° \sin 81°$$

$$= (\sin 9° \cos 9°)(\sin 18° \cos 18°)$$

$$\times (\sin 27° \cos 27°)(\sin 36° \cos 36°) \sin 45°$$

$$= \left(\frac{1}{4}\right)^{2} \cdot \frac{\sqrt{2}}{2} \sin 18° \sin 36° \sin 54° \sin 72°$$

$$= \left(\frac{1}{4}\right)^{3} \cdot \frac{\sqrt{2}}{2} \sin 36° \sin 72°.$$

Suppose

$$x = \sin 36° \sin 72°,$$

$$\because \cos 36° \cos 72° = \frac{1}{4}$$

$$\therefore \begin{cases} x + \dfrac{1}{4} = \cos 36° \\[2mm] x - \dfrac{1}{4} = \cos 72° \end{cases}$$

$$\therefore x^2 - \frac{1}{16} = \cos 36° \cos 72° = \frac{1}{4}, \quad x^2 = \frac{5}{16},$$

$$\therefore x = \frac{\sqrt{5}}{4}$$

$$\therefore \sin 1° \sin 2° \sin 3° \cdots \sin 89° = \left(\frac{1}{4}\right)^{45} \cdot 6\sqrt{10}.$$

□

11. Find $\cos \dfrac{2\pi}{5} + \cos \dfrac{4\pi}{5}$.

Solution 1:

$$\cos\frac{2\pi}{5} + \cos\frac{4\pi}{5} = \frac{2\sin\frac{\pi}{5}\left(\cos\frac{2}{5}\pi + \cos\frac{4}{5}\pi\right)}{2\sin\frac{\pi}{5}}$$

$$= \frac{1}{2\sin\frac{\pi}{5}}\left(\sin\frac{3}{5}\pi - \sin\frac{\pi}{5} + \sin\pi - \sin\frac{3}{5}\pi\right)$$

$$= \frac{1}{2\sin\frac{\pi}{5}}\cdot\left(-\sin\frac{\pi}{5}\right)$$

$$= -\frac{1}{2}.$$

Solution 2:

Suppose $x = \cos\frac{2}{5}\pi + \cos\frac{4}{5}\pi, y = \cos\frac{2}{5}\pi - \cos\frac{4}{5}\pi$

$$\therefore xy = \cos^2\frac{2}{5}\pi - \cos^2\frac{4}{5}\pi = \frac{1}{2}\left(1 + \cos\frac{4}{5}\pi\right)$$

$$-\frac{1}{2}\left(1 + \cos\frac{8}{5}\pi\right)$$

$$= \frac{1}{2}\left(\cos\frac{4}{5}\pi - \cos\frac{8}{5}\pi\right) = -\frac{1}{2}y$$

$$\because y \neq 0 \quad \therefore x = -\frac{1}{2}.$$

Solution 3:

Suppose: $\cos x + \cos 2x = \cos\frac{2}{5}\pi + \cos\frac{4}{5}\pi$

$$\therefore x = \frac{2}{5}\pi, \frac{4}{5}\pi$$

$$\because 2\cos^2 x + \cos x - \left(1 + \cos\frac{2}{5}\pi + \cos\frac{4}{5}\pi\right) = 0$$

$$\therefore \cos\frac{2}{5}\pi \quad \text{and} \quad \cos\frac{4}{5}\pi \text{ are two roots of equation}$$

$$2y^2 + y - \left(1 + \cos \frac{2}{5}\pi + \cos \frac{4}{5}\pi\right) = 0$$

$$\therefore \ \cos \frac{2}{5}\pi + \cos \frac{4}{5}\pi = -\frac{1}{2}.$$

Comment: This solution uses sines of subtraction of $\frac{4}{5}\pi$ and $\frac{2}{5}\pi$, and then eliminate several items after product-to-sum. Finally we get the result. Generally,

$$\sin(\alpha + \beta) + \sin(\alpha + 2\beta) + \cdots + \sin(\alpha + n\beta)$$

$$= \frac{\sin \frac{n\beta}{2} \sin \left(\alpha + \frac{n+1}{2}\beta\right)}{\sin \frac{\beta}{2}}$$

$$\cos(\alpha + \beta) + \cos(\alpha + 2\beta) + \cdots + \cos(\alpha + n\beta)$$

$$= \frac{\cos \frac{n\beta}{2} \cos \left(\alpha + \frac{n+1}{2}\beta\right)}{\cos \frac{\beta}{2}}.$$

12. If $\frac{\cos^4 A}{\cos^2 B} + \frac{\sin^4 A}{\sin^2 B} = 1$, prove $\frac{\cos^4 B}{\cos^2 A} + \frac{\sin^4 B}{\sin^2 A} = 1$.

Analysis: Exchange the A and B in condition, which is the conclusion to proof, and so consider to prove $\cos^2 A = \cos^2 B$, $\sin^2 A = \sin^2 B$. Note the conditions belong to the type $a^2 + b^2 = 1$, so use substitution method.

Proof: Suppose

$$\frac{\cos^2 A}{\cos B} = \cos \varphi, \ \frac{\sin^2 A}{\sin B} = \sin \varphi$$

$$\therefore \ \cos^2 A = \cos B \cos \varphi, \tag{a}$$

$$\sin^2 A = \sin B \sin \varphi \tag{b}$$

(a) + (b) $\Rightarrow 1 = \cos B \cos \varphi + \sin B \sin \varphi = \cos(B - \varphi)$

$$\therefore \ B - \varphi = 2k\pi, \quad (k \in Z)$$

$$\therefore \ \cos B = \cos \varphi, \quad \sin B = \sin \varphi$$

$$\therefore \ \cos^2 A = \cos^2 B, \sin^2 A = \sin^2 B$$

$$\therefore \ \frac{\cos^4 B}{\cos^2 A} + \frac{\sin^4 B}{\sin^2 A} = \cos^2 B + \sin^2 B = 1.$$

□

13. Prove $2 \sin^4 x + \frac{3}{4} \sin^2 2x + 5 \cos^4 x - \cos 3x \cos x = 2(1 + \cos^2 x)$.

Analysis: This problem is complex on the surface. There are angles x, $2x$, $3x$ and functions sines, cosines. Consider translate function of angle x and $2x$ to $\cos x$ and $\sin x$. In addition, it can be considered to be unified into $2x$ to achieve the purpose of reducing the power.

Proof 1:

$$\text{left} = 2\sin^4 x + 5\cos^4 x + \frac{3}{4}(2\sin x \cos x)^2$$

$$-(4\cos^3 x - 3\cos x)\cos x$$

$$= 2\sin^4 x + 5\cos^4 x + 3\sin^2 x \cos^2 x - 4\cos^4 x + 3\cos^3 x$$

$$= 2\sin^4 x - 2\cos^4 x \cos + 6\cos^2 x$$

$$= 2(1 - 2\cos^2 x) + 6\cos^2 x$$

$$= 2(1 + \cos^2 x) = \text{right}.$$

□

Proof 2:

$$\text{left} = 2\left(\frac{1 - \cos 2x}{2}\right)^2 + \frac{3}{4}(1 - \cos^2 2x) + 5\left(\frac{1 + \cos 2x}{2}\right)^2$$

$$-\frac{1}{2}(2\cos^2 2x - 1 + \cos 2x)$$

$$= 3 + \cos 2x = 2(1 + \cos^2 x) = \text{right}.$$

□

14. Prove $\tan \alpha + 2 \tan 2\alpha + 4 \tan 4\alpha + 8 \cot 8\alpha = \cot \alpha$.

Proof:

$$\because \ \tan 2\alpha = \frac{2\tan\alpha}{1-\tan^2\alpha}$$

$$\therefore \ \cot 2\alpha = \frac{1}{2}(\cot\alpha - \tan\alpha)$$

$$\therefore \ \tan\alpha = \cot\alpha - 2\cot 2\alpha \tag{9}$$

$$\therefore \ \tan 2\alpha = \cot 2\alpha - 2\cot 4\alpha \tag{10}$$

$$\therefore \ \tan 4\alpha = \cot 4\alpha - 2\cot 8\alpha \tag{11}$$

$$(9) + (10)\times 2 + (11)\times 4 \Rightarrow \tan\alpha + 2\tan 2\alpha$$

$$+\, 4\tan 4\alpha = \cot\alpha - 8\cot 8\alpha.$$

$$\square$$

Comment: This method can be used to prove more general conclusions:
$\tan x + 2\tan 2x + 2^2\tan 2^2 x + \cdots + 2^n\tan 2^n x = \cot x - 2^{n+1}\tan 2^{n+1}x.$

15. If $\frac{\sin^4 x}{a} + \frac{\cos^4 x}{b} = \frac{1}{a+b}(a>0, b>0)$, prove

$$\frac{\sin^8 x}{a^3} + \frac{\cos^8 x}{b^3} = \frac{1}{(a+b)^3} \quad (a>0, b>0).$$

Proof 1: Suppose $\sin^2 x = u, \cos^2 x = v$,

$$\therefore \begin{cases} u + v = 1 & (12) \\[2mm] \dfrac{u^2}{a} + \dfrac{v^2}{b} = \dfrac{1}{a+b} & (13) \end{cases}$$

$(12) \Rightarrow v = 1 - u$; put into (13), we get

$$\therefore \ \frac{u^2}{a} + \frac{(1-u)^2}{b} = \frac{1}{a+b}$$

$$\therefore (a+b)^2 u^2 - 2a(a+b)u + a^2 = 0$$

$$\therefore u = \frac{a}{a+b}$$

$$\therefore v = 1 - u = \frac{b}{a+b}$$

$$\therefore \frac{\sin^8 x}{a^3} + \frac{\cos^8 x}{b^3} = \frac{1}{a^3}\left(\frac{a}{a+b}\right)^4 + \frac{1}{b^3}\left(\frac{b}{a+b}\right)^4 = \frac{1}{(a+b)^3}.$$

□

Proof 2:

$$(a+b)\left(\frac{\sin^4 x}{a} + \frac{\cos^4 x}{b}\right) \geq (\sin^2 x + \cos^2 x)^2$$

$$= 1 \Rightarrow \left(\frac{\sin^4 x}{a} + \frac{\cos^4 x}{b}\right) \geq \frac{1}{a+b}$$

(1) $\because \dfrac{\sin^4 x}{a} + \dfrac{\cos^4 x}{b} = \dfrac{1}{a+b}$ $\quad (a>0, b>0),$

$$\therefore \frac{a}{\sin^2 x} = \frac{b}{\cos^2 x} = k, \quad \therefore a+b = k,$$

$$\therefore \text{left} = \frac{\sin^8 x}{k^3 \sin^6 x} + \frac{\cos^8 x}{k^3 \cos^6 x}$$

$$= \frac{\sin^2 x + \cos^2 x}{k^3} = \frac{1}{k^3} = \frac{1}{(a+b)^3}.$$

□

16. Suppose α, β are acute angles and $\sin^2 \alpha + \sin^2 \beta = \sin(\alpha + \beta)$, prove $\alpha + \beta = \dfrac{\pi}{2}$.

Proof 1:

$$\sin^2 \alpha - \sin \alpha \cos \beta + \sin^2 \beta - \cos \alpha \sin \beta = 0,$$

$$\sin \alpha(\sin \alpha - \cos \beta) + \sin \beta(\sin \beta - \cos \alpha) = 0$$

$$\frac{\sin\alpha(\sin^2\alpha - \cos^2\beta)}{\sin\alpha + \cos\beta} + \frac{\sin\beta(\sin^2\beta - \cos^2\alpha)}{\sin\beta + \cos\alpha} = 0,$$

$$\because \sin^2\alpha - \cos^2\beta = \sin^2\beta - \cos^2\alpha$$

$$\therefore (\sin^2\alpha - \cos^2\beta)\left(\frac{\sin\alpha}{\sin\alpha + \cos\beta} + \frac{\sin\beta}{\sin\beta + \cos\alpha}\right) = 0$$

$$\because \frac{\sin\alpha}{\sin\alpha + \cos\beta} + \frac{\sin\beta}{\sin\beta + \cos\alpha} \neq 0, \therefore \sin^2\alpha - \cos^2\beta = 0,$$

$$\therefore \sin\alpha - \cos\beta = 0, \sin\alpha = \sin\left(\frac{\pi}{2} - \beta\right)$$

$$\therefore \alpha + \beta = \frac{\pi}{2} \qquad\qquad\qquad \square$$

Proof 2:

$\because \alpha, \beta$ are acute angles,

$\therefore \cos(\alpha - \beta) > 0$

$$\sin(\alpha + \beta) = \sin^2\alpha + \sin^2\beta = \frac{1 - \cos 2\alpha}{2} + \frac{1 - \cos 2\beta}{2}$$

$$= 1 - \cos(\alpha + \beta)\cos(\alpha - \beta) \qquad\qquad (14)$$

$\because 0 < \sin(\alpha + \beta) \leq 1 \therefore 0 \leq \cos(\alpha + \beta) \cdot \cos(\alpha - \beta) < 1$

①+③$\Rightarrow \cos(\alpha + \beta) \geq 0$

$\because 0 < \alpha + \beta \leq \dfrac{\pi}{2} \quad \therefore 0 \leq |\alpha - \beta| < \alpha + \beta \leq \dfrac{\pi}{2}$

$\because 0 \leq \cos(\alpha + \beta) < \cos(\alpha + \beta).$

Substituting into (14)

$$\Rightarrow 0 \leq \sin(\alpha + \beta) \leq 1 - \cos^2(\alpha + \beta) = \sin^2(\alpha + \beta)$$

$\therefore \sin(\alpha + \beta) \geq 1$

Only $\sin(\alpha + \beta) = 1$

$$\therefore\ \alpha + \beta = \frac{\pi}{2}. \qquad\qquad\qquad \square$$

Exercises 2

1. Suppose $\sin\alpha + \cos\alpha = -\frac{1}{5}$, $\alpha \in (\frac{3\pi}{2}, 2\pi)$, find
   $$\frac{2\sin\alpha}{\cos\alpha - \sin\alpha - \cos 3\alpha + \sin 3\alpha}.$$

2. In $\triangle ABC$, $\cos A = \frac{12}{13}$, $\sin B = \frac{3}{5}$, find $\sin C$.

3. Suppose $\sin\alpha = \frac{12}{13}, \sin(\alpha + \beta) = \frac{4}{5}, \alpha, \beta \in (0, \frac{\pi}{2})$, find $\cos\frac{\beta}{2}$.

4. Find $\cos^2 x + \cos^2(x + \frac{2\pi}{3}) + \cos^2(x + \frac{4\pi}{3})$.

5. If $\cos\left(a + \frac{\pi}{4}\right) = \frac{3}{5}, \frac{\pi}{2} \le a \le \frac{3\pi}{2}$, find $\cos\left(2a + \frac{\pi}{4}\right)$.

6. If $\tan(\alpha - \beta) = \frac{1}{2}, \tan\beta = -\frac{1}{7}$, $\alpha, \beta \in (0, \pi)$, find $2\alpha - \beta$.

7. In $\triangle ABC$, $AB = \frac{4\sqrt{6}}{3}$, $\cos B = \frac{\sqrt{6}}{6}$, $BD = \sqrt{5}$, which is the median of the triangle. Find $\sin A$.

8. Find period of function $f(x) = \sin\left(x + \frac{\pi}{3}\right)\sin\left(x + \frac{\pi}{2}\right)$.

9. If $\frac{3\pi}{4} < \alpha < \pi, \tan\alpha + \cot\alpha = -\frac{10}{3}$, find
   $$\frac{5\sin^2\frac{\alpha}{2} + 8\sin\frac{\alpha}{2}\cos\frac{\alpha}{2} + 11\cos^2\frac{\alpha}{2} - 8}{\sqrt{2}\sin(\alpha - \frac{\pi}{2})}.$$

10. In $\triangle ABC$, $\cot A + \cot B + \cot C = \sqrt{3}$, judge the shape of $\triangle ABC$.

11. In isosceles trapezoid $ABCD$, $AD//BC$, height is 5, lower bottom $BC = 5$, radius of $\odot O$ is 1, which remains a tangent to the bottom-up position and two equal sides, the base angles are ϕ. Find $\sin\phi$. (See Fig. 2.2.)

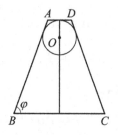

Fig. 2.2.

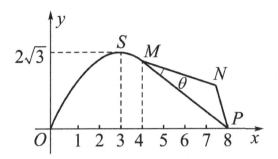

Fig. 2.3.

12. Find the maximum and minimum of function $y = 7 - 4\sin x \cos x + 4\cos^2 x - 4\cos^4 x$.

13. If $-\frac{\pi}{2} \le x \le \frac{\pi}{2}$, find the maximum and minimum of $f(x) = \sqrt{3}\sin x + \cos x$.

14. As shown in Fig. 2.3, the city plans to build a sports lane on road OP of the length of $8\,\mathrm{km}$. The former part of the lane is the curved line OSM, which is the image for the function $y = A\sin\omega x (A > 0, \omega > 0)$, $x \in [0, 4]$. The top of the image is $S\,(3, 2\sqrt{3})$. The latter part of the lane is the folding line MNP. To ensure the athlete's safety, limit the angle $MNP = 120°$.

(1) Find the distance between M and P and the value of A, ω.
(2) How to design to make the length of the folding line MNP to its utmost.

15. Prove $\sum_{k=0}^{88} \frac{1}{\cos k° \cos(k+1)°} = \frac{\cos 1°}{\sin^2 1°}$.

16. Suppose $n \in N^*$, prove $\prod_{k=1}^{n} \cos \frac{k\pi}{2n+1} = \frac{1}{2^n}$.

17. Prove: $\sum_{k=1}^{n} \left(\frac{1}{2^k} \tan \frac{x}{2^k} \right) = \frac{1}{2^n} \cot \frac{x}{2^n} - \cot x$.

18. Suppose integers a and b meet $\sqrt{9-8\sin 50°} = a + b\csc 50°$, find a and b.

19. In $\triangle ABC$, A, B, C are the corresponding angles of side a, b, c. If $a^2 + b^2 - c^2 = ab$, $\frac{\tan A - \tan B}{\tan A + \tan B} = \frac{c-b}{c}$, prove $\triangle ABC$ is equilateral triangle.

20. If $a, b \in N^*$, and $a > b$, $\sin\theta = \frac{2ab}{a^2+b^2} (\theta \in (0, \frac{\pi}{2}))$, $A_n = (a^2 + b^2)^n \sin n\theta$. Prove, to all positive integers n, A_n is integer.

Chapter 3

Trigonometric Functions

3.1. Graphs and Properties of Trigonometric Functions

Graphs and properties of sine, cosine, tangent and cotangent functions:

1. Definition: $y = \sin x$, $x \in R$, is named sine function; $y = \cos x$, $x \in R$, is named cosine function. $y = \tan x$, $x \neq k\pi + \frac{\pi}{2}(k \in Z)$, is named tangent function; and $y = \cot x$, $x \neq k\pi$ $(k \in Z)$, is named cotangent function.

Sine, cosine, tangent and cotangent functions are all named trigonometric functions.

2. Primary properties of sine, cosine, tangent and cotangent functions

Please see the table in next page.

Exercise

1. Find the domain of function $y = \sqrt{-\tan x - 1} + \frac{\sqrt{16-x^2}}{1 - \log_{\frac{\sqrt{3}}{2}} \sin x}$.

Solution:

$$\begin{cases} -\tan x - 1 \geq 0, & (1) \\ 16 - x^2 \geq 0, & (2) \\ 1 - \log_{\frac{\sqrt{3}}{2}} \sin x \neq 0, & (3) \\ \sin x > 0. & (4) \end{cases}$$

$$\because \ (1) \Rightarrow k\pi + \frac{\pi}{2} < x \leq k\pi + \frac{3}{4}\pi, k \in Z$$

$$(2) \Rightarrow -4 \leq x \leq 4$$

Function	$y = \sin x$	$y = \cos x$	$y = \tan x$	$y = \cot x$
Domain	R	R	$\{x \mid x \in R$ and $x \neq k\pi + \frac{\pi}{2},\ k \in Z\}$	$\{x \mid x \in R$ and $x \neq k\pi,\ k \in Z\}$
Range	$\{y \mid \|y\| \leq 1\}$	$\{y \mid \|y\| \leq 1\}$	R	R
Parity	Odd function	Even function	Odd function	Odd function
Monotonicity	$[2k\pi - \frac{\pi}{2}, 2k\pi + \frac{\pi}{2}]$ monotone increasing $[2k\pi + \frac{\pi}{2}, 2k\pi + \frac{3\pi}{2}]$ monotone decreasing $(k \in Z)$	$[2k\pi - \pi, 2k\pi]$ monotone increasing $[2k\pi, 2k\pi + \pi]$ monotone decreasing $(k \in Z)$	$(k\pi - \frac{\pi}{2}, k\pi + \frac{\pi}{2})$ monotone increasing $(k \in Z)$	$(k\pi, k\pi + \pi)$ monotone decreasing $(k \in Z)$
Periodicity	$T = 2\pi$	$T = 2\pi$	$T = \pi$	$T = \pi$
Graph				

(3) $\Rightarrow x \neq 2k\pi + \dfrac{\pi}{3}$ and $x \neq 2k\pi + \dfrac{2}{3}\pi, k \in Z$

(4) $\Rightarrow 2k\pi < x < (2k+1)\pi, k \in Z$

$$\therefore \quad x \in \left[-4, -\dfrac{5}{4}\pi\right] \cup \left(\dfrac{\pi}{2}, \dfrac{2}{3}\pi\right) \cup \left(\dfrac{2}{3}\pi, \dfrac{3}{4}\pi\right].$$

Comment: Draw all solutions on the number axis, and find intersection sets by combination of quantities and spatial forms.

2. If $f(x) = \dfrac{\sin x}{|\sin x|} + \dfrac{\cos x}{|\cos x|} + \dfrac{\tan x}{|\tan x|} + \dfrac{\cot x}{|\cot x|} + \dfrac{\sin x}{|\csc x|} + \dfrac{\cos x}{|\sec x|}$, find the range of $f(x)$.

Solution: According to the condition $x \neq k\pi, x \neq k\pi + \frac{\pi}{2}$ $(k \in Z)$

$$x \in \mathrm{I} \Rightarrow y = f(x) = 5, \quad x \in \mathrm{II} \Rightarrow y = f(x) \in (-3, -1)$$

$$x \in \mathrm{III} \Rightarrow y = f(x) = -1, \quad x \in \mathrm{IV} \Rightarrow y = f(x) \in (-3, -1)$$

$$y = f(x) = (-3, -1] \cup \{5\}.$$

3. If $y = \sin x + \sqrt{1 + \cos^2 x}$, find the maximum and minimum.

Analysis: Use the change element method according to

$$(\sin x)^2 + (\sqrt{1 + \cos^2 x})^2 = 2 \Rightarrow \left(\dfrac{\sin x}{\sqrt{2}}\right)^2 + \left(\dfrac{\sqrt{1 + \cos^2 x}}{\sqrt{2}}\right)^2 = 1.$$

Solution 1: Suppose $\sin x = \sqrt{2}\cos\theta$, $\sqrt{1 + \cos^2 x} = \sqrt{2}\sin\theta$ $(\frac{\pi}{4} \leq \theta \leq \frac{3}{4}\pi)$,

$$\therefore \quad y = \sqrt{2}\cos\theta + \sqrt{2}\sin\theta = 2\sin\left(\theta + \dfrac{\pi}{4}\right)$$

$$\because \ \dfrac{\pi}{4} \leq \theta \leq \dfrac{3}{4}\pi \quad \therefore \ \dfrac{\pi}{2} \leq \theta + \dfrac{\pi}{4} \leq \pi \quad \therefore \ 0 \leq \sin\left(\theta + \dfrac{\pi}{4}\right) \leq 1$$

$$\theta = \dfrac{3}{4}\pi \Rightarrow x = 2k\pi - \dfrac{\pi}{2}(k \in Z) \quad \therefore \ y_{\min} = 0$$

$$\theta = \dfrac{1}{4}\pi \Rightarrow x = 2k\pi + \dfrac{\pi}{2} \ (k \in Z) \quad \therefore \ y_{\max} = 2.$$

Solution 2:

$$y = \sin x + \sqrt{1 + \cos^2 x} \leq \sqrt{2[\sin^2 x + (\sqrt{1 + \cos^2 x})^2]} = 2$$

$$\text{and } |\sin x| \leq 1 \leq \sqrt{1 + \cos^2 x}, \therefore 0 \leq \sin x + \sqrt{1 + \cos^2 x} \leq 2$$

$$\therefore \sqrt{1 + \cos^2 x} = \sin x \Rightarrow x = 2k\pi + \frac{\pi}{2}, \ k \in Z, y_{\max} = 2$$

$$\sqrt{1 + \cos^2 x} = -\sin x \Rightarrow x = 2k\pi - \frac{\pi}{2}, \ k \in Z, y_{\min} = 0.$$

4. (1) If the minimum of $f(x) = \cos 2x - 2a(1 + \cos x)$ is $-\frac{1}{2}$, find the range of values of a.

(2) If $f(x) = \sin^4 x - \sin x \cos x + \cos^4 x$, find the range of $f(x)$.

Solution: (1)

$$f(x) = 2\cos^2 x - 1 - 2a - 2a\cos x$$

$$= 2\left(\cos x - \frac{a}{2}\right)^2 - \frac{1}{2}a^2 - 2a - 1.$$

(i) When $a > 2$, $f(x)_{\min} = 1 - 4a$, where $\cos x = 1$.

(ii) When $a < -2$, $f(x)_{\min} = 1$, where $\cos x = -1$.

(iii) When $-2 \leq a \leq 2$, $f(x)_{\min} = -\frac{1}{2}a^2 - 2a - 1$, where $\cos x = \frac{a}{2}$.

When $a > 2$ or $a < -2$, $f(x)_{\min} \neq -\frac{1}{2}$,

$$\therefore -\frac{1}{2}a^2 - 2a - 1 = -\frac{1}{2},$$

$$\therefore a = -2 + \sqrt{3}, \ a = -2 - \sqrt{3} \text{ (Rejection)}.$$

(2) $f(x) = \sin^4 x - \sin x \cos x + \cos^4 x = 1 - \frac{1}{2}\sin 2x - \frac{1}{2}\sin^2 2x$.

Suppose $t = \sin 2x$, then

$$f(x) = g(t) = 1 - \frac{1}{2}t - \frac{1}{2}t^2 = \frac{9}{8} - \frac{1}{2}\left(t + \frac{1}{2}\right)^2.$$

$$\min_{-1\leq t\leq 1} g(t) = g(1) = \frac{9}{8} - \frac{1}{2}\cdot\frac{9}{4} = 0, \quad \max_{-1\leq t\leq 1} g(t)$$

$$= g\left(-\frac{1}{2}\right) = \frac{9}{8} - \frac{1}{2}\cdot 0 = \frac{9}{8}$$

$$\therefore 0 \leq f(x) \leq \frac{9}{8}.$$

5. If

$$x, y \in \left[-\frac{\pi}{4}, \frac{\pi}{4}\right], a \in R, \quad \begin{cases} x^3 + \sin x - 2a = 0, \\ 4y^3 + \sin y \cos y + a = 0, \end{cases}$$

find $\cos(x + 2y)$.

Solution: Eliminate $a \Rightarrow x^3 + \sin x = -8y^3 - 2\sin y \cos y$

$$\Rightarrow x^3 + \sin x = (-2y)^3 + \sin(-2y).$$

Suppose $f(t) = t^3 + \sin t$

$\therefore f(t)$ is an increasing function in $\left[-\frac{\pi}{2}, \frac{\pi}{2}\right]$

$$\because x \in \left[-\frac{\pi}{4}, \frac{\pi}{4}\right], \quad -2y \in \left[-\frac{\pi}{2}, \frac{\pi}{2}\right], \quad f(x) = f(-2y)$$

$$\therefore x = -2y \Rightarrow x + 2y = 0$$

$$\therefore \cos(x + 2y) = 1.$$

Graphs and properties of function: $y = A\sin(\omega x + \varphi)$
$(A > 0, \omega > 0, x \in R)$

1. Five Point Method

Using "Five Point Method" to plot the graph of $y = A\sin(\omega x + \varphi)(A > 0, \omega > 0, X \in R)$

(1) confirm periodicity $T = \frac{2\pi}{\omega}$;

(2) let $\omega x + \varphi = 0,\ \frac{\pi}{2},\ \pi,\ \frac{3\pi}{2},\ 2\pi$; then

$$x = \frac{-\varphi}{\omega},\ \frac{1}{\omega}\left(\frac{\pi}{2} - \varphi\right),\ \frac{1}{\omega}(\pi - \varphi),\ \frac{1}{\omega}\left(\frac{3\pi}{2} - \varphi\right),\ \frac{1}{\omega}(2\pi - \varphi).$$

Find five points:

$$\left(-\frac{\varphi}{\omega}, 0\right),\ \left(\frac{1}{\omega}\left(\frac{\pi}{2} - \varphi\right), 1\right),\ \left(\frac{1}{\omega}(\pi - \varphi), 0\right),$$

$$\left(\frac{1}{\omega}\left(\frac{3\pi}{2} - \varphi\right), -1\right),\ \left(\frac{1}{\omega}(2\pi - \varphi),\ 0\right).$$

(3) Charting, plot the graph of function in one period, then according to the periodicity, will get the graph of function $y = A\sin(\omega x + \varphi)(A > 0, \omega > 0, X \in R)$.

2. Translation Method

We can get the graph of the function $y = A\sin(\omega x + \varphi)$ $(A > 0, \omega > 0, X \in R)$ as follows.

By the graph of function $y = \sin x$, the graph of function $y = \sin(\omega x + \varphi)$ has the following ways to transform:

(1) Transform, and then translation: graph of $y = \sin x$

The abscissa of all points be extended $(0 < \omega < 1)$ or narrowed $(\omega > 1)$ by $\frac{1}{\omega}$ times the graph of $y = \sin(\omega x)$ left $(\varphi > 0)$ or right $(\varphi < 0)$ translation $|\frac{\varphi}{\omega}|$ unit, the graph of $y = \sin(\omega x + \phi)$

The ordinates of all points be extended or narrowed $(0 < A < 1)$ by A times the graph of $y = A\sin(\omega x + \phi)$

(2) Translation, and then transform. The graph of $y = \sin x$ left $(\varphi > 0)$ or right $(\varphi < 0)$ translation $|\frac{\varphi}{\omega}|$ unit, the graph of $y = \sin(x + \phi)$

The abscissa of all points be extended $(0 < \omega < 1)$ or narrowed $(\omega > 1)$ by $\frac{1}{\omega}$ times the graph of $y = \sin(\omega x + \phi)$

The ordinates of all points be extended or narrowed $(0 < A < 1)$ by A times the graph of $y = A\sin(\omega x + \phi)$.

3. Primary Properties of $y = A\sin(\omega x + \varphi)$ $(A > 0, \omega > 0)(k \in Z)$

Domain	Range	Periodicity	Parity	Monotone increasing interval	Monotone decreasing interval
R	$[-A, A]$	$\frac{2\pi}{\omega}$	When $\varphi \neq \frac{k\pi}{2}$, none when $\varphi = k\pi - \frac{\pi}{2}$, even function When $\varphi = k\pi$, odd function	$\left[\frac{1}{2\omega}(4k\pi - \pi - 2\varphi), \frac{1}{2\omega}(4k\pi + \pi - 2\varphi)\right]$	$\left[\frac{1}{2\omega}(4k\pi + \pi - 2\varphi), \frac{1}{2\omega}(4k\pi + 3\pi - 2\varphi)\right]$

When $y = A\sin(\omega x + \varphi)$ means simple harmonic oscillation, A is named amplitude, which means maximum distance from the equilibrium position; $f = \frac{1}{T} = \frac{\omega}{2\pi}$ named frequency, which means number of times per unit time of reciprocating vibration; $\omega x + \varphi$ is named phase. When $x = 0$, φ is named primary phase.

The graph of $y = A\sin(\omega x + \varphi)$ is both axisymmetric graph and central symmetry graph. Its symmetric axis is $x = \frac{1}{\omega}\left(k\pi + \frac{\pi}{2} - \varphi\right)$ $(k \in Z)$, and its center of symmetry is $\left(\frac{1}{\omega}(k\pi - \varphi), 0\right)$.

Similarly, we can research *primary properties* of function of $y = A\cos(\omega x + \varphi)(A > 0, \omega > 0, x \in R)$ and $y = A\tan(\omega x + \varphi)(A > 0, \omega > 0, x \in R)$ $(k \in Z)$.

Exercise

6. Find monotone interval of $f(x) = 2\sin(\frac{\pi}{3} - 2x)$.

Solution:

$$\because f(x) = 2\sin\left(\frac{\pi}{3} - 2x\right) = -2\sin\left(2x - \frac{\pi}{3}\right),$$

$\because f(x)$ is monotone increasing function

$\therefore \sin\left(2x - \frac{\pi}{3}\right)$ is monotone decreasing function

$$\because 2k\pi + \frac{\pi}{2} \le 2x - \frac{\pi}{3} \le 2k + \pi$$

$$\therefore k\pi + \frac{5}{12}\pi \le x \le k\pi + \frac{2}{3}\pi \quad (k \in Z)$$

$\because f(x)$ is monotone decreasing function

$\therefore \sin\left(2x - \frac{\pi}{3}\right)$ is monotone increasing function

$$\because 2k\pi - \frac{\pi}{2} \le 2x - \frac{\pi}{3} \le 2k\pi + \frac{\pi}{2}$$

$$\therefore k\pi - \frac{1}{12}\pi \le x \le k\pi + \frac{5}{12}\pi \quad (k \in Z)$$

$\therefore$ decreasing interval is $\left[k\pi - \frac{\pi}{12}, \ k\pi + \frac{5}{12}\pi\right]$,

increasing interval is $\left[k\pi + \frac{5}{12}\pi, k\pi + \frac{2}{3}\pi\right]$.

Comment: Translate function into standard form $y = A\sin(\omega x + \varphi)(A > 0, \omega > 0)$, and then find the solution.

7. If $x \in \left[-\frac{5\pi}{12}, -\frac{\pi}{3}\right]$, find maximum of $y = \tan(x + \frac{2\pi}{3}) - \tan(x + \frac{\pi}{6}) + \cos(x + \frac{\pi}{6})$.

Solution:

$$y = \tan\left(x + \frac{2\pi}{3}\right) + \cot\left(x + \frac{2\pi}{3}\right) + \cos\left(x + \frac{\pi}{6}\right)$$

$$= \frac{1}{\cos(x + \frac{2\pi}{3})\sin(x + \frac{2\pi}{3})} + \cos\left(x + \frac{\pi}{6}\right)$$

$$= \frac{2}{\sin(2x + \frac{4\pi}{3})} + \cos\left(x + \frac{\pi}{6}\right)$$

$$\because x \in \left[-\frac{5\pi}{12}, -\frac{\pi}{3}\right],$$

$$\therefore 2x + \frac{4\pi}{3} \in \left[\frac{\pi}{2}, \frac{2\pi}{3}\right], \quad x + \frac{\pi}{6} \in \left[-\frac{\pi}{4}, -\frac{\pi}{6}\right],$$

$$\therefore \quad \frac{2}{\sin(2x + \frac{4\pi}{3})} \text{ and}$$

$\cos\left(x + \dfrac{\pi}{6}\right)$ are increasing functions in $\left[-\dfrac{5\pi}{12}, -\dfrac{\pi}{3}\right]$.

$\therefore$ The maximum of function is $f\left(-\dfrac{\pi}{3}\right) = \dfrac{11}{6}\sqrt{3}$.

Comment: If it is difficult to translate the function, consider the use of the monotonous of function. To prove that monotonous of two functions are the same in the same interval, the maximum (minimum) value is obtained at the same point.

8. Let $f(x) = \dfrac{\sin(\pi x) - \cos(\pi x) + 2}{\sqrt{x}}$ $(\frac{1}{4} \le x \le \frac{5}{4})$, find minimum of $f(x)$.

Solution:

$$f(x) = \frac{\sqrt{2}\sin(\pi x - \frac{\pi}{4}) + 2}{\sqrt{x}} \quad \left(\frac{1}{4} \le x \le \frac{5}{4}\right),$$

Suppose $g(x) = \sqrt{2}\sin(\pi x - \frac{\pi}{4})(\frac{1}{4} \le x \le \frac{5}{4})$, then $g(x) \ge 0$. So $g(x)$ is increasing function in $[\frac{1}{4}, \frac{3}{4}]$, and decreasing function in $[\frac{3}{4}, \frac{5}{4}]$, and the symmetric axis of $y = g(x)$ is $x = \frac{3}{4}$, so for all $x_1 \in [\frac{1}{4}, \frac{3}{4}]$, there exists $x_2 \in [\frac{3}{4}, \frac{5}{4}]$, and mange $g(x_2) = g(x_1)$ (see Fig. 3.1).

so

$$f(x_1) = \frac{g(x_1) + 2}{\sqrt{x_1}} = \frac{g(x_2) + 2}{\sqrt{x_1}} \ge \frac{g(x_2) + 2}{\sqrt{x_2}} = f(x_2).$$

$\because$ $f(x)$ is decreasing function in $\left[\dfrac{3}{4}, \dfrac{5}{4}\right]$,

$\therefore$ $4 = f\left(\dfrac{1}{4}\right) > f\left(\dfrac{3}{4}\right)$,

$\therefore$ $f(x) \ge f\left(\dfrac{5}{4}\right) = \dfrac{4\sqrt{5}}{5}$,

$\therefore$ minimum of $f(x)$ is $\dfrac{4\sqrt{5}}{5}$ in $\left[\dfrac{1}{4}, \dfrac{5}{4}\right]$.

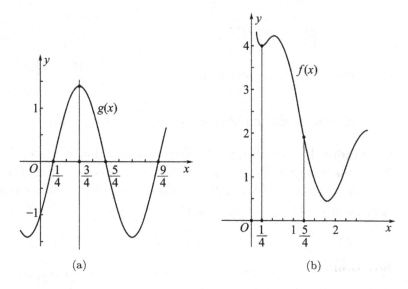

(a)   (b)

Fig. 3.1.

Comment: If it is difficult or impossible to solve the problem of non-conventional functions, we can use outflanking tactics like monotonous and symmetries of function and also use the calculus to solve.

9. If $f(x)$ is the decreasing function in $(-\infty, 1]$. To $x \in R$, $f(k - \sin x) \geq f(k^2 - \sin^2 x)$ is always satisfied, find k.
   Solution:

$$\Leftrightarrow \begin{cases} k - \sin x \leq 1 \\ k^2 - \sin^2 x \leq 1 \\ k - \sin x \leq k^2 - \sin^2 x \end{cases} \Leftrightarrow \begin{cases} k^2 \leq 1 + \sin^2 x & (1) \\ k^2 - k + \frac{1}{4} \geq \left(\sin x - \frac{1}{2} \right)^2. & (2) \end{cases}$$

According to (1)

$$k^2 \leq \min_{x \in R} |1 + \sin^2 x| = 1. \qquad (3)$$

According to (2):

$$k^2 - k + \frac{1}{4} \geq \min_{x \in R} \left| \left(\sin x - \frac{1}{2} \right)^2 \right| = \frac{9}{4}. \tag{4}$$

According to (3), (4): $k = -1$.

Comment: Using the monotone property of the function, take the problem to inequality problem. The inequality problem is often using the separation variable method. After separation of variables, the following conclusions are often used.

$\forall x \in f_D$, inequality $a \leq f(x)$ always holds $\Leftrightarrow a \leq \min_{x \in f_D} f(x)$.

$\forall x \in f_D$, inequality $a \geq f(x)$ always holds $\Leftrightarrow a \geq \max_{x \in f_D} f(x)$.

$\exists x \in f_D$ inequality $a \geq f(x)$ always holds $\Leftrightarrow a \geq \min_{x \in f_D} f(x)$.

$\exists x \in f_D$ inequality $a \leq f(x)$ always holds $\Leftrightarrow a \leq \max_{x \in f_D} f(x)$.

10. Suppose $0 \leq \theta \leq \frac{\pi}{2}$, inequality $\sin^2 \theta + 3m \cos \theta - 6m - 4 < 0$ is satisfied, find the range of m.

Solution: $\because 3m > \frac{3 + \cos^2 \theta}{\cos \theta - 2}$ $(0 \leq \theta \leq \frac{\pi}{2})$ is always satisfied,

$$\therefore 3m > \max_{\theta \in [0, \frac{\pi}{2}]} \frac{3 + \cos^2 \theta}{\cos \theta - 2}.$$

Let $\cos \theta - 2 = t, -2 \leq t \leq -1, u = \frac{3 + \cos \theta}{\cos \theta - 2} = \frac{3 + (t+2)^2}{t} = t + \frac{7}{t} + 4$.
When $t = -2$, $u_{\max} = -2 + \left(-\frac{7}{2} \right) + 4 = -\frac{3}{2}$.

$$\therefore 3m > -\frac{3}{2}, \quad m > -\frac{1}{2}.$$

11. If $(x + 3 + 2 \sin \theta \cos \theta)^2 + (x + a \sin \theta + a \cos \theta)^2 \geq \frac{1}{8}$ is always satisfied to all $x \in R$ and all $\theta \in \left[0, \frac{\pi}{2} \right]$, find the range value of a.

Solution 1: Using

$$\sqrt{\frac{a^2 + b^2}{2}} \geq \frac{a + b}{2} \Rightarrow a^2 + b^2 \geq 2 \left(\frac{a + b}{2} \right)^2,$$

$$(x + 3 + 2 \sin \theta \cos \theta)^2 + (x + a \sin \theta + a \cos \theta)^2$$

$$= [x + 3 + 2\sin\theta\cos\theta)]^2 + [-x - a\sin\theta - a\cos\theta]^2$$

$$\geq 2\left(\frac{x + 3 + 2\sin\theta\cos\theta - x - a\sin\theta - a\cos\theta}{2}\right)^2.$$

$$\because (x + 3 + 2\sin\theta\cos\theta)^2 + (x + a\sin\theta + a\cos\theta)^2$$

$$\geq \frac{1}{8} \text{ is always satisfied.}$$

$$\therefore [(x + 3 + 2\sin\theta\cos\theta)^2 + (x + a\sin\theta + a\cos\theta)^2]_{\min}$$

$$\geq 2\left(\frac{x + 3 + 2\sin\theta\cos\theta - x - a\sin\theta - a\cos\theta}{2}\right)^2$$

$$\geq \frac{1}{8} \text{ is always satisfied.}$$

$$\Leftrightarrow (3 + 2\sin\theta\cos\theta - a\sin\theta - a\cos\theta)^2 \geq \frac{1}{4},$$

$$\Leftrightarrow a \geq \frac{3 + 2\sin\theta\cos\theta + \frac{1}{2}}{\sin\theta + \cos\theta} \quad \text{or}$$

$$\Leftrightarrow a \leq \frac{3 + 2\sin\theta\cos\theta - \frac{1}{2}}{\sin\theta + \cos\theta}, \ \theta \in \left[0, \frac{\pi}{2}\right].$$

Let $t = \sin\theta + \cos\theta$, 则 $t \in [1, \sqrt{2}]$, $\sin\theta\cos\theta = \dfrac{t^2 - 1}{2}$.

$$\therefore \Leftrightarrow a \geq \max_{\theta \in [0, \frac{\pi}{2}]} \frac{3 + 2\sin\theta\cos\theta + \frac{1}{2}}{\sin\theta + \cos\theta}$$

$$= \max_{t \in [1, \sqrt{2}]} \left\{ t + \frac{5}{2t} \right\} = 1 + \frac{5}{2} = \frac{7}{2}$$

$$\text{or } a \leq \min_{\theta \in [0, \frac{\pi}{2}]} \frac{3 + 2\sin\theta\cos\theta - \frac{1}{2}}{\sin\theta + \cos\theta}$$

$$= \min_{t \in [1,\sqrt{2}]} \left\{ t + \frac{3}{2t} \right\} = 2\sqrt{\frac{3}{2}} = \sqrt{6},$$

$$\therefore a \geq \frac{7}{2} \text{ or } a \leq \sqrt{6}.$$

Solution 2: Let $y = x$,

$$\therefore \begin{cases} x - y = 0 \\ [x - (-3 - 2\sin\theta\cos\theta)]^2 + [y - (-a\sin\theta - a\cos\theta)]^2 \geq \frac{1}{8} \end{cases}$$

$\Leftrightarrow$ The square of distance of point

$$P(-3 - 2\sin\theta\cos\theta, -a\sin\theta - a\cos\theta)$$

and point $Q(x,y) \geq \dfrac{1}{8}$

$\Leftrightarrow$ The minimum of square of distance of point

$$P(-3 - 2\sin\theta\cos\theta, -a\sin\theta - a\cos\theta)$$

and point $Q(x,y) \geq \dfrac{1}{8}$

$\Leftrightarrow$ The square of distance of point

$$P(-3 - 2\sin\theta\cos\theta, -a\sin\theta - a\cos\theta)$$

to the line $x - y = 0$ is $\geq \dfrac{1}{8}$ (see Fig. 3.2),

$$\therefore \left[\frac{|-3 - 2\sin\theta\cos\theta + a\sin\theta + a\cos\theta|}{\sqrt{1+1}} \right]^2 \geq \frac{1}{8}.$$

12. Suppose $f(x) = 3\sin x + 2\cos x + 1$, $a, b, c \in R$, and $af(x) + bf(x - c) = 1$ is always satisfied for all $x \in R$, find $\frac{b\cos c}{a}$.

Solution: Let $c = \pi$, $\therefore$ for all $x \in R$, $f(x) + f(x - c) = 2$ is satisfied, $\therefore$ let $a = b = \frac{1}{2}$, $c = \pi$,

$\therefore$ for all $x \in R$, $af(x) + bf(x - c) = 1$, $\therefore \frac{b\cos c}{a} = -1$.

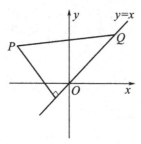

Fig. 3.2.

Generally, $\because$ $f(x) = \sqrt{13}\sin(x + \varphi) + 1$, $f(x - c) = \sqrt{13}\sin(x + \varphi - c) + 1$ $(0 < \varphi < \frac{\pi}{2}$ and $\tan\varphi = \frac{2}{3})$,

$\therefore$ $af(x) + bf(x - c) = 1$ is *equivalent to*

$$\sqrt{13}a\sin(x + \varphi) + \sqrt{13}b\sin(x + \varphi - c) + a + b = 1, \text{ namely}$$

$$\sqrt{13}\,a\sin(x + \varphi) + \sqrt{13}b\sin(x + \varphi)\cos c$$

$$-\sqrt{13}b\sin c\cos(x + \varphi) + (a + b - 1) = 0,$$

$$\therefore\ \sqrt{13}(a + b\cos c)\sin(x + \varphi)$$

$$-\sqrt{13}b\sin c\cos(x + \varphi) + (a + b - 1) = 0.$$

According to known condition, for all $x \in R$, we have

$$\begin{cases} a + b\cos c = 0, & \text{(1)} \\ b\sin c = 0, & \text{(2)} \\ a + b - 1 = 0. & \text{(3)} \end{cases}$$

If $b = 0$, then according to (1), $a = 0$, which does not meet (3), $\therefore$ $b \neq 0$. $\therefore$ According to (2), $\sin c = 0$, $\therefore$ $c = 2k\pi + \pi$ or $c = 2k\pi$ $(k \in Z)$.
When $c = 2k\pi$, $\cos c = 1$, then (1) and (3) are contradiction.

$\therefore$ $c = 2k\pi + \pi(k \in Z)$, $\cos c = -1$.

According to (1) and (3), $a = b = \frac{1}{2}$, $\frac{b\cos c}{a} = -1$.

Comment: First guess: the use of special values, extreme values to explore, discover the law of the problem. Solve the problem by the method of undetermined coefficient method. First guess is an important mean and method to promote the development of mathematics.

13. (1) If $y = A \sin \omega x (A > 0, \omega > 0)$ has at least 50 maximums in $[0,1]$, find the maximum of ω.
   (2) If $y = 3 \sin(\omega x)(\omega > 0)$ has at least 50 maximums in $[a, a + 1]$, find the maximum of ω.
   (3) In plane rectangular coordinate system xoy, find the area of the image formed by the closed graph of $f(x) = a \sin ax + \cos ax$ $(a > 0)$ and $g(x) = \sqrt{a^2 + 1}$.
   (4) The large of interval $(m, n), [m, n], (m, n], [m, n)$ is defined $n - m(n > m)$, the sum of interval of the solution set which meets $\sin x \cos x + \sqrt{3} \cos^2 x + b > 0, x \in [0, \pi]$ **exceeded** $\frac{\pi}{3}$, find the range interval of b.

Solution: (1) Minimum occurs once in a period, but minimum occurs twice in a period and $\frac{3}{4}$ period. Moreover, $49\frac{3}{4}$ period can guarantee the emergence of 50 minimum.
If y has at least 50 maximums in $[0,1]$

$$\therefore \text{ at least include } \left(49 + \frac{3}{4}\right) \text{ periods,}$$

$$\therefore \begin{cases} 49\frac{3}{4}T \leq 1, \\ T = \dfrac{2\pi}{\omega}. \end{cases}$$

$$\therefore \omega \geq \frac{199}{2}\pi, \quad \therefore \omega_{\min} = \frac{199}{2}\pi.$$

$$(2) \begin{cases} 50T \leq 1, \\ T = \dfrac{2\pi}{\omega}. \end{cases} \Rightarrow \omega \geq 100\pi, \quad \therefore \omega_{\min} = 100\pi.$$

(3) $f(x) = \sqrt{a^2 + 1}\sin(ax + \varphi)$, where $\varphi = \arctan\frac{1}{a}$, its period is $\frac{2\pi}{a}$, *amplitude is* $\sqrt{a^2 + 1}$. The image formed by the closed graph of $f(x)$ and $g(x)$ is shown in image 1, which has the symmetry $(S_1 = S_2, S_3 = S_{44})$. This can be cut growth for graphics as *rectangle ABCD, where* length is $\frac{2\pi}{a}$, and width is $\sqrt{a^2 + 1}$ as the shown in image 2, $\therefore$ its area is $\frac{2\pi}{a}\sqrt{a^2 + 1}$.

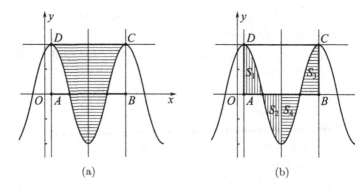

(a)   (b)

Fig. 3.3.

(4) $\because \sin x \cos x + \sqrt{3}\cos^2 x + b = \sin\left(2x + \frac{\pi}{3}\right) + b + \frac{\sqrt{3}}{2}$.

Suppose $f(x) = \sin(2x + \frac{\pi}{3})$, $\sin x \cos x + \sqrt{3}\cos^2 x + b > 0$, $x \in [0, \pi]$, is equivalent to $f(x) > -b - \frac{\sqrt{3}}{2}$, $x \in [0, \pi]$.

When $x \in [0, \frac{\pi}{6}]$, its large of the solution set is equal to $\frac{\pi}{6}$, short of $\frac{\pi}{6}$;

$\therefore$ left and right of the solution set are short of $\frac{\pi}{12}$.

When $x \in [\frac{\pi}{4}, \frac{11\pi}{12}]$, $-b - \frac{\sqrt{3}}{2} < f(x) = \sin(2 \cdot \frac{\pi}{4} + \frac{\pi}{3}) = \frac{1}{2}$, $\therefore b > -\frac{\sqrt{3}+1}{2}$.

Comment: By means of combining numbers with shapes, the problem is solved by using the method of periodic, symmetry and cut-paste algorithm.

Maximum and minimum of trigonometric function

Find the maximum of triangular function by means of the appropriate trigonometric transformation or algebraic

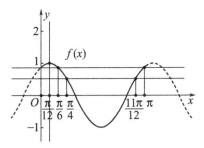

Fig. 3.4.

transformation. Turn to essential trigonometric function or algebraic function, and then using the boundedness of trigonometric function or common method, find the maximum of function.

1. The type of $y = a\sin x + b$ (or $y = a\cos x + b$), using $|\sin x| \leq 1$ (or $|\cos x| \leq 1$) to solve.

2. The type of $y = a\sin x + b\cos x$, introducing auxiliary angle ϕ, translate into $y = \sqrt{a^2 + b^2}\sin(x + \phi)$, using $|\sin(x + \phi)| \leq 1$ to solve.

3. The type of $y = a\sin^2 x + b\sin x + c$ (or $y = a\cos^2 x + b\cos + c$), let $t = \sin x$ (or $t = \cos x, |t| \leq 1$), and then translate it into the most value problem of quadratic function in the closed interval.

4. The type of $y = \frac{a\sin x + b}{c\sin x + d}$ (or $y = \frac{a\cos x + b}{c\cos x + d}$), get $\sin x$ (or $\cos x$), using $|\sin x| \leq 1$ (or $|\cos x| \leq 1$ to solve, or using separation of constant method to solve.

5. The type of $y = \frac{a\sin x + b}{c\cos x + d}$ (or $y = \frac{a\cos x + b}{c\sin x + d}$), and then translate it into the $\sin(x + \phi) = g(y)$.

6. To the most problems of function which include $\sin x \pm \cos x$ or $\sin x \cos x$, let $\sin x \pm \cos x = t, |t| \leq \sqrt{2}$, and then translate $\sin x \cos x$ into the relational expression about t.

7. In the solution of the problem with the maximum value of trigonometric functions, the parameters are should be discussed.

8. Can be solved by calculus.

Exercise:

14. If $f(x) = 2\cos^2 x + \sqrt{3}\sin 2x + a$, and $x \in \left[0, \frac{\pi}{2}\right], |f(x)| < 2$, find the range value of a.

Solution:

$$f(x) = \cos 2x + \sqrt{3}\sin 2x + a + 1 = 2\sin\left(2x + \frac{\pi}{6}\right) + a + 1$$

$$\because 0 \le x \le \frac{\pi}{2} \quad \therefore \frac{\pi}{6} \le 2x + \frac{\pi}{6} \le \frac{7}{6}\pi$$

$$\therefore a \le f(x) \le a + 3$$

$$\because |f(x)| < 2, \therefore [a, a+3] \subset (-2, 2)$$

$$\therefore \begin{cases} a > -2 \\ a + 3 < 2 \end{cases}, \quad \therefore -2 < a < -1.$$

15. If $\alpha, \beta \in \left(0, \frac{\pi}{2}\right), \alpha + \beta \ne \frac{\pi}{2}$, and α, β meet $\sin\beta = \sin\alpha\cos(\alpha+\beta)$, (1) use $\tan\alpha$ to represent $\tan\beta$; (2) find maximum of $\tan\beta$.

Solution:

(1) $\sin\beta = \sin\alpha \cdot \cos\alpha \cdot \cos\beta - \sin^2\alpha \cdot \sin\beta$,

$\sin\beta(1 + \sin^2\alpha) = \sin\alpha \cdot \cos\alpha \cdot \cos\beta$

$\because \cos\beta \ne 0, \ \tan\beta(1 + \sin^2\alpha) = \sin\alpha\cos\alpha$

$$\therefore \tan\beta = \frac{\sin\alpha \cdot \cos\alpha}{1 + \sin^2\alpha} = \frac{\sin\alpha \cdot \cos\alpha}{2\sin^2\alpha + \cos^2\alpha} = \frac{\tan\alpha}{2\tan^2\alpha + 1}.$$

(2) $\because x = \tan\alpha, \ (x > 0)$

$$\therefore \tan\beta = y = \frac{x}{2x^2 + 1}, \quad \therefore 2yx^2 - x + y = 0$$

$$\because \Delta = 1 - 8y^2 \ge 0, \ y > 0$$

$$\therefore 0 < y \le \frac{\sqrt{2}}{4}$$

$$\therefore \tan\beta \text{ maximum is } \frac{\sqrt{2}}{4}.$$

16. Suppose $f(x, y, z) = \sin^2(x - y) + \sin^2(y - z) + \sin^2(z - x)$, $x, y, z \in R$, find the maximum of $f(x, y, z)$.

Solution:

$$f(x, y, z) = \sin^2(x - y) + \sin^2(y - z) + \sin^2(z - x)$$

$$= \frac{1}{2}[1 - \cos 2(x - y) + 1 - \cos 2(y - z) + 1 - \cos 2(z - x)]$$

$$= \frac{3}{2} - \frac{1}{2}[(\cos 2x \cos 2y + \sin 2x \sin 2y)$$

$$+ (\cos 2y \cos 2z + \sin 2y \sin 2z)]$$

$$- \frac{1}{2}(\cos 2z \cos 2x + \sin 2z \sin 2x)$$

$$= \frac{3}{2} - \frac{1}{4}[(\cos 2x + \cos 2y + \cos 2z)^2$$

$$+ (\sin 2x + \sin 2y + \sin 2z)^2 - 3]$$

$$\leq \frac{3}{2} + \frac{3}{4} = \frac{9}{4}.$$

When $x = \frac{\pi}{3}$, $y = \frac{2\pi}{3}$, $z = \frac{3\pi}{3}$ the maximum of $f(x, y, z)$ is $\frac{9}{4}$.

17. The maximum of $F(x) = |\cos^2 x + 2 \sin x \cos x - \sin^2 x + Ax + B|$ is M in $0 \leq x \leq \frac{3\pi}{2}$, which is related to A and B. What about the value of A, B, M is minimum? Prove your result.

Solution 1:

$$F(x) = \left| \sqrt{2} \sin \left(2x + \frac{\pi}{4} \right) + Ax + B \right|$$

If $A = B = 0$, $F(x) = \sqrt{2} \left| \sin \left(2x + \frac{\pi}{4} \right) \right|$,

when $x = \frac{\pi}{8}$, $x = \frac{5}{8}\pi$, $x = \frac{9}{8}\pi$, $F(x)_{\min} = M = \sqrt{2}$.

Proof: $\forall A, B, M \geq \sqrt{2}$,

if $A, B, M < \sqrt{2}$, $\therefore F \left(\frac{\pi}{8} \right)$, $F \left(\frac{5\pi}{8} \right)$, $F \left(\frac{9\pi}{8} \right) < \sqrt{2}$

$$\sqrt{2} + \frac{\pi}{8}A + B < \sqrt{2} \tag{10}$$

$$-\sqrt{2} + \frac{5\pi}{8}A + B < -\sqrt{2} \tag{11}$$

$$\sqrt{2} + \frac{9\pi}{8}A + B < \sqrt{2} \tag{12}$$

$(10) + (12) - 2 \times (11) \Rightarrow 0 < 0$ which is a contradiction.

$\because A = B = 0$

$\therefore M \geq \sqrt{2}$, $\therefore A = B = 0$, $\therefore M = \sqrt{2}$.

$\square$

Solution 2: $F(x) = |\sqrt{2}\sin(2x + \frac{\pi}{4}) + Ax + B|$, when $A = B = 0$, $F(x)_{\max} = M = \sqrt{2}$, to prove $\sqrt{2}$ is the minimum value of M, i.e., implies to prove: when $A \neq 0$, $B \neq 0$, $M > \sqrt{2}$.

(i) when $A > 0, B > 0$, take $x = \frac{\pi}{8}$, $M \geq F\left(\frac{\pi}{8}\right) = \left|\sqrt{2} + \frac{1}{8}A\pi + B\right| > \sqrt{2}$;

(ii) $A < 0$, $B < 0$, take $x = \frac{5\pi}{8}$, $M \geq F\left(\frac{5\pi}{8}\right) = \left|-\sqrt{2} - (-A\frac{5\pi}{8} - B)\right| > \sqrt{2}$;

(iii) when $A > 0$, $B < 0$,

- take $A \cdot \frac{9\pi}{8} > -B$, $x = \frac{9\pi}{8}$, $M \geq F\left(\frac{9\pi}{8}\right) = \left|\sqrt{2} + \frac{9}{8}A\pi - (-B)\right| > \sqrt{2}$,
- take $A \cdot \frac{9\pi}{8} < -B$, $x = \frac{5\pi}{8}$, $M \geq F\left(\frac{5\pi}{8}\right) = \left|-\sqrt{2} - (-B - \frac{9}{8}A\pi)\right| > \sqrt{2}$,

(iv) $A < 0$, $B > 0$,

- take $-A \cdot \frac{5\pi}{8} > B$, $x = \frac{5\pi}{8}$, $M \geq F\left(\frac{5\pi}{8}\right) = \left|-\sqrt{2} - (-\frac{5}{8}A\pi - B)\right| > \sqrt{2}$;
- take $-A \cdot \frac{9\pi}{8} < B$, $x = \frac{\pi}{8}$, $M \geq F\left(\frac{\pi}{8}\right) = \left|\sqrt{2} + B - (-\frac{1}{8}A\pi)\right| > \sqrt{2}$;

$\therefore$ when $A = 0$, $B = 0$, $M_{\min} = \sqrt{2}$.

18. If $f(x) = -\frac{1}{2} + \frac{\sin\frac{5}{2}x}{\sin\frac{1}{2}x}$ $(0 < x < \pi)$.

(1) represent $f(x)$ as polynomials about $\cos x$;
(2) find the range value of $f(x)$;
(3) if $\cos x$ has two different sign in $f(x) = k(\cos x - 2)$, find the range value of k.

Solution:

(1) $f(x) = -\dfrac{1}{2} + \dfrac{\sin\frac{5}{2}x}{\sin\frac{1}{2}x} = \dfrac{\sin\frac{5}{2}x - \sin\frac{x}{2}}{\sin\frac{1}{2}x} + \dfrac{1}{2}$

$\qquad = \dfrac{2\cos\frac{3}{2}x\sin x}{\sin\frac{1}{2}x} + \dfrac{1}{2} = \dfrac{4\cos\frac{3}{2}x\cos\frac{x}{2}\sin\frac{x}{2}}{\sin\frac{x}{2}} + \dfrac{1}{2}$

$\qquad = 4\cos\dfrac{3x}{2}\cos\dfrac{x}{2} + \dfrac{1}{2} = 4\cos^2 x + 2\cos x - \dfrac{3}{2}$

$\qquad = 4\left(\cos x + \dfrac{1}{4}\right)^2 - \dfrac{7}{4}.$

(2) $f(x) \in \left[-\dfrac{7}{4}, \dfrac{9}{2}\right).$

(3) $4\cos^2 x + 2\cos x - \dfrac{3}{2} = k(\cos x - 2),$

let $t = \cos x$, $t \in [-1, 1]$, $g(t) = 4t^2 + (2 - k)t + 2k - \frac{3}{2}$, one root in $[-1, 0)$, another root in $(0, 1]$

$$\begin{cases} g(-1) \geq 0 \\ f(1) < 0 \\ g(0) \geq 0 \end{cases} \Rightarrow -\dfrac{1}{6} \leq k < \dfrac{3}{4}.$$

19. If the minimum of $f(x) = (\sin x + 4\sin\theta + 4)^2 + (\cos x - 5\cos\theta)^2$ is $g(\theta)$, find the maximum of $g(\theta)$ (see Fig. 3.5).

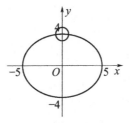

Fig. 3.5.

Solution 1:

$$f(x) = 8(1 + \sin\theta)\sin x - 10\cos\theta\cos x$$

$$-9\sin^2\theta + 32\sin\theta + 42$$

$$= \sqrt{64(1 + \sin\theta)^2 + 100\cos^2\theta}\sin(x + \theta)$$

$$-9\sin^2\theta + 32\sin\theta + 42$$

$$\therefore g(\theta) = -\sqrt{64(1 + \sin\theta)^2 + 100\cos^2\theta}$$

$$-9\sin^2\theta + 32\sin\theta + 42$$

$$= -2\sqrt{-9\sin^2\theta + 32\sin\theta + 41} - 9\sin^2\theta + 32\sin\theta + 42;$$

let $t = \sqrt{-9\sin^2\theta + 32\sin\theta + 41}, t \in [0, 8]$,

$$\therefore g(\theta) = t^2 - 2t + 1 = (t - 1)^2, \ t \in [0, 8],$$

$$\therefore g(\theta)_{\max} = (8 - 1)^2 = 49.$$

Solution 2: In the cartesian coordinate plane, let

$$A(\cos x, 4 + \sin x), \ B(5\cos\theta, -4\sin\theta)$$

$$\therefore |AB|^2 = (\cos x - 5\cos\theta)^2 + (\sin x + 4\sin\theta + 4)^2$$

$$f(x) = |AB|^2, A \text{ in the circle} : x^2 + (y - 4)^2 = 1$$

B in the ellipse : $\dfrac{x^2}{25} + \dfrac{y^2}{16} = 1$.

For all $B, g(\theta) = (|BC| - 1)^2$, $C(0, 4)$ is the center of a circle

$$\therefore g(\theta) = (\sqrt{(5\cos\theta)^2 + (4\sin\theta + 4)^2} - 1)^2$$

ellipse $\dfrac{x^2}{25} + \dfrac{y^2}{16} = 1$ in circle $x^2 + (y - 4)^2 = 64$

$\therefore$ the maximum distance between $C(0, 4)$

and ellipse $\dfrac{x^2}{25} + \dfrac{y^2}{16} = 1$ is 8

$\therefore\ g(\theta) \le (8-1)^2 = 49, g(\theta)_{\max} = 49\ \left(x = -\dfrac{\pi}{2}, \theta = \dfrac{\pi}{2}\right).$

20. $a, b, A, B \in R$ and $f(\theta) = 1 - a\cos\theta - b\sin\theta - A\cos 2\theta - B\sin 2\theta$. Prove: To all $\theta \in R$, $f(\theta) \ge 0$ is satisfied, then $a^2 + b^2 \le 2$ and $A^2 + B^2 \le 1$.

Proof: Using *auxiliary angle formula*:

$$f(\theta) = 1 - a\cos\theta - b\sin\theta - A\cos 2\theta - B\sin 2\theta$$

$$= 1 - \sqrt{a^2 + b^2}\cos(\theta - \alpha) - \sqrt{A^2 + B^2}\cos(\theta - \beta).$$

Let $r = \sqrt{a^2 + b^2}$, $R = \sqrt{A^2 + B^2}$; then

$$f(\theta) = 1 - r\cos(\theta - \alpha) - R\cos(\theta - \beta).$$

First prove : $a^2 + b^2 \le 2$

Suppose : $\theta = a + 45°$, $\theta = a - 45°$

$$f(a + 45°) = 1 - \frac{\gamma}{\sqrt{2}} - R\cos 2(\alpha - \beta + 45°) \ge 0 \qquad (13)$$

$$f(a - 45°) = 1 - \frac{\gamma}{\sqrt{2}} - R\cos 2(\alpha - \beta - 45°) \qquad (14)$$

$$= 1 - \frac{\gamma}{\sqrt{2}} + R\cos 2(\alpha - \beta + 45°) \ge 0$$

$(13) + (14) \Rightarrow 2\left(1 - \dfrac{\gamma}{\sqrt{2}}\right) \ge 0, \therefore\ \gamma^2 = a^2 + b^2 \le 2.$

Then prove: $A^2 + B^2 \le 1$.

Let $\theta = \beta$, $\theta = \beta + \pi$,

$$f(\beta) = 1 - \gamma \cos(\beta - \alpha) - R\cos 0 = 1 - \gamma\cos(\beta - a) - R \geq 0 \tag{15}$$

$$f(\beta + \pi) = 1 - \gamma\cos(\beta - a + \pi) - R\cos 2\pi$$

$$= 1 + \gamma\cos(\beta - a) - R \geq 0 \tag{16}$$

$$(15) + (16) \Rightarrow R^2 = A^2 + B^2 \leq 1.$$

$\square$

21. Suppose M is the maximum of $f(x) = 1 + \frac{2x + \sin x}{x^4 + x^2 + \cos x}$, and m is the minimum. Find $M + m$.

Solution: Suppose $F(x) = f(x) - 1$, we know $F(x)$ is odd function, and its graph symmetrizes origin

$$\because F_{\min}(x) = m - 1, \quad F_{\max}(x) = M - 1,$$

$$\therefore -(m - 1) = M - 1,$$

$$\therefore M + m = 2.$$

22. If the graph of $f(x) = |\sin x|$ has only three intersection points with line $y = kx$ $(k > 0)$, α is the maximum of the abscissa of intersection points. Prove:

$$\frac{\cos \alpha}{\sin \alpha + \sin 3\alpha} = \frac{1 + \alpha^2}{4\alpha}.$$

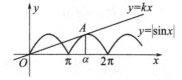

Fig. 3.6.

Proof: As shown in Fig. 3.6, two graph are tangency in $\left(\pi, \frac{3\pi}{2}\right)$.

Suppose point of tangency $A(\alpha, -\sin\alpha)$, $\alpha \in (\pi, \frac{3\pi}{2})$.

$$\because \ f'(x) = -\cos x, \ x \in \left(\pi, \frac{3}{2}\pi\right),$$

$$\therefore \ -\cos\alpha = -\frac{\sin\alpha}{\alpha}, \ \therefore \ \alpha = \tan\alpha.$$

$$\therefore \ \frac{\cos\alpha}{\sin\alpha + \sin 3\alpha} = \frac{\cos\alpha}{2\sin 2\alpha \cos\alpha}$$

$$= \frac{1}{4\sin\alpha\cos\alpha}$$

$$= \frac{\cos^2\alpha + \sin^2\alpha}{4\sin\alpha\cos\alpha}$$

$$= \frac{1 + \tan^2\alpha}{4\tan\alpha}$$

$$= \frac{1 + \alpha^2}{4\alpha}.$$

Comment: Solving the problem about the tangent of the curve, calculus can be a simple solution to this problem. □

23. Suppose $f(x) = \frac{\sin x}{2 + \cos x}$.

(1) Find the monotony interval of $f(x)$.

(2) If $f(x) \le ax$ is satisfied to all $x \ge 0$, find the range value of a.

Solution:

(1) $f'(x) = \dfrac{(2 + \cos x)\cos x - \sin x(-\sin x)}{(2 + \cos x)^2} = \dfrac{2\cos x + 1}{(2 + \cos x)^2}.$

- When $2k\pi - \frac{2\pi}{3} < x < 2k\pi + \frac{4\pi}{3}$ ($k \in \mathbf{Z}$), $\cos x > -\frac{1}{2}$, $\therefore \ f'(x) > 0.$
- When $2k\pi + \frac{2\pi}{3} < x < 2k\pi + \frac{4\pi}{3}$ ($k \in \mathbf{Z}$), $\cos x < -\frac{1}{2}$, $\therefore \ f'(x) < 0.$

∴ $f(x)$ is monotony increasing function in every interval

$$\left(2k\pi - \frac{2\pi}{3},\ 2k\pi + \frac{4\pi}{3}\right)\ (k \in \mathbf{Z}),$$

$f(x)$ is monotony decreasing function in every interval

$$\left(2k\pi + \frac{2\pi}{3}, 2k\pi + \frac{4\pi}{3}\right)\ (k \in \mathbf{Z}).$$

(2) By convention, variables separation: $a \geq \max_{x\geq 0} \frac{f(x)}{x} = \max_{x\geq 0} \frac{\sin x}{2(2+\cos x)}$, it is unable to solve whether the derivation or the other methods. Only to find the necessary condition of $f(x) \leq ax$, and then improve gradually.

Let $g(x) = ax - f(x)$,

$$\therefore g'(x) = a - \frac{2\cos x + 1}{(2 + \cos x)^2}$$

$$= a - \frac{2}{2 + \cos x} + \frac{3}{(2 + \cos x)^2}$$

$$= 3\left(\frac{1}{2 + \cos x} - \frac{1}{3}\right)^2 + a - \frac{1}{3}.$$

∴ When $a \geq \frac{1}{3}$, $g'(x) \geq 0$.

∵ $g(0) = 0$, ∴ when $x \geq 0$, $g(x) \geq g(0) = 0$, ∴ $f(x) \leq ax$.

Let $h(x) = \sin x - 3ax$, $h'(x) = \cos x - 3a$.

If $h'(x) > 0 \Leftrightarrow \begin{cases} \cos x > 3a \\ x \geq 0 \end{cases} \Leftrightarrow 0 \leq x < \arccos 3a,$

∴ $0 < 3a < 1 \Leftrightarrow 0 < a < \frac{1}{3}.$

When $0 < a < \frac{1}{3}$, let $h(x) = \sin x - 3ax$, ∴ $h'(x) = \cos x - 3a$.

∴ When $x \in [0, \arccos 3a), h'(x) > 0$.

∴ $h(x)$ is monotone increasing in $[0, \arccos 3a)$.

∴ When $x \in (0, \arccos 3a), h(x) > h(0) = 0, \therefore \sin x > 3ax$.

∴ When $x \in (0, \arccos 3a), f(x) = \dfrac{\sin x}{2 + \cos x} > \dfrac{\sin x}{3} > ax$.

∴ When $a \le 0, f\left(\dfrac{\pi}{2}\right) = \dfrac{1}{2} > 0 \ge a \cdot \dfrac{\pi}{2}$.

∴ $a \in \left[\dfrac{1}{3}, +\infty\right)$.

Comment: In the process of solving problem, if it is difficult, retreat to the simple without losing nature, gradually weave, detour step by step in circuitous process. In the course of the detour, we have used the method of strengthening the proposition to solve the problem by the differential calculus.

24. Suppose $a, b \in R_+ n \in N, n \ge 1$.
Find the minimum of $y = \dfrac{a}{\sin^n \theta} + \dfrac{b}{\cos^n \theta}$ $(\theta \in (0, \frac{\pi}{2}))$.

Solution: List:

$\dfrac{a}{\sin^n \theta}$	$\dfrac{a}{\sin^n \theta}$	$\sin^2\theta$	$\sin^2\theta$	□	$\sin^2\theta$
$\dfrac{b}{\cos^n \theta}$	$\dfrac{b}{\cos^n \theta}$	$\cos^2\theta$	$\cos^2\theta$	□	$\cos^2\theta$

$$\therefore y^2 = \left(\frac{a}{\sin^n \theta} + \frac{b}{\cos^n \theta}\right)\left(\frac{a}{\sin^n \theta} + \frac{b}{\cos^n \theta}\right)$$

$$\underbrace{(\sin^2 \theta + \cos^2 \theta)(\sin^2 \theta + \cos^2 \theta) \cdots (\sin^2 \theta + \cos^2 \theta)}_{n\Uparrow}$$

$$\ge \left\{\left[\left(\frac{a}{\sin^n \theta}\right)^2 (\sin^2 \theta)^n\right]^{\frac{1}{n+2}} + \left[\left(\frac{a}{\cos^n \theta}\right)^2 (\cos^2 \theta)^n\right]^{\frac{1}{n+2}}\right\}^{n+2}$$

$$= (a^{\frac{2}{n+2}} + b^{\frac{2}{n+2}})^{n+2}$$

$$\therefore y \geq (a^{\frac{2}{n+2}} + b^{\frac{2}{n+2}})^{\frac{n+2}{2}};$$

when $\dfrac{\frac{a}{\sin^n \theta}}{\frac{b}{\cos^n \theta}} = \dfrac{\sin^2 \theta}{\cos^2 \theta} \Leftrightarrow \tan^{n+2} \theta = \frac{a}{b} \Leftrightarrow \theta = \arctan \left(\frac{a}{b}\right)^{\frac{1}{n+2}}$, equal sign is satisfied;

$$\therefore \text{ when } \theta = \arctan \left(\frac{a}{b}\right)^{\frac{1}{n+2}}, \ y_{\min} = (a^{\frac{2}{n+2}} + b^{\frac{2}{n+2}})^{\frac{n+2}{2}}.$$

25. Suppose $p, q \in R, x \in (0, \frac{\pi}{2})$. Find the minimum of $f(x) = \dfrac{p}{\sqrt{\sin x}} + \dfrac{q}{\sqrt{\cos x}}$.

Solution 1: Let $\alpha = \frac{5}{4}, \beta = 5 \Rightarrow \frac{1}{\alpha} + \frac{1}{\beta} = 1$.

According to Hölder's inequality: $a_i, b_i \in R \ (1 \leq i \leq n)$, $\alpha + \beta = 1, \ \alpha\beta > 0$; then

$$\sum_{i=1}^{n} a_i^{\alpha} b_i^{\beta} \leq \left(\sum_{i=1}^{n} a_i\right)^{\alpha} \left(\sum_{i=1}^{n} b_i\right)^{\beta}$$

$$(\text{when } i = 2, \ a_1^{\alpha} b_1^{\beta} + a_2^{\alpha} b_2^{\beta} \leq (a_1 + a_2)^{\alpha} (b_1 + b_2)^{\beta}).$$

$$\therefore p^{\frac{5}{4}} + q^{\frac{5}{4}} = \frac{p^{\frac{5}{4}}}{(\sin x)^{\frac{2}{5}}} (\sin x)^{\frac{2}{5}} + \frac{q^{\frac{5}{4}}}{(\cos x)^{\frac{2}{5}}} (\cos x)^{\frac{2}{5}}$$

$$= \left(\frac{p}{\sqrt{\sin x}}\right)^{\frac{5}{4}} (\sin^2 x)^{\frac{1}{5}} + \left(\frac{q}{\sqrt{\cos x}}\right)^{\frac{5}{4}} (\cos^2)^{\frac{1}{5}}$$

$$\leq \left(\frac{p}{\sqrt{\sin x}} + \frac{q}{\sqrt{\cos x}}\right)^{\frac{4}{5}} (\sin^2 x + \cos^2 x)^{\frac{1}{5}}$$

$$\therefore f(x) = \frac{p}{\sqrt{\sin x}} + \frac{q}{\sqrt{\cos x}} \geq (p^{\frac{4}{5}} + q^{\frac{4}{5}})^{\frac{5}{4}}.$$

Solution 2: Carlson inequality:

$\dfrac{p}{\sqrt{\sin x}}$	$\dfrac{p}{\sqrt{\sin x}}$	$\dfrac{p}{\sqrt{\sin x}}$	$\dfrac{p}{\sqrt{\sin x}}$	$\sin^2 x$
$\dfrac{q}{\sqrt{\cos x}}$	$\dfrac{q}{\sqrt{\cos x}}$	$\dfrac{q}{\sqrt{\cos x}}$	$\dfrac{q}{\sqrt{\cos x}}$	$\cos^2 x$

According to Carlson's inequality,

$$[f(x)]^{\frac{4}{5}} = \left[\left(\frac{p}{\sqrt{\sin x}} + \frac{q}{\sqrt{\cos x}}\right)\left(\frac{p}{\sqrt{\sin x}} + \frac{q}{\sqrt{\cos x}}\right)\right.$$

$$\left. \times \left(\frac{p}{\sqrt{\sin x}} + \frac{q}{\sqrt{\cos x}}\right)\left(\frac{p}{\sqrt{\sin x}} + \frac{q}{\sqrt{\cos x}}\right)\right]^{\frac{1}{5}}$$

$$\times (\sin^2 x + \cos^2 x)^{\frac{1}{5}} \geq p^{\frac{4}{5}} + q^{\frac{4}{5}},$$

$$\therefore \ f(x) \geq (p^{\frac{4}{5}} + q^{\frac{4}{5}})^{\frac{5}{4}};$$

when

$$\frac{\frac{p}{\sqrt{\sin x}}}{\frac{q}{\sqrt{\cos x}}} = \frac{\sin^2 x}{\cos^2 x} \Rightarrow \tan x = \left(\frac{p}{q}\right)^{\frac{2}{5}},$$

equal sign is satisfied.

Solution 3:
Suppose $t = \dfrac{p}{\sqrt{\sin x}} + \dfrac{q}{\sqrt{\cos x}}$,

$$\therefore \ 1 = \frac{p}{f\sqrt{\sin x}} + \frac{q}{f\sqrt{\cos x}}, \quad \text{associate } 1 = \sin^2 x + \cos^2 x,$$

$$5 = 4 + 1 = 4\left(\frac{p}{f\sqrt{\sin x}} + \frac{q}{f\sqrt{\cos x}}\right) + \sin^2 x + \cos^2$$

$$= \left(4 \cdot \frac{p}{f\sqrt{\sin x}} + \sin^2 x\right) + \left(4 \cdot \frac{q}{f\sqrt{\cos x}} + \cos^2 x\right)$$

$$\geq 5\frac{\left(\sqrt[5]{p^4} + \sqrt[5]{q^4}\right)}{\sqrt[5]{f^4}}$$

$$\therefore \ f(x) \geq (p^{\frac{4}{5}} + q^{\frac{4}{5}})^{\frac{5}{4}},$$

when

$$\frac{\frac{p}{\sqrt{\sin x}}}{\frac{q}{\sqrt{\cos x}}} = \frac{\sin^2 x}{\cos^2 x} \Rightarrow \tan x = \left(\frac{p}{q}\right)^{\frac{2}{5}},$$

equal sign is satisfied.

Exercise

1. If the graph of $y = 3\cos(2x + \phi)$ is centrally symmetrical about point $\frac{4\pi}{3}, 0)$, find the minimum of $|\varphi|$.

2. If the graph of $f(x) = \sqrt{3}\sin\omega x + \cos\omega x$ $(\omega > 0)$ has two *adjacent intersections with line* $y = 2$, and distance of these two points is π. Find the monotony increasing interval of $f(x)$.

3. If there is an overlap between graph of $y = \tan\left(\omega x + \frac{\pi}{4}\right)(\omega > 0)$ to the right translation 6 unit and graph of $y = \tan\left(\omega x + \frac{\pi}{6}\right)$, find the minimum of ω.

4. When $0 \le x \le 1$, inequation $\sin\frac{\pi x}{2} \ge kx$ is satisfied. Find the range value of k.

5. Find the monotony decreasing interval of $f(x) = \sqrt{\log_2 \frac{1}{\sin x} - 1}$.

6. Find the maximum and minimum of $y = \frac{1-\sin x \cos x}{1+\sin x \cos x}$ in $[0, \pi]$.

7. If the graph of $f(x) = \sin(\omega x + \varphi), (\omega > 0)$ has first point $N(6,0)$ on the origin of the right with x-axis, and $f(2 + x) = f(2 - x)$, $f(0) < 0$. Find analytic form of $f(x)$.

8. If $f(x) = 2a\sin^2 x - 2\sqrt{3} - a\sin x \cos x + a + b - 1$, $(a, b \in R$, $a < 0)$, which domain is $[0, \frac{\pi}{2}]$ and range is $[-3, 1]$, find $a + b$.

9. If the graph of $f(x) = A\sin(\omega x + \varphi), x \in R$ $(A > 0, \omega > 0, 0 < \varphi < \frac{\pi}{2})$ has two intersection points with x-axis, and distance of these two points is $\frac{\pi}{2}$, the lowest point is $M(\frac{2\pi}{3}, -2)$.

   (1) Find analytic form of $f(x)$.
   (2) When $x \in [\frac{\pi}{12}, \frac{\pi}{2}]$, find the range of $f(x)$.

10. Suppose $f(x) = \sin(\frac{\pi x}{4} - \frac{\pi}{6}) - 2\cos^2\frac{\pi x}{8} + 1$.

    (1) Find the period of $f(x)$.
    (2) If the graph of $y = g(x)$ and $y = f(x)$ is symmetrical about straight line $x = 1$, find the maximum of $y = g(x)$ $(x \in [0, \frac{4}{3}])$.

11. $f(x) = \cos^2(x + \frac{\pi}{12}), g(x) = 1 + \frac{1}{2}\sin 2x$.

    (1) Suppose $x = x_0$ is symmetric axis of $y = f(x)$. Find $g(x_0)$.
    (2) Find the monotony increasing interval of $h(x) = f(x) + g(x)$.

12. Prove, there are only real number pairs (c, d) in $(0, \frac{\pi}{2})$, $c, d \in (0, \frac{\pi}{2})$, and $c < d$, let $\sin(\cos c) = c, \cos(\sin d) = d$.

13. $f(x) = \frac{m-2\sin x}{\cos x}$ is monotone decreasing in $(0, \frac{\pi}{2})$, find the range of m.

14. For all a, θ constant $F(a, \theta) = \frac{a^2+2a\sin\theta+2}{a^2+2a\cos\theta+2}$ $(a, \theta \in R, a \neq 0)$, find the range of $F(a, \theta)$

15. $\alpha, \beta, \gamma \in (0, \frac{\pi}{2})$, $\cos\alpha + \cos\beta + \cos\gamma = 1$, find the minimum of $\tan^2 = \alpha + \tan^2 = \beta + 8\tan^2 = \gamma$.

16. Suppose x_1, x_2, x_3, x_4 are positive real numbers and $x_1 + x_2 + x_3 + x_4 = \pi$, find the minimum of

$$f(x_1, x_2, x_3, x_4) = \prod_{i=1}^{4}\left(2\sin^2 x_i + \frac{1}{\sin^2 x_i}\right).$$

17. The equation $x^3\sin\theta - (\sin\theta - 2)x^2 + 6x - 4 = 0$ has three positive real roots, find the minimum of $u = \frac{9\sin^2\theta+3-4\sin\theta}{(1-\cos\theta)(2\cos\theta-6\sin\theta-3\sin 2\theta+2)}$.

18. Whether there is a function $f : R \to R$, that makes the real numbers x, y, $f(x + f(y)) = f(x) + \sin y$ constant?

19. $-\frac{\pi}{2} < f(x) + g(x) < \frac{\pi}{2}$ and $-\frac{\pi}{2} < f(x) + g(x) < \frac{\pi}{2}$, prove for all $x \in R$, $\cos f(x) > \sin g(x)$, and thus prove for all $x \in R$, $\cos f(x) > \sin g(x)$.

20. Set the function $f(x)$ for all real x are satisfied $f(x+2\pi) = f(x)$. Prove there are four functions $f_i(i = 1, 2, 3, 4)$:

(1) $i = 1, 2, 3, 4$, $f(x)$ is even function, and $f_i(x + \pi) = f_i(x)$;
(2) $f(x) = f_1(x) + f_2(x)\cos x + f_3(x)\sin x + f_4(x)\sin 2x$.

21. Suppose $a_1, a_2, \ldots, a_n$ are constant real numbers, x is variable number, and

$$f(x) = \cos(a_1 + x) + \frac{1}{2}\cos(a_2 + x) + \frac{1}{4}\cos(a_3 + x)$$

$$+ \cdots + \frac{1}{2^{n-1}}\cos(a_n + x),$$

$$f(x_1) = f(x_2) = 0,$$

prove $x_2 - x_1 = m\pi$, and m is an integer.

22. Find the range of a, if

$$\sin 2\theta - (2\sqrt{2} + \sqrt{2}a)\sin\left(\theta + \frac{\pi}{4}\right) - \frac{2\sqrt{2}}{\cos(\theta - \frac{\pi}{4})} > -3 - 2a$$

always holds in $\theta \in \left[0, \frac{\pi}{2}\right]$.

23. $\theta_i \in \left(0, \frac{\pi}{2}\right), \tan\theta_1 \cdot \tan\theta_2 \cdots \tan\theta_n = 2^{\frac{n}{2}}, n \in N^*$, there is $\cos\theta_1 + \cos\theta_2 + \cdots + \cos\theta_n \le \lambda$, If the above conditions are met for $\theta_1, \theta_2, \ldots, \theta_n$, find the minimum of λ.

Chapter 4

Inverse Trigonometric Functions and Trigonometric Equations

4.1. Inverse Trigonometric Functions

Inverse sine function, inverse cosine function and inverse tangent function

1. The inverse function of $y = \sin x \left(x \in \left[-\frac{\pi}{2}, \frac{\pi}{2} \right] \right)$ is called inverse sine function, which is defined as $y = \arcsin x, x \in [-1, 1]$.
2. The inverse function of $y = \cos x$ $(x \in [0, \pi])$ is called inverse cosine function, which is defined as $y = \arccos x, x \in [-1, 1]$.
3. The inverse function of $y = \tan x \left(x \in \left(-\frac{\pi}{2}, \frac{\pi}{2} \right) \right)$ is called inverse tangent function, which is defined as $y = \arctan x$.

Primary properties of inverse trigonometric functions

Function	$y = \arcsin x$	$y = \arccos x$	$y = \arctan x$
Domain	$[-1, 1]$	$[-1, 1]$	$(-\infty, +\infty)$
Range	$\left[-\frac{\pi}{2}, \frac{\pi}{2} \right]$	$[0, \pi]$	$\left(-\frac{\pi}{2}, \frac{\pi}{2} \right)$
Parity	Odd function	None	Odd function
Monotonicity	Increasing function	Decreasing function	Increasing function
Graph			

Operation of inverse trigonometric functions

1. **Trigonometric operation of inverse trigonometric functions:**
   $\sin(\arcsin x) = x \quad (|x| \leq 1),$
   $\cos(\arccos x) = x \quad (|x| \leq 1),$
   $\tan(\arctan x) = x \quad (x \in R).$

2. **Negative arguments:**
   $\arcsin(-x) = -\arcsin x,$
   $\arccos(-x) = \pi - \arccos x,$
   $\arctan(-x) = -\arctan x.$

3. **Inverse trigonometric operation of trigonometric functions:**
   $\arcsin(\sin x) = x, \quad x \in \left[-\dfrac{\pi}{2}, \dfrac{\pi}{2}\right],$

   $\arccos(\cos x) = x, \quad x \in [0, \pi],$

   $\arctan(\tan x) = x, \quad x \in \left(-\dfrac{\pi}{2}, \dfrac{\pi}{2}\right).$

   The calculation of the inverse trigonometric function can be obtained by setting the angle. Get trigonometric function value of the known angle and the range of the angle, so that the problem of inverse trigonometric function is converted into the problem of trigonometric function value.

Exercise

1. If $\sin x = \frac{1}{3}$,

   (1) when $x \in \left[-\frac{\pi}{2}, \frac{\pi}{2}\right]$, find x;
   (2) when $x \in [0, 2\pi]$, find x;
   (3) when $x \in R$, find the set of value of x.

   Analysis: Find α if trigonometric function value of α is known. The angle has more than one, so the number of angles should be determined according to the range of the angle. The method can be divided into the following steps:

   (1) Determine the quadrant of angle α.

(2) If the function value is positive, find the corresponding acute angle.

(3) If the function value is negative, according to its quadrant, find the corresponding form.

(4) If the angle is out of $[0, 2\pi]$, express this angle with the form of coterminal angle.

Solution:

(1) $x \in \left[-\dfrac{\pi}{2}, \dfrac{\pi}{2}\right]$, $\sin x$ is increasing function;

$\therefore$ one x satisfies $\sin x = \dfrac{1}{3}$

$\therefore x = \arcsin \dfrac{1}{3}$.

(2) The value of $\sin x$ is positive in the first quadrant and second quadrant;

$\therefore$ two x satisfy $\sin x = \frac{1}{3}$ in $(0, 2\pi]$, which are $x = \arcsin \frac{1}{3}$ and $x = \pi - \arcsin \frac{1}{3}$.

(3) When $x \in R$, according to the periodicity of sine function, $x \in \{x \mid x = \arcsin \frac{1}{3} + 2k\pi$ or $x = (2k+1)\pi - \arcsin \frac{1}{3}, k \in Z\}$.

Comment: Known trigonometric function values for angle, the general idea is to find trigonometric functions corresponding to the absolute value of the acute angle, and then use induction formula for the districts corresponding to the angle. In the end, we use the periodicity of the trigonometric function for the general solution.

2. Express following x with inverse trigonometric:

(1) $\sin x = \dfrac{1}{3}, x \in \left[\dfrac{5\pi}{2}, \dfrac{7\pi}{2}\right]$;

(2) $\cos x = a, a \in [-1, 1], x \in [-4\pi, -3\pi]$;

(3) $\sin 2x = a, a \in [-1, 1], x \in \left[\dfrac{\pi}{2}, \dfrac{3\pi}{2}\right]$.

Solution:

(1) $\because x \in \left[\dfrac{5\pi}{2}, \dfrac{7\pi}{2}\right]$, $\quad \therefore x - 3\pi \in \left[-\dfrac{\pi}{2}, \dfrac{\pi}{2}\right]$

$\because \sin(x - 3\pi) = -\sin x = -\dfrac{1}{3}, \quad \therefore x - 3\pi = \arcsin\left(-\dfrac{1}{3}\right)$

$\therefore x = 3\pi - \arcsin\left(\dfrac{1}{3}\right).$

(2) $\because x \in [-4\pi, -3\pi],$

$\therefore x + 4\pi \in [0, \pi], \cos(x + 4\pi) = \cos x = a$

$\therefore x = -4\pi + \arccos a.$

(3) $\because x \in \left[\dfrac{\pi}{2}, \dfrac{3}{2}\pi\right], \quad \therefore 2x \in [\pi, 3\pi];$

$\because$ when $2x \in \left[\pi, \dfrac{3}{2}\pi\right], \quad \therefore 2x - \pi \in \left[0, \dfrac{\pi}{2}\right],$

$\sin(2x - \pi) = -\sin 2x = -a$

$\therefore 2x - \pi = \arcsin(-a), \quad x = \dfrac{\pi}{2} - \dfrac{1}{2}\arcsin a;$

$\because$ when $2x \in \left[\dfrac{3\pi}{2}, \dfrac{5}{2}\pi\right], \quad 2x - 2\pi \in \left[-\dfrac{\pi}{2}, \dfrac{\pi}{2}\right],$

$\sin(2x - 2\pi) = \sin 2x = a$

$\therefore 2x - 2\pi = \arcsin a, x = \pi + \dfrac{1}{2}\arcsin a;$

$\because$ when $2x \in \left[\dfrac{5\pi}{2}, 3\pi\right], \quad 2x - 3\pi \in \left[-\dfrac{\pi}{2}, 0\right],$

$\sin(2x - 3\pi) = -\sin 2x = -a$

$\therefore 2x - 3\pi = \arcsin(-a), \quad x = \dfrac{3\pi}{2} - \dfrac{1}{2}\arcsin a.$

Comment: $\sin x = a$, so $x \in \left[-\dfrac{\pi}{2}, \dfrac{\pi}{2}\right]$ is the condition of $x = \arcsin x$. If x is not in $\left[-\dfrac{\pi}{2}, \dfrac{\pi}{2}\right]$, make a compound angle of x in $\left[-\dfrac{\pi}{2}, \dfrac{\pi}{2}\right]$ by induction formula, then carry out the transformation, and therefore pay special attention to the interval of the angle. Other trigonometric functions are similar to the corresponding transformation.

3. If $f(x) = x^2 - \pi x, \alpha = \arcsin \frac{1}{3}, \beta = \operatorname{arccot} \frac{5}{4}, \gamma =$ $\arccos \left(-\frac{1}{3}\right), \delta = \operatorname{arccot} \left(-\frac{5}{4}\right)$, try to compare the size of $f(\alpha), f(\beta), f(\gamma), f(\delta)$.

Solution 1: $f(x)$ symmetric $x = \frac{\pi}{2}$, monotone increasing in $\left(-\infty, \frac{\pi}{2}\right)$, monotone decreasing in $\left(\frac{\pi}{2}, +\infty\right)$, and $\left|x_1 - \frac{\pi}{2}\right| > \left|x_2 - \frac{\pi}{2}\right|, f(x_1) > f(x_2)$.

$\because 0 < \alpha < \dfrac{\pi}{6}, \quad \dfrac{\pi}{4} < \beta \dfrac{\pi}{3}, \quad \dfrac{\pi}{2} < \gamma < \dfrac{2}{3}\pi, \quad \dfrac{3\pi}{4} < \delta < \dfrac{5}{6}\pi$

$\therefore 0 < \left|\gamma - \dfrac{\pi}{2}\right| < \dfrac{\pi}{6} < \left|\beta - \dfrac{\pi}{2}\right| < \dfrac{\pi}{4},$

$\dfrac{\pi}{4} < \left|\delta - \dfrac{\pi}{2}\right| < \dfrac{\pi}{3} < \left|\alpha - \dfrac{\pi}{2}\right| < \dfrac{\pi}{2}$

$\therefore f(\alpha) > f(\delta) > f(\beta) > f(\gamma).$

Solution 2:

$$\because \alpha = \operatorname{Arg}(2\sqrt{2} + i), \quad \beta = \operatorname{Arg}(4 + 5i),$$

$$\gamma = \operatorname{Arg}(-1 + 2\sqrt{2}i), \quad \delta = \operatorname{Arg}(-5 + 4i)$$

$$\therefore \alpha = \operatorname{arctg} \frac{\sqrt{2}}{4}, \quad \beta = \operatorname{arctg} \frac{5}{4}$$

$$\gamma = \pi + \operatorname{arctg}(-2\sqrt{2}) = \pi - \operatorname{arctg} 2\sqrt{2}$$

$$\delta = \pi + \operatorname{arctg}\left(-\frac{4}{5}\right) = \pi - \operatorname{arctg} \frac{4}{5}$$

$$\because f(x) \text{ symmetric } x = \frac{\pi}{2}$$

$$\therefore f(x) \text{ monotone decreasing in } \left[0, \frac{\pi}{2}\right]$$

$$\because f(x) = f(\pi - x)$$

$$f(\alpha) = f\left(\operatorname{arctg} \frac{\sqrt{2}}{4}\right), \quad f(\beta) = f\left(\operatorname{arctg} \frac{5}{4}\right)$$

$$\therefore f(\gamma) = f(\pi - \text{arctg } 2\sqrt{2}) = f(\text{arctg } 2\sqrt{2})$$

$$f(\delta) = f\left(\pi - \text{arctg } \frac{4}{5}\right) = f\left(\text{arctg} \frac{2}{5}\right)$$

$$\because \frac{\sqrt{2}}{4} < \frac{4}{5} < \frac{5}{4} < 2\sqrt{2}$$

$$\therefore 0 < \text{arctg } \frac{\sqrt{2}}{4} < \text{arctg } \frac{4}{5} < \text{arctg } \frac{5}{4} < \text{arctg } 2\sqrt{2}.$$

Comment: Using the geometric meaning as well, make α, β, γ, δ and $\pi - \alpha$, $\pi - \delta$ to lie in the complex plane, and then we have $f(\alpha) > f(\beta) > f(\gamma) > f(\delta)$ because $f(x)$ is decreasing function in $\left[0, \frac{\pi}{2}\right]$.

4. Find the domain and range of the following function:

(1) $y = \arcsin \frac{1-x^2}{1+x^2}$;

(2) $y = \arccos(x^2 - x)$:

(3) $y = \arctan \left(x + \frac{1}{x}\right)$;

(4) $\ln[\arccos(e^x - 1)]$;

(5) $y = \arctan x + \arcsin x$.

Solution:

(1) $\because -1 \leq \dfrac{1-x^2}{1+x^2} \leq 1, \quad \therefore x \in R$

$$\because -1 \leq \frac{1-x^2}{1+x^2} = \frac{2}{1+x^2} - 1 \leq 1$$

$$\therefore -\frac{\pi}{2} \leq \arcsin \frac{1-x^2}{1+x^2} \leq \frac{\pi}{2}$$

$\therefore$ domain: R, range: $\left(-\dfrac{\pi}{2}, \dfrac{\pi}{2}\right]$.

(2) $\because -1 \leq x^2 - x \leq 1, \quad \therefore \dfrac{1-\sqrt{5}}{2} \leq x \leq \dfrac{1+\sqrt{5}}{2}$

$$\because x^2 - x = \left(x - \frac{1}{2}\right)^2 - \frac{1}{4} \geq -\frac{1}{4}$$

$$\therefore -\frac{1}{4} \leq x^2 - x \leq 1$$

$$\therefore \ 0 \le y \le \pi - \arccos \frac{1}{4}$$

$\therefore$ domain: $\left[\dfrac{1-\sqrt{5}}{2}, \dfrac{1+\sqrt{5}}{2}\right]$,   range: $\left[0, \pi - \arccos \dfrac{1}{4}\right]$.

(3) $\because$ Inverse trigonometric functions domain: R.
So we only need $x \ne 0$.

$$\because \ \left| x + \frac{1}{x}\right| = |x| + \frac{1}{|x|} \ge 2$$

$$\therefore \ x + \frac{1}{x} \le -2 \quad \text{or} \quad x + \frac{1}{x} \ge 2$$

$$\therefore \ -\frac{\pi}{2} < y \le -\operatorname{arctg} 2 \ \text{or} \ \operatorname{arctg} 2 \le y < \frac{\pi}{2}$$

$\therefore$ domain:$(-\infty, 0) \cup (0, +\infty)$

range: $\left(-\dfrac{\pi}{2}, -\operatorname{arctg} 2\right] \cup \left[\operatorname{arctg}2, \dfrac{\pi}{2}\right)$.

(4) $\because$ $\begin{cases} \arccos(e^x - 1) > 0 \\ -1 \le e^x - 1 \le 1 \end{cases}$

$\therefore \ -1 \le e^x - 1 < 1, \quad e^x < 2$

$\therefore \ x < \ln 2$

$\because \ -1 < e^x - 1 < 1,$

$\therefore \ 0 < \arccos(e^x - 1) < \pi, \ln\lfloor \arccos(e^x - 1)\rfloor < \ln \pi$

$\therefore$ domain:$(-\infty, \ln 2) \cup (-\infty, \ln \pi)$.

(5) Domain of function is $[-1, 1]$,

$\because \ y = \arctan x \quad$ and

$y = \arcsin x$ are increasing functions in $[-1, 1]$

$\therefore \ f(-1) \le y \le f(1)$,

$\therefore$ domain is $[-1, 1]$,

range is $[-\arctan 1 - \arcsin 1, \arctan 1 + \arcsin 1]$,

namely $\left[-\dfrac{\pi}{4} - \arcsin 1, \dfrac{\pi}{4} + \arcsin 1\right]$.

Comment: Pay attention to that range of inner function of compound function which is the definition domain of outer function.

5. In $\triangle ABC$, $y = \arccos(\sin A) + \arccos(\sin B) + \arccos(\sin C)$, find the range of y.

Solution:

(i) If $\triangle ABC$ is an acute triangle, then

$$\because \text{arc}\cos(\sin A) = \arccos\left[\cos\left(\frac{\pi}{2} - A\right)\right] = \frac{\pi}{2} - A > 0$$

$$\therefore y = \left(\frac{\pi}{2} - A\right) + \left(\frac{\pi}{2} - B\right) + \left(\frac{\pi}{2} - C\right) = \frac{\pi}{2}.$$

(ii) if $\triangle ABC$ is a non-acute triangle, then

$$\because \arccos(\sin A) = \arccos\left[\cos\left(\frac{\pi}{2} - A\right)\right]$$

$$= \arccos\left[\cos\left(A - \frac{\pi}{2}\right)\right] = A - \frac{\pi}{2} \geq 0$$

$$\therefore y = \left(A - \frac{\pi}{2}\right) + \left(B - \frac{\pi}{2}\right) + \left(\frac{\pi}{2} - C\right)$$

$$= 2A - \frac{\pi}{2} \in \left[\frac{\pi}{2}, \frac{3}{2}\pi\right)$$

$$\therefore y \in \left[\frac{\pi}{2}, \frac{3}{2}\pi\right).$$

6. Find the value of the following expression:

(1) $\arcsin[\sin(-\frac{13\pi}{4})]$;

(2) $\arccos(\cos\frac{11\pi}{6})$;

(3) $\arctan\frac{1}{7} + 2\arcsin\frac{1}{\sqrt{10}}$;

(4) $\arcsin(\sin 2000°)$.

Solution:

(1) $\arcsin\left[\sin\left(-\frac{13}{4}\pi\right)\right] = \arcsin\left[\sin\left(-3\pi - \frac{\pi}{4}\right)\right]$

$$= \arcsin\left(\sin\frac{\pi}{4}\right) = \tfrac{\pi}{4}.$$

$$(2)\ \arccos\left(\cos\frac{11}{6}\pi\right) = \arccos\left[\cos\left(2\pi - \frac{\pi}{6}\right)\right]$$

$$= \arccos\left(\cos\frac{\pi}{6}\right) = \frac{\pi}{6}.$$

$$(3)\ \because\ \tan\left(2\arcsin\frac{1}{\sqrt{10}}\right) = \frac{2\tan\left(\arcsin\dfrac{1}{\sqrt{10}}\right)}{1 - \tan^2\left(\arcsin\dfrac{1}{\sqrt{10}}\right)}$$

$$= \frac{2\cdot\frac{1}{3}}{1-\left(\frac{1}{3}\right)^2} = \frac{3}{4}$$

$$\therefore\ \tan\left(\arcsin\frac{1}{7} + 2\arcsin\frac{1}{\sqrt{10}}\right)$$

$$= \frac{\tan\left(\arcsin\dfrac{1}{7}\right) + \tan\left(2\arcsin\dfrac{1}{\sqrt{10}}\right)}{1 - \tan\left(\arcsin\dfrac{1}{7}\right)\tan\left(2\arcsin\dfrac{1}{\sqrt{10}}\right)}$$

$$= \frac{\frac{1}{7}+\frac{3}{4}}{1-\frac{1}{7}\cdot\frac{3}{4}} = 1$$

$$\because\ 0 < \arcsin\frac{1}{\sqrt{10}} < \arcsin\frac{\sqrt{2}}{2} = \frac{\pi}{4},$$

$$0 < \arcsin\frac{1}{7} < \arcsin 1 = \frac{\pi}{4},$$

$$\therefore 0 < \arcsin\frac{1}{7} + 2\arcsin\frac{1}{\sqrt{10}} < \frac{3\pi}{4},$$

$$\therefore \arcsin\frac{1}{7} + 2\arcsin\frac{1}{\sqrt{10}} = \frac{\pi}{4}.$$

$$(4)\ \arcsin(\sin 2000°) = \arcsin\left[\sin(5\times360° + 200°)\right]$$

$$= \arcsin[\sin(200°)] = -\arcsin[\sin(20°)] = -20°.$$

Comment: In this problem, to pay special attention, the method of finding angle is similar to the method in function problem, but the range of the angle is more subtle.

7. Solve the inequality $\arccos x > \arccos x^2$.

    Solution:

$$\begin{cases} -1 \le x \le 1, \\ -1 \le x^2 \le 1, \\ x < x^2 \end{cases}$$

$$\therefore -1 \le x < 0$$

8. If

$$\begin{cases} \arctan x + \dfrac{1}{2}(2^x - 2^{-x}) - 1 = 0, \\ \arctan 3y + \dfrac{1}{2}(8^y - 8^{-y}) + 1 = 0, \end{cases}$$

find the value of $\arctan(x + 3y)$.

    Solution:

$$\arctan 3y + \frac{1}{2}(8^y - 8^{-y}) + 1 = 0$$

$$\Rightarrow \arctan(-3y) + \frac{1}{2}(2^{-3y} - 2^{3y}) - 1 = 0.$$

$$\because f(x) = \arctan x + \frac{1}{2}(2^x - 2^{-x}) - 1 \text{ is increasing function}$$

$$\therefore f(x) = f(-3y) \Rightarrow x = 3y$$

$$\therefore \arctan(x + 3y) = 0.$$

9. Prove $\arcsin(\cos x) = \arccos(\sin x), 0 < x < \dfrac{\pi}{2}$.

    Proof 1:

$$\because 0 < x < \frac{\pi}{2}$$

$$\therefore 0 < \frac{\pi}{2} - x < \frac{\pi}{2}$$

$$\therefore \arcsin(\cos x)$$

$$= \arcsin\left[\sin\left(\frac{\pi}{2} - x\right)\right] = \frac{\pi}{2} - x$$

$$\therefore \arccos(\sin x) = \frac{\pi}{2} - x$$

$$\therefore \arcsin(\cos x) = \arccos(\sin x).$$

Proof 2: Suppose $A = \arcsin(\cos x)$.

$$\because 0 < x < \frac{\pi}{2}$$

$$\therefore 0 < A < \frac{\pi}{2}$$

According to Fig. 4.1, we can know $AC = \sin x$.

$$\therefore \cos A = \sin x, \quad A = \arccos(\sin x).$$

Therefore, the original inequality is established.

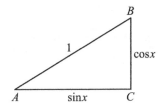

Fig. 4.1.

10. If a, b, c are three sides of a right triangle (where c is hypotenuse), prove

$$\operatorname{arccot}\sqrt{\frac{c+a}{c-a}} + \operatorname{arccot}\sqrt{\frac{c+b}{c-b}} = \frac{\pi}{4}.$$

Proof: Suppose $\alpha \left(0 < \alpha < \frac{\pi}{2}\right)$.

$$\therefore a = c\sin\alpha, b = c\cos\alpha$$

$$\therefore \text{left} = \operatorname{arccot}\sqrt{\frac{1+\sin\alpha}{1-\sin\alpha}} + \operatorname{arccot}\sqrt{\frac{1+\cos\alpha}{1-\cos\alpha}}$$

$$= \operatorname{arccot}\sqrt{\frac{1+\cos\left(\frac{\pi}{2}-\alpha\right)}{1-\cos\left(\frac{\pi}{2}-\alpha\right)}} + \operatorname{arccot}\sqrt{\frac{1+\cos\alpha}{1-\cos\alpha}}$$

$$= \text{arccot}\left|\cot\left(\frac{\pi}{4} - \frac{\alpha}{2}\right)\right| + \text{arccot}\left|\cot\frac{\alpha}{2}\right|$$

$$\because 0 < \alpha < \frac{\pi}{2}, \quad \therefore 0 < \frac{\pi}{4} - \frac{\alpha}{2} < \frac{\pi}{4}$$

$$\therefore \cot\left(\frac{\pi}{4} - \frac{\alpha}{2}\right) > 0, \quad \cot\frac{\alpha}{2} > 0$$

$$\therefore \text{left} = \text{arccot} \cdot \cot\left(\frac{\pi}{4} - \frac{\alpha}{2}\right) + \text{arccot} \cdot \cot\frac{\alpha}{2}$$

$$= \frac{\pi}{4} - \frac{\alpha}{2} + \frac{\alpha}{2} = \frac{\pi}{4} = \text{right.}$$

11. Find the value of $\arctan\frac{1}{3} + \arctan\frac{1}{5} + \arctan\frac{1}{7} + \arctan\frac{1}{8}$.

Solution: Suppose $\alpha = \arctan\frac{1}{3}$, $\beta = \arctan\frac{1}{5}$, $\gamma = \arctan\frac{1}{7}$, $\delta = \arctan\frac{1}{8}$.

$$\therefore \tan\alpha = \frac{1}{3}, \tan\beta = \frac{1}{5}, \tan\gamma = \frac{1}{7}, \tan\delta = \frac{1}{8},$$

$$0 < \alpha, \beta, \gamma, \delta < \frac{\pi}{4},$$

$$\therefore \tan(\alpha + \beta) = \frac{4}{7}, \tan(\gamma + \delta) = \frac{3}{11},$$

$$\tan(\alpha + \beta + \gamma + \delta) = 1, 0 < \alpha + \beta + \gamma + \delta < \frac{\pi}{4},$$

$$\therefore \arctan\frac{1}{3} + \arctan\frac{1}{5} + \arctan\frac{1}{7} + \arctan\frac{1}{8} = \frac{\pi}{4}.$$

12. If $0 < x < 1$, find the value of $2\arctan\frac{1+x}{1-x} + \arcsin\frac{1-x^2}{1+x^2}$.

Solution: Suppose $x = \tan\alpha$, $0 < \alpha < \frac{\pi}{4}$.

$$\therefore 2\arctan\frac{1+x}{1-x} + \arcsin\frac{1-x^2}{1+x^2}$$

$$= 2\arctan\frac{1+\tan\alpha}{1-\tan\alpha} + \arcsin\frac{1-\tan^2\alpha}{1+\tan^2\alpha}$$

$$= 2 \arctan \tan \left(\frac{\pi}{4} + \alpha \right) + \arcsin \cos 2\alpha$$

$$= 2 \arctan \tan \left(\frac{\pi}{4} + \alpha \right) + \arcsin \sin \left(\frac{\pi}{2} - 2\alpha \right)$$

$$\because 0 < \alpha < \frac{\pi}{4}, \quad \therefore \frac{\pi}{4} < \alpha + \frac{\pi}{4} < \frac{\pi}{2}, \ 0 < \frac{\pi}{2} - 2\alpha < \frac{\pi}{2}$$

$$\therefore = 2 \left(\alpha + \frac{\pi}{4} \right) + \left(\frac{\pi}{2} - 2\alpha \right) = \pi.$$

13. Find the value of $\arctan \frac{1}{2} + \arctan \frac{1}{8} + \cdots + \arctan \frac{1}{2n^2}$.

Solution:

$$\frac{1}{2n^2} \overline{\overline{\text{constructed by tangents of subtraction of two angles}}} \frac{2}{4n^2}$$

$$= \frac{(2n+1) - (2n-1)}{(2n+1)(2n-1) + 1} \overline{\overline{\text{expressed by inverse trigonometric}}}$$

$$\overline{\overline{\text{functions}}}$$

$$\frac{\tan[\arctan(2n+1) - \tan[\arctan(2n-1)]}{1 + \tan[\arctan(2n+1)] \cdot \tan[\arctan(2n-1)]}$$

$$= \tan[\arctan(2n+1) - \arctan(2n+1)].$$

At the same time on both sides of the inverse tangent:

$$\arctan \frac{1}{2n^2} = \arctan(2n+1) - \arctan(2n-1)$$

$$= (\arctan 3 - \arctan 1) + (\arctan 5 - \arctan 3) + \cdots$$

$$+ [\arctan(2n+1)] - [\arctan(2n-1)]$$

$$= \arctan(2n+1) - \arctan 1 = \arctan \frac{n}{n+1}.$$

14. Find the value of

$$\arcsin \frac{\sqrt{3}}{2} + \arcsin \frac{\sqrt{8} - \sqrt{6}}{6} + \arcsin \frac{\sqrt{15} - \sqrt{8}}{12} + \cdots$$

$$+ \arcsin \frac{\sqrt{(n+1)^2 - 1} - \sqrt{n^2 - 1}}{n(n+1)}.$$

Solution:

According to the sine operation of inverse sine function:

$$\frac{\sqrt{(n+1)^2 - 1} - \sqrt{n^2 - 1}}{n(n+1)},$$

let $\frac{1}{n} = \sin \alpha_n \left(\alpha_n \in \left(0, \frac{\pi}{2}\right]\right).$

15. Draw the graph of $y = \arctan x + \arctan \frac{1-x}{1+x}$.

Solution:

Suppose $x = \tan \alpha \left(-\frac{\pi}{2} < \alpha < \frac{\pi}{2}, \alpha \neq -\frac{\pi}{4}\right)$ (see Fig. 4.2).

$$\therefore y = \arctan \tan \alpha + \arctan \frac{1 - \tan \alpha}{1 + \tan \alpha}$$

$$= \arctan \tan \alpha + \arctan \tan \left(\frac{\pi}{4} - \alpha\right)$$

$$\because -\frac{\pi}{2} < \alpha < \frac{\pi}{2}, \quad \therefore -\frac{\pi}{4} < \frac{\pi}{4} - \alpha < \frac{3}{4}\pi, \quad \frac{\pi}{4} - \alpha \neq \frac{\pi}{2}$$

$$\therefore -\frac{\pi}{4} < \alpha < \frac{\pi}{2}, -\frac{\pi}{4} < \frac{\pi}{4} - \alpha < \frac{1}{2}\pi \ (x > -1), \ y = \alpha + \frac{\pi}{4} - \alpha = \frac{\pi}{4};$$

$$\therefore \text{when} \ -\frac{\pi}{2} < \alpha < -\frac{\pi}{4}, \frac{\pi}{2} < \frac{\pi}{4} - \alpha < \frac{3}{4}\pi \ (x < -1)$$

$$\because \tan \left(\frac{\pi}{4} - \alpha\right) = \tan \left(\frac{\pi}{4} - \alpha - \pi\right), \ -\frac{\pi}{2} < \frac{\pi}{4} - \alpha - \pi < -\frac{\pi}{4},$$

$$\therefore y = \alpha + \frac{\pi}{4} - \alpha - \pi = -\frac{3}{4}\pi;$$

$$\therefore y = \begin{cases} \dfrac{\pi}{4} & (x > -1) \\ -\dfrac{3}{4}\pi & (x < -1) \end{cases} \quad \text{(see Fig. 4.2).}$$

4.2. Trigonometric Equations

A trigonometric equation is any equation that contains a trigonometric function.

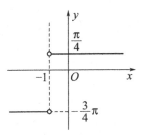

Fig. 4.2.

1. Solution set of simplest trigonometric equation:

Simplest trigonometric equation		Solution set
$\sin x = a$	$\|a\| \leq 1$	$\{x \mid x = k\pi + (-1)^k \arcsin a, k \in Z\}$
	$\|a\| > 1$	$\emptyset$
$\cos x = a$	$\|a\| \leq 1$	$\{x \mid x = 2k\pi \pm \arccos a, k \in Z\}$
	$\|a\| > 1$	$\emptyset$
$\tan x = a$		$\{x \mid x = k\pi + \arctan a, k \in Z\}$

2. Most trigonometric equations can be divided into one of these types:

(1) Simplest type: as $a\sin(wx + \varphi) + b = 0$.
   Method: Transform to $\sin x = A$, and solve the equation with *solution set of simplest trigonometric equation*.

(2) Substitution type: as $a \sin^2 x + b \sin x + c = 0$.
   Method: Transform to $(A_1 \sin x - B_1)(A_2 \sin x - B_2) = 0$ and then transform to type (1).

(3) Type of $a \sin x + b \cos x + c = 0$.
   Method: Transform to type (1) with the auxiliary angle formula.

(4) The same equivalent triangle ratio type: as $\sin(w_1 x + \varphi_1) = \sin(w_2 x + \varphi_2)$

$\sin A = \sin B$	$\cos A = \cos B$	$\tan A = \tan B$	$\cot A = \cot B$
$A = k\pi + (-1)^k B$	$A = 2k\pi \pm B$	$A = k\pi + B$	$A = k\pi + B$

(5) Homogeneous type: as $a\sin^2 x + b\sin x \cos x + c\cos^2 x = 0$.
Method: Transform to $a\tan^2 x + b\tan x + c = 0$

3. Attention in solving trigonometric equations:

(1) For a range of solutions of the equation, first find out the general form of the equation. Then according to the problem, determine the value range of k, and finally calculate the equation in the specified range. It can avoid the omission of the solution of equations.

(2) Using the formula to the denominator, square, square root. The use of tangent, cotangent may generate extraneous roots or root lost. Therefore in the deformation process it should pay attention to whether the deformation of each step of the solution is the same strain; attention should be paid to the inspection.

(3) The expression of general solution of trigonometric equation form is not the only form. Do not get excited over a little thing, if there are different expressions.

4. Method of solving trigonometric equations: keep simple, namely

Trigonometric Equations $\xrightarrow{\text{Substitution or Transform}}$ Simplest

Trigonometric Equation $\xrightarrow{\text{Inverse Trigonometric Function}}$ Solution Set.

Exercise

1. Find the value range of k, if equation $(2 - \cos x)k = 2 + \cos x$ has real solution.

Solution: The original equation transforms into
$(k + 1) \cos x = 2k - 2$.

- When $k = -1$, the equation has no real solution.
- When $k \neq -1$, $\cos x = \frac{2k-2}{k+1}$.

If the equation has real solution, $\left| \frac{2k-2}{k+1} \right| \leq 1, \therefore \frac{1}{3} \leq k \leq 3$.

Comment: When solving equation $\sin x = a$, $\cos x = a$, pay attention to the range of a. Equation has solution if and only if $|a| \leq 1$.

2. Solve equations:

(1) $2\sin(3x + \frac{\pi}{4}) = \sqrt{3}$;

(2) $\sin(3x + \frac{\pi}{3}) = \sin 2x$;

(3) $\tan(3x + \frac{\pi}{3}) = 5$;

(4) $5\cos(3x + \frac{\pi}{4}) = 1$;

(5) $\sin x + \cos x + \tan x + \cot x + \sec x + \csc x + 2 = 0$;

(6) $\tan(x + \frac{\pi}{4}) + \tan(x - \frac{\pi}{4}) = 2\cot x$.

Solution:

(1) $3x + \dfrac{\pi}{4} = k\pi + (-1)^k \dfrac{\pi}{3}, \quad \therefore x = \dfrac{k\pi}{3} - \dfrac{\pi}{12} + (-1)^k \dfrac{\pi}{9}, k \in Z$.

(2) $3x + \dfrac{\pi}{3} = k\pi + (-1)^k 2x$,

$$\therefore x = \begin{cases} \dfrac{2n\pi}{3} - \dfrac{4\pi}{9}, k = 2n - 1, n \in Z, \\ \\ 2n\pi - \dfrac{\pi}{3}, k = 2n, n \in Z. \end{cases}$$

(3) $x = \dfrac{k\pi}{3} + \dfrac{1}{3}\arctan 5 - \dfrac{\pi}{9}, \quad k \in Z$.

(4) $x = \dfrac{2k\pi}{3} \pm \dfrac{1}{3}\arccos\dfrac{1}{5}, \quad k \in Z$.

(5) $\sin x + \cos x + \dfrac{1}{\sin x \cos x} + \dfrac{\sin x + \cos x}{\sin x \cos x} + 2 = 0$

$\because y = \sin x + \cos x, \quad \therefore \sin x \cos x = \dfrac{y^2 - 1}{2}$

$\therefore y + \dfrac{2}{y^2 - 1} + \dfrac{2y}{y^2 - 1} + 2 = 0$

$\therefore y(y+1)^2 = 0, \quad \therefore y = 0 \quad \text{or} \quad y = -1$

$\because \sin x \neq 0, \cos x \neq 0, \quad \therefore \dfrac{y^2 - 1}{2} = \sin x \cos x \neq 0$

$\therefore y \neq -1$

$\therefore y = 0, \sin x + \cos x = 0$

$\therefore \tan x = -1, x = k\pi - \dfrac{\pi}{4}, \quad k \in Z$

$\therefore \left\{ x \,\middle|\, x = k\pi - \dfrac{\pi}{4}, \ k \in Z \right\}.$

(6) $\dfrac{\sin\left(x + \frac{\pi}{4}\right)}{\cos\left(x + \frac{\pi}{4}\right)} + \dfrac{\sin\left(x - \frac{\pi}{4}\right)}{\cos\left(x - \frac{\pi}{4}\right)} = \dfrac{2\cos x}{\sin x}$

Go to the denominator: $\sin 2x \sin x$

$= 2\cos\left(x + \dfrac{\pi}{4}\right)\cos\left(x - \dfrac{\pi}{4}\right)\cos x$

$\therefore \sin 2x \sin x = \cos 2x \cos x, \quad \cos 3x = 0$

$\therefore x = \dfrac{k\pi}{3} + \dfrac{\pi}{6}, \quad k \in Z.$

3. Solve the equation

$$\arccos\left|\dfrac{x^2-1}{x^2+1}\right| + \arcsin\left|\dfrac{2x}{x^2+1}\right| + \text{arccot}\left|\dfrac{x^2-1}{2x}\right| = \pi.$$

Solution: Let $|x| = \tan \alpha, \alpha \in \left(0, \frac{\pi}{2}\right)$

$$\arccos \left| \frac{\tan^2 \alpha - 1}{\tan^2 \alpha + 1} \right| + \arcsin \left| \frac{2 \tan \alpha}{\tan^2 \alpha + 1} \right|$$

$$+ \operatorname{arccot} \left| \frac{1 - \tan^2 \alpha}{2 \tan \alpha} \right| = \pi.$$

When $\tan \alpha \geq 1 \Rightarrow \dfrac{\pi}{2} > \alpha \geq \dfrac{\pi}{4}$, $\tan^2 \alpha - 1 \geq 0, \pi > 2\alpha \geq \dfrac{\pi}{2}$,

$\therefore \arccos(-\cos 2\alpha) + \arcsin \sin 2\alpha + \operatorname{arccot}(-\cot 2\alpha) = \pi$

$\therefore (\pi - 2\alpha) + (\pi - 2\alpha) + (\pi - 2\alpha) = \pi$

$\therefore \alpha = \dfrac{\pi}{3}, \; x = \pm\sqrt{3}.$

When $\tan \alpha < 1 \Rightarrow 0 < \alpha < \dfrac{\pi}{4}$,

$\therefore \tan^2 \alpha - 1 < 0$

$\therefore \arccos \cos 2\alpha + \arcsin \sin 2\alpha + \operatorname{arccot} \cot 2\alpha = \pi$

$\therefore 6\alpha = \pi, \; \alpha = \dfrac{\pi}{6}, \; x = \pm\dfrac{\sqrt{3}}{3}.$

Comment: Pay attention to application of universal replacement formula.

4. Solve the equation $\arctan x + \arcsin \dfrac{x}{\sqrt{x^2 + \frac{9}{4}}} = \dfrac{\pi}{4}$.

Solution:

$$\tan \left(\arcsin \frac{x}{\sqrt{x^2 + \frac{9}{4}}} \right) = \frac{\sqrt{x^2 + \frac{9}{4}}}{\sqrt{1 - \left(\dfrac{x}{\sqrt{x^2 + \frac{9}{4}}} \right)^2}} = \frac{2x}{3}.$$

Take two tangents:

$$\tan\left(\operatorname{arctg} x + \arcsin \frac{x}{\sqrt{x^2 + \frac{9}{4}}}\right) = 1 \Rightarrow \frac{x + \frac{2}{3}x}{1 - x \cdot \frac{2}{3}x} = 1,$$

$$\frac{5}{3}x = 1 - \frac{2}{3}x^2$$

$$\therefore x = \frac{1}{2} \quad \text{or} \quad x = -3.$$

Comment: When solving the inverse sine function equation and inverse cosine function equation, we need to check the root. Because domain of these two function is $[-1,1]$, the range of x may be expanded during solving process.

5. Find x in $[0, 2\pi]$, which satisfies the following condition:

$$2\cos x \le \sqrt{1 + \sin 2x} + \sqrt{1 - \sin 2x} \le \sqrt{2}.$$

Solution: Let $y = \sqrt{1 + \sin 2x} + \sqrt{1 - \sin 2x}$

$$\therefore y^2 = 2 + 2\sqrt{1 - \sin^2 2x} = 2 + 2|\cos 2x|$$

$$\because \text{right} = 2 + 2|\cos 2x| \le 2$$

$$\therefore \cos 2x = 0, \quad x = \frac{k\pi}{2} + \frac{\pi}{4}, \ k \in Z$$

$$\because x \in [0, 2\pi], \ \therefore x = \frac{\pi}{4}, \frac{3}{4}\pi, \frac{5}{4}\pi, \frac{7}{4}\pi$$

$$\therefore x \Rightarrow \left\{\frac{\pi}{4}, \frac{3}{4}\pi, \frac{5}{4}\pi, \frac{7}{4}\pi\right\}.$$

6. (1) Solve the equation $16\sin \pi x \cos \pi x = 16x + \frac{1}{x}$.

Solution: When $x > 0$, $16x + \frac{1}{x} \ge 8$. (When $x = \frac{1}{4}$, equal sign is satisfied.)

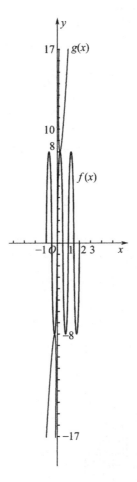

$\because 16\sin \pi x \cos \pi x = 8\sin 2\pi x \le 8 \quad \left(x = \dfrac{1}{4} + k, k \in \mathbb{Z} \right).$

When $x > 0, \quad x = \dfrac{1}{4}.$

$\because$ It is an odd function.

$\therefore x = -\dfrac{1}{4}$

∴ Solution set is $\left\{\dfrac{1}{4}, -\dfrac{1}{4}\right\}$.

(2) How many real roots satisfy equation $\sqrt{x}\sin(x^2) - 2 = 0$ in $[0, 20]$?

Solution: Suppose $y = \sqrt{x}\sin(x^2)$.

When $y = 0$, the solution of equation is $x = \sqrt{k\pi} \in [0, 20]$, $k \in Z$.

$$\therefore x = \sqrt{k\pi}\,(k = 0, 1, 2, \ldots, 127) \text{ has } 128 \text{ roots.}$$

When $\sqrt{x} \geq 2$, the graph of $y = \sqrt{x}\sin(x^2)$ can intersect to the graph of $y = 2$, $\therefore x \geq 4$ $\therefore \sqrt{5\pi} < 4$, $\sqrt{6\pi} > 4$.

∴ When $x \in [\sqrt{6\pi}, \sqrt{7\pi}]$, $\sqrt{x}\sin(x^2) \geq 0$.

∴ The function $y = \sqrt{x}\sin(x^2)$ is increasing from $(\sqrt{6\pi}, 0)$, and then descending to $(\sqrt{7\pi}, 0)$,

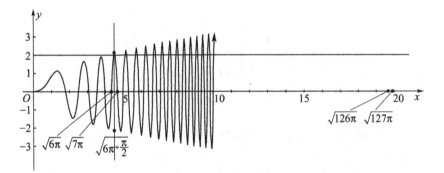

$$y_{\max} \geq \sqrt{\sqrt{6\pi + \frac{\pi}{2}} \cdot \sin\left[\left(\sqrt{6\pi + \frac{\pi}{2}}\right)^2\right]} = \sqrt{\sqrt{6\pi + \frac{\pi}{2}}} > 2.$$

∴ equation $\sqrt{x}\sin(x^2) - 2 = 0$ has two roots in $[\sqrt{6\pi}, \sqrt{7\pi}]$.

Similarly, the equation $\sqrt{x}\sin(x^2) - 2 = 0$ has two roots in $[\sqrt{8\pi}, \sqrt{9\pi}]$, $[\sqrt{10\pi}, \sqrt{11\pi}], \ldots, [\sqrt{126\pi}, \sqrt{127\pi}]$. Therefore, there are 122 roots.

Exercise 4

1. Solve the inequality
$$\begin{cases} \dfrac{x^2 + x - 2}{x^2 + x - 6} < 0, \\[2mm] \cos 5x + \cos x > 2\cos 2x. \end{cases}$$

2. Solve the equation $\log_{\sin 3x}(\cos x - \cos 2x) = 1$.

3. In $\triangle ABC$, if $\cos 3A + \cos 3B + \cos 3C = 1$, prove one angle in $\triangle ABC$ is constant value.

4. In $\triangle ABC$, $\tan \angle CAB = \frac{22}{7}$, draw a vertical line of BC from A, the BC is divided into 3 and 17 two line segments. Find the area of $\triangle ABC$.

5. If $\arcsin(\sin \alpha + \sin \beta) + \arcsin(\sin \alpha - \sin \beta) = \frac{\pi}{2}$, find the value of $\sin^2 \alpha + \sin^2 \beta$.

6. Find the range of $y = \arctan x + \arctan \frac{1-x}{1+x}$.

7. There is an inscribed square in the right-angled triangle $\triangle ABC$, and one of the side is hypotenuse BC of $\triangle ABC$.

   (1) Suppose $AB = a$, $\angle ABC = \theta$, express S_1 which is area of $\triangle ABC$ and S_2 which is area of square with a and θ.

   (2) If a is constant, and θ is variable, find the θ when $\frac{S_1}{S_2}$ is minimum.

8. Find $\arctan \frac{1}{3} + \arctan \frac{1}{7} + \arctan \frac{1}{13} + \cdots + \arctan \frac{1}{1+n+n^2}$.

9. Find all constants c, let $f(x) = c + \arctan \frac{2-2x}{1+4x}$ is an odd function in $\left(-\frac{1}{4}, \frac{1}{4}\right)$.

10. Solve $\sin^3 x + \cos^5 x = 1$.

11. If x_1, x_2 are two roots of $x^2 - 6x + 7 = 0$, find $\arctan x_1 + \arctan x_2$.

12. Find $\sum_{k=1}^{n} \arcsin \frac{\sqrt{(k+1)^2-1}-\sqrt{k^2-1}}{k(k+1)}$.

13. Let $a \in R$. Discussion about the solutions of the equation $\cos 2x + 2\sin x + 2a - 3 = 0$ in $[0, 2\pi]$.

14. Solve $\cos^n x - \sin^n x = a$, where n is arbitrarily positive integer.

15. Solve $\cos^2 x + \cos^2 2x + \cos^2 3x = 1$.

Chapter 5

Solutions of Triangles

5.1. Solutions of Triangles

Law of sines and law of cosines, area of triangles

1. Laws of Sines

In $\triangle ABC$, $\frac{a}{\sin A} = \frac{b}{\sin B} = \frac{c}{\sin C} = 2R$ (R is the radius of circumcircle of $\triangle ABC$)

Laws of Sines reveal the quantitative relationship between three sides of the triangle and its diagonal sine. There are four elements, if any three, for fourth.

Function of law of sines:

(1) known two angles and arbitrary side of triangle, for other sides and an angle (one solution);

(2) known two sides and the diagonal of one side, for the other side and two angles (one solution or two solutions).

2. Law of Cosines

In $\triangle ABC$,

$$a^2 = b^2 + c^2 - 2bc \cos A,$$

$$b^2 = c^2 + a^2 - 2ca \cos B,$$

$$c^2 = a^2 + b^2 - 2ab \cos C.$$

Laws of Cosines reveal the quantitative relationship between three sides of the triangle and one side diagonal cosine. There are four known elements, if three, for fourth.

Function of law of cosines:

(1) known three sides, for three angles (one solution);

(2) known two sides and their included angle, for the third side and other angles (one solution).

3. Formula of Triangles Area

In $\triangle ABC$,

$$S_{\triangle ABC} = \tfrac{1}{2}ab\sin C = \tfrac{1}{2}bc\sin A$$
$$= \tfrac{1}{2}ca\sin B.$$

4. Study Relationship of Trigonometric Ratios in the Triangle, Accumulating Some Familiar Knowledge Blocks

$A + B + C = \pi, A = \pi - (B + C), B + C = \pi - A,$

$\frac{A}{2} = \frac{\pi}{2} - \frac{B+C}{2}, \sin A = \sin(B + C), \cos A = -\cos(B + C),$

$\tan A = -\tan(B + C),$

$\sin \frac{A}{2} = \cos \frac{B+C}{2}, \quad \cos \frac{A}{2} = \sin \frac{B+C}{2}, \tan \frac{A}{2} = \cot \frac{B+C}{2}.$

5.
The shape of the triangle is determined from the trigonometric equation, often by Laws of Sines, translate side to the sine of one angle or vice versa. By Laws of Cosines, translate three sides to cosine of one angle or translate cosine of one angle to relationship of three sides.

6.
By using Laws of Sines or formula of triangle area $S_\triangle = \tfrac{1}{2}ab\sin C$ find the angle, and finally calculate the angle by its sine. Because angle belongs to $(0, \pi)$, two supplemental non-right angles are an arbitrary acute angle and an obtuse angle, their sines must be the same. Control the range of the angle in order to judge whether which angle is correct.

In $\triangle ABC$, if $a = 20\sqrt{3}, b = 20, A = 60°$, find B by law of sines,

$$\frac{a}{\sin A} = \frac{b}{\sin B} \Rightarrow \frac{20\sqrt{3}}{\sin 60°} = \frac{20}{\sin B} \Rightarrow \sin B = \frac{1}{2},$$

$\because B = 30°$ or $B = 150°$,

$\because A = 60°$ by triangular sum theorem

$B \in (0°, 120°), B = 30°,$

Commonly used formula in $\triangle ABC$

1. In $\triangle ABC$, $A > B \Leftrightarrow \sin A > \sin B$.
2. In $\triangle ABC$, $\sin^2 A + \sin^2 B + \sin^2 C = 2(1 + \cos A \cos B \cos C)$,

$$\cos^2 A + \cos^2 B + \cos^2 C = 1 - 2\cos A \cos B \cos C.$$

3. In non-right-angled triangle $\triangle ABC$, $\tan A + \tan B + \tan C = \tan A \tan B \tan C$.

4. In $\triangle ABC$, $\tan \frac{A}{2} \tan \frac{B}{2} + \tan \frac{B}{2} \tan \frac{C}{2} + \tan \frac{C}{2} \tan \frac{A}{2} = 1$.

5. $\sin A + \sin B + \sin C = 4 \cos \frac{A}{2} \cos \frac{B}{2} \cos \frac{C}{2}$.

6. $\cos A + \cos B + \cos C = 4 \sin \frac{A}{2} \sin \frac{B}{2} \sin \frac{C}{2} + 1$.

7. $\sin \frac{A}{2} + \sin \frac{B}{2} + \sin \frac{C}{2} = 1 + 4 \sin \frac{\pi - A}{4} \sin \frac{\pi - B}{4} \sin \frac{\pi - C}{4}$.

8. $\cos \frac{A}{2} + \cos \frac{B}{2} + \cos \frac{C}{2} = 4 \cos \frac{A+B}{4} \cos \frac{B+C}{4} \cos \frac{C+A}{4}$.

Exercise

1. In $\triangle ABC$, angles A, B, C correspond to the side a, b, c.
If $a^2 + b^2 - c^2 = ab$, and $\frac{\tan A - \tan B}{\tan A + \tan B} = \frac{c-b}{c}$, prove $\triangle ABC$ is a equilateral triangle.

Analysis: $\because a^2 + b^2 - c^2 = ab$, by Laws of Cosines, $C = 60^0$;

$\therefore \dfrac{\tan A - \tan B}{\tan A + \tan B} = \dfrac{c-b}{c}$ is a hybrid containing sides and angles.

By Laws of Sines and Cosines, transform relationship of angles to sides first or relationship of sides to angles, so as to achieve the purpose to solve this problem.

Solution 1:

$$\because a^2 + b^2 - c^2 = 2ab \cos C = ab,$$

$$\cos C = \frac{1}{2}, 0 < C < 180°, C = 60°.$$

$$\frac{\tan A - \tan B}{\tan A + \tan B} = \frac{\frac{\sin A}{\cos A} - \frac{\sin B}{\cos B}}{\frac{\sin A}{\cos A} + \frac{\sin B}{\cos B}} = \frac{\sin A \cos B - \cos A \sin B}{\sin A \cos B + \cos A \sin B}$$

$$= \frac{\sin(A-B)}{\sin(A+B)} = \frac{\sin(A-B)}{\sin C},$$

$$\because \frac{c-b}{c} = \frac{\sin C - \sin B}{\sin C},$$

$$\therefore \frac{\sin C - \sin B}{\sin C} = \frac{\sin(A-B)}{\sin C},$$

$\therefore\ \sin C - \sin B = \sin(A - B),$

$\sin(A + B) - \sin B = \sin(A - B),$

$\sin A \cos B + \cos A \sin B - \sin B = \sin A \cos B - \cos A \sin B,$ which implies $\sin B = 2 \cos A \sin B,$

$\because \sin B > 0, \therefore\ \cos A = \frac{1}{2}, \because\ 0 < A < 180°, \therefore\ A = 60°,$

$\therefore\ \triangle ABC$ is an equilateral triangle.

Solution 2: $C = 60°.$

$$\because\ \frac{\tan A - \tan B}{\tan A + \tan B} = \frac{c - b}{c}$$

$$\because\ \frac{\tan A}{\tan B} = \frac{2c - b}{b}$$

$$\because\ \frac{\sin A \cos B}{\cos A \sin B} = \frac{2 \sin C - \sin B}{\sin B}$$

$$\therefore\ \sin A \cos B = 2 \sin C \cos A - \cos A \sin B$$

$$\therefore\ \sin(A + B) = \sin C = 2 \sin C \cos A$$

$$\therefore\ \cos A = \frac{1}{2}, \quad 0 < A < \pi$$

$$\therefore\ A = 60° \quad \therefore\ A = B = C = 60°.$$

Solution 3: $C = 60°.$

$$\frac{\tan A - \tan B}{\tan A + \tan B} = \frac{\frac{\sin A}{\cos A} - \frac{\sin B}{\cos B}}{\frac{\sin A}{\cos A} + \frac{\sin B}{\cos B}} = \frac{\sin A\ \cos B - \cos A\ \sin B}{\sin A\ \cos B + \cos A\ \sin B}$$

$$= \frac{\frac{a}{2R} \cdot \frac{a^2+c^2-b^2}{2ac} - \frac{b}{2R} \cdot \frac{b^2+c^2-a^2}{2bc}}{\frac{a}{2R} \cdot \frac{a^2+c^2-b^2}{2ac} + \frac{b}{2R} \cdot \frac{b^2+c^2-a^2}{2bc}} = \frac{a^2 - b^2}{c^2}.$$

(R is the radius of circumcircle of $\triangle ABC$)

$\therefore\ \frac{a^2-b^2}{c^2} = \frac{c-b}{c},$ which implies $b^2 + c^2 - a^2 = bc.$

$\therefore\ \cos A = \frac{1}{2}, \because\ 0 < A < 180°, \therefore\ A = 60°.$

$\therefore\ \triangle ABC$ is a equilateral triangle.

Comment:

(1) If relationship between the sides and angles is known in the triangle, determine the shape of a triangle. Solution 1 translates sides into angles and Solution 2 translates angles into sides. No matter what kind of solutions, cosine law is the bridge of realizing the transformation.

(2) When solving the questions about triangles, some of the more complex problems often need to using sine law and cosine law alternately, so as to achieve the purpose of simplification. Sometimes, some conclusions and formulas are derived from the sine and cosine theorems.

(3) This case that involves the basic mathematics thinking method is a method of equivalent translation, such as side into angle, angle into edge, tangent into sine (cosine). In the transformation process, but also penetrate the elimination (reduce the number of unknowns) the basic problem solving method.

2. In $\triangle ABC$, if $B = 30°, c = 2\sqrt{3}, b = 2$, find the area of $\triangle ABC$.

Analysis: This problem belongs to the problem to solve triangle which is known as the diagonal of the two sides and one opposite angles. There are many ways to deal with it.

Solution 1: First find angles, and find $\sin C$ by sine law. Then find the area.

Solution 2: First find sides, and find a by cosine law. Then find the area.

Solution 1: According to law of sines: $\frac{c}{\sin C} = \frac{b}{\sin B}$,

$\therefore \frac{2\sqrt{3}}{\sin C} = \frac{2}{\sin 30°}, \therefore \sin C = \frac{\sqrt{3}}{2}.$

(1) When C is an acute angle, $C = 60°, A = 90°$,

$\therefore S_{\triangle ABC} = \frac{1}{2}bc = \frac{1}{2} \times 2 \times 2\sqrt{3} = 2\sqrt{3}.$

(2) When C is an obtuse angle, $C = 120°, A = 30°$,

$\therefore S_{\triangle ABC} = \frac{1}{2}bc \sin 30° = \frac{1}{2} \times 2 \times 2\sqrt{3} \times \frac{1}{2} = \sqrt{3}.$

Solution 2: According to law of cosines, $\cos B = \frac{a^2+c^2-b^2}{2ac}$.

$$\therefore \cos 30° = \frac{a^2 + (2\sqrt{3})^2 - 2^2}{2a \times 2\sqrt{3}}, \quad \therefore a^2 - 6a + 8 = 0,$$

$$\therefore a = 2 \quad \text{or} \quad a = 4.$$

$$\because S_{\triangle ABC} = \frac{1}{2} ac \sin B, \quad \therefore S_{\triangle ABC} = \sqrt{3} \quad \text{or} \quad 2\sqrt{3}.$$

Comment: When solving triangle where two sides and opposite angle are known, the discussion on the triangle is a difficult point in the application of the sine law to solve the triangle.

3. In $\triangle ABC$, angles A, B, C correspond to sides a, b, c, which are geometric sequence. Find the range of $y = \frac{1+\sin 2B}{\sin B + \cos B}$.

Analysis: Because the function is expressed in the form of the triangle form of angle B, the condition $b^2 = ac$ can be transformed to the form of the triangle of angle B.

Solution:

$$\because b^2 = ac$$

$$\therefore \cos B = \frac{a^2 + c^2 - b^2}{2ac} = \frac{a^2 + c^2 - ac}{2ac} = \frac{1}{2}\left(\frac{a}{c} + \frac{c}{a}\right) - \frac{1}{2} \geq \frac{1}{2}$$

$$\therefore 0 < B \leq \frac{\pi}{3}$$

$$y = \frac{1 + \sin 2B}{\sin B + \cos B} = \frac{(\sin B + \cos B)^2}{\sin B + \cos B} = \sin B + \cos B$$

$$= \sqrt{2}\sin\left(B + \frac{\pi}{4}\right)$$

$$\because \frac{\pi}{4} < B + \frac{\pi}{4} \leq \frac{7\pi}{12}$$

$$\therefore \frac{\sqrt{2}}{2} < \sin\left(B + \frac{\pi}{4}\right) \leq 1,$$

$$\therefore 1 < y \leq \sqrt{2}.$$

Comment: This question is used knowledge of inequality, cosine law, cosines and sines of addition and subtraction of two angles. Translate it into general function by auxiliary formula, and then find the range of trigonometric. The range of angle B in $[0, \frac{\pi}{3}]$ is always ignored in the problem solving, leading to the wrong answer $y \in [-\sqrt{2}, \sqrt{2}]$.

4. In $\triangle ABC$, angles A, B, C correspond to sides a, b, c. If $c - a$ is equal to the height of side AC, find the value of $\sin \frac{C-A}{2} + \cos \frac{C+A}{2}$.

   Solution: In the right-angled triangle $\triangle ABC$, let $\angle C = 90°$, $\angle A = 30°$, $AB = 2$

   $\because c - a = 2 - 1 = 1 = h$

   $\therefore \sin \frac{C-A}{2} + \cos \frac{C+A}{2} = \sin 30° + \cos 60° = 1.$

5. In $\triangle ABC$, angles A, B, C correspond to sides a, b, c. If $9a^2 + 9b^2 - 19c^2 = 0$, find the value of $\frac{\cot C}{\cot A + \cot B}$.

   Solution 1:
   According to the condition,

   $$a^2 + b^2 = \frac{19}{9}c^2. \tag{1}$$

   According to the law of cosines,

   $$\cos C = \frac{a^2 + b^2 + c^2}{2ab}. \tag{2}$$

   Putting (1) into (2) implies

   $\cos C = \frac{5}{9} \cdot \frac{c^2}{ab} = \frac{5}{9} \cdot \frac{\sin^2 C}{\sin A \sin B}$

   $\therefore \cot C = \frac{5}{9} \cdot \frac{\sin C}{\sin A \sin B} = \frac{5}{9} \cdot \frac{\sin(A+B)}{\sin A \sin B}$

   $= \frac{5}{9} \cdot \frac{\sin A \cos B + \cos A \sin B}{\sin A \sin B} = \frac{5}{9}(\cot A + \cot B)$

   $\therefore \frac{\cot C}{\cot A + \cot B} = \frac{5}{9}.$

Solution 2:

$$\frac{\cot C}{\cot A + \cot B} = \frac{\frac{\cos C}{\sin C}}{\frac{\cos A}{\sin A} + \frac{\cos B}{\sin B}} = \frac{\cos C}{\sin C} \cdot \frac{\sin A \cdot \sin B}{\sin(A+B)}$$

$$= \frac{\sin A \cdot \sin B \cdot \cos C}{\sin^2 C} = \frac{ab}{c^2} \cdot \frac{a^2 + b^2 - c^2}{2ab}$$

$$= \left(\frac{19}{9}c^2 - c^2\right) \cdot \frac{1}{2c^2} = \frac{5}{9}.$$

Comment: Solution 1 uses sine law, cosine law and the mutual conversion of the side and angle relations, which shows the flexibility of thinking. Solution 2 uses sine law and cosine law from required type, which is concise and lively.

6. In $\triangle ABC$, angles A, B, C correspond to sides a, b, c, which are geometric sequence. Find the range of $\frac{\sin A \cot C + \cos A}{\sin B \cot C + \cos B}$.

Solution: Let common ratio of a, b, c is q, so $b = aq, c = aq^2$,

$$\therefore \frac{\sin A \cot C + \cos A}{\sin B \cot C + \cos B} = \frac{\sin A \cos C + \cos A \sin C}{\sin B \cos C + \cos B \sin C}$$

$$= \frac{\sin(A+C)}{\sin(B+C)} = \frac{\sin(\pi - B)}{\sin(\pi - A)} = \frac{\sin B}{\sin A} = \frac{b}{a} = q.$$

$\therefore$ Only find the range of q.

$\because$ a, b, c are geometric sequence; the maximum side is a or c.

$\therefore$ If a, b, c are sides of triangle, $a + b > c$ and $b + c > a$.

$$\therefore \begin{cases} a + aq > aq^2, \\ aq + aq^2 > a, \end{cases} \quad \therefore \begin{cases} q^2 - q - 1 < 0, \\ q^2 + q - 1 > 0. \end{cases}$$

$$\therefore \begin{cases} \dfrac{1 - \sqrt{5}}{2} < q < \dfrac{\sqrt{5}+1}{2}, \\ q > \dfrac{\sqrt{5}-1}{2} \text{ or } < -\dfrac{\sqrt{5}+1}{2}. \end{cases}$$

$$\therefore \frac{\sqrt{5}-1}{2} < q < \frac{\sqrt{5}+1}{2}.$$

$$\therefore \frac{\sin A \cot C + \cos A}{\sin B \cot C + \cos B} \in \left(\frac{\sqrt{5}-1}{2}, \frac{\sqrt{5}+1}{2}\right).$$

Comment: When solving problems, we often use two implied conditions: "the sum of the two sides of the triangle is more than the third one", "the difference between the two sides of the triangle is less than the third one".

7. If $\triangle ABC$ is inscribed in the unit circle, three bisectors of angles A, B, C extended to A_1, B_1, C_1 intersection of this circle, find the value of

$$\frac{AA_1 \cdot \cos \frac{A}{2} + BB_1 \cdot \cos \frac{B}{2} + CC_1 \cdot \cos \frac{C}{2}}{\sin A + \sin B + \sin C}.$$

Solution: As shown in the figure, connect BA_1, $\frac{AA_1}{\sin(B+\frac{A}{2})}=2R=2$

$$AA_1 = 2\sin\left(B + \frac{A}{2}\right) = 2\sin\left(\frac{A+B+C}{2} + \frac{B}{2} - \frac{C}{2}\right)$$

$$= 2\cos\left(\frac{B}{2} - \frac{C}{2}\right).$$

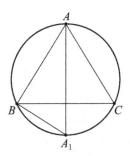

$$\therefore \; AA_1 \cos \frac{A}{2} = 2\cos\left(\frac{B}{2} - \frac{C}{2}\right)\cos \frac{A}{2} = \cos \frac{A+B-C}{2}$$

$$+ \cos \frac{A+C-B}{2} = \cos\left(\frac{\pi}{2} - C\right) + \cos\left(\frac{\pi}{2} - B\right)$$

$$= \sin C + \sin B \; \therefore \; BB_1 \cos \frac{B}{2} = \sin A + \sin C,$$

$$CC_1 \cos \frac{C}{2} = \sin A + \sin B, \quad \therefore \quad AA_1 \cos \frac{A}{2} + BB_1$$

$$\cos \frac{B}{2} + CC_1 \cos \frac{C}{2} = 2(\sin A + \sin B + \sin C),$$

$$\therefore \frac{AA_1 \cdot \cos \frac{A}{2} + BB_1 \cdot \cos \frac{B}{2} + CC_1 \cdot \cos \frac{C}{2}}{\sin A + \sin B + \sin C}$$

$$= \frac{2(\sin A + \sin B + \sin C)}{\sin A + \sin B + \sin C} = 2.$$

8. In $\triangle ABC$, if $a^2 + b^2 = 6c^2$, find the value of $(\cot A + \cot B) \tan C$.

Solution:

$$(\cot A + \cot B) \tan C = \frac{\sin(A + B)}{\sin A \cdot \sin B} \cdot \frac{\sin C}{\cos C}$$

$$= \frac{\sin^2 C}{\sin A \cdot \sin B} \cdot \frac{1}{\cos C} = \frac{c^2}{ab} \cdot \frac{2ab}{a^2 + b^2 - c^2}$$

$$= \frac{2c^2}{a^2 + b^2 - c^2} = \frac{2c^2}{6c^2 - c^2} = \frac{2}{5}.$$

We always meet the inequality about the side (a, b, c) of triangle, angle (A, B, C), height (h_a, h_b, h_c), midline (m_a, m_b, m_c), angular bisector (t_a, t_b, t_c), semi-perimeter (p), radius of inscribed circle (r), radius of circumcircle (R), radius of escribed circle (r_a, r_b, r_c) and area (Δ). These inequalities are called trigonometric inequalities. Proof of trigonometric inequalities always solved by uniform replacement as following type $(*)$. First prove the following conclusion.

Sufficient and necessary condition, that a, b, c are three sides of triangle, is that there exist three positive numbers x, y, z where $a = y + z, b = z + x, c = x + y$ and type $(*)$ holds at the same time.

Proof: First prove the necessity. If there exist positive numbers x, y, z that meet type $(*)$, then $a + b > c, b + c > a, c + a > b$. So a, b, c are three sides of triangle. $\square$

And then prove sufficiency. If a, b, c are three sides of triangle, let $x = \frac{1}{2}(b + c - a)$, $y = \frac{1}{2}(a + c - b)$, $z = \frac{1}{2}(a + b - c)$. So type $(*)$ holds and $x, y, z \in R^+$.

The relationship of a, b, c, x, y, z as following figure. In type $(*)$ replacement:

$$p = \frac{1}{2}(a + b + c) = x + y + z,$$

$$p - a = \frac{1}{2}(b + c - a) = x; \quad p - b = \frac{1}{2}(c + a - b) = y,$$

$$p - c = \frac{1}{2}(a + b - c) = z; \quad \Delta = \sqrt{(x + y + z)xyz},$$

$$r = \frac{\Delta}{p} = \sqrt{\frac{(p - a)(p - b)(p - c)}{p}} = \sqrt{\frac{xyz}{x + y + z}},$$

$$R = \frac{abc}{4\Delta} = \frac{abc}{4 \cdot \sqrt{p(p - a)(p - b)(p - c)}} = \frac{(x + y)(y + z)(z + x)}{4 \cdot \sqrt{xyz(x + y + z)}},$$

$$\sin A = \frac{a}{2R} = \frac{2\sqrt{(x + y + z)xyz}}{(z + x)(x + y)},$$

$$\cos A = \frac{b^2 + c^2 - a^2}{2bc} = \frac{x(x + y + z) - yz}{x(x + y + z) + yz},$$

$$\tan A = \frac{\sin A}{\cos A} = \frac{2\sqrt{(x + y + z)xyz}}{x(x + y + z) - yz},$$

$$\sin \frac{A}{2} = \sqrt{\frac{1 - \cos A}{2}} = \sqrt{\frac{yz}{(z + x)(x + y)}},$$

$$\cos \frac{A}{2} = \sqrt{\frac{1 + \cos A}{2}} = \sqrt{\frac{x(x + y + z)}{(z + x)(x + y)}},$$

$$\tan \frac{A}{2} = \frac{r}{x} = \sqrt{\frac{yz}{x(x + y + z)}} = \frac{yz}{\sqrt{(x + y + z)xyz}},$$

$$t_a = \frac{2\Delta}{(b + c) \sin \frac{A}{2}} = \frac{2\sqrt{x(z + x)(x + y)(x + y + z)}}{2x + y + z},$$

$$h_a = \frac{2}{a} = \frac{2\sqrt{(x + y + z)xyz}}{y + z},$$

$$m_a = \frac{1}{2}\sqrt{2(b^2 + c^2) - a^2},$$

$$= \frac{1}{2}\sqrt{(4x^2 + y^2 + z^2) + 2(2zx + 2xy - yz)}$$

$$r_a = \frac{\Delta}{x} = \frac{\sqrt{(x+y+z)xyz}}{x}.$$

$\therefore$ Through type $(*)$ replacement, trigonometric inequality can be transformed into algebraic inequality containing x, y, z, and then make the proof algebra.

Exercise

9. In the convex quadrilateral $ABCD$, the diagonal line AC bisects $\angle BAD$, and point E is on the extended line of CD. G is intersection of BE and AC. Extend DG to the extended line of CB to F. Prove $\angle BAF = \angle DAE$.

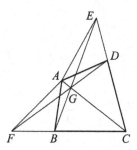

Analysis: This problem involves many angles. If we use the method of pure plane geometry, it may be more annoying. We use of triangular method to prove: first set the parameters, and then calculation.

Proof: Suppose $\angle BAF = \alpha$, $\angle DAE = \beta$. $\angle BAC = \angle DAC = \theta$.

In $\triangle AFB$ and $\triangle ABC$, according to the Laws of Sines:

$$\frac{FB}{\sin \alpha} = \frac{AF}{\sin \angle ABF}, \quad \frac{BC}{\sin \theta} = \frac{AC}{\sin \angle ABC},$$

$\because \sin \angle ABF = \sin \angle ABC,$

$$\therefore \frac{FB}{BC} \cdot \frac{\sin \theta}{\sin \alpha} = \frac{AF}{AC} \tag{3}$$

$$\frac{DE}{BC} \cdot \frac{\sin \theta}{\sin \beta} = \frac{AE}{AC}, \tag{4}$$

Suppose O is the intersection of EF and AC,

so $\frac{1}{2}AF \times AO \sin (\alpha + \theta) = S_{\triangle AFO}$

$$\frac{1}{2}AE \times AO \sin(\beta + \theta) = S_{\triangle AEO}.$$

$$\therefore \frac{AF}{AE} \cdot \frac{\sin(\alpha + \theta)}{\sin(\beta + \theta)} = \frac{S_{\triangle AFO}}{S_{\triangle AEO}} = \frac{OF}{OE}. \tag{5}$$

In $\triangle EFC$, according to Ceva's theorem:

$$\frac{FB}{BC} \times \frac{CD}{DE} \times \frac{EO}{OF} = 1. \tag{6}$$

Therefore, by (3)–(6)

$$\frac{\sin \alpha}{\sin \beta} = \frac{\sin(\alpha + \theta)}{\sin(\beta + \theta)} \Rightarrow \sin \theta \sin(\alpha - \beta) = 0,$$

$$\because |\alpha - \beta| < \pi, \theta \in \left(0, \frac{\pi}{2}\right).$$

$$\therefore \alpha = \beta$$

$$\therefore \angle BAF = \angle DAE.$$

10. Vertex angle of isosceles triangle ABC is $\pi/7$, and D is the point on the waist AB. $CD = \sqrt{2}AD$, Prove $AD = BC$.

Analysis: AD and BC are in a relatively dispersed position. $A = \frac{\pi}{7}$ should be a breakthrough, and prove it by the relationship of sides and angles of triangle.

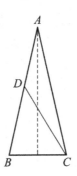

Proof: $\because$ $A = \pi/7$, suppose $\alpha = \pi/14$.

$\therefore$ $A = 2\alpha$, and $7\alpha = \pi/2$.

$\therefore$ $3\alpha = \pi/2 - 4\alpha$.

$$\therefore\ \sin 3\alpha = \sin\left(\frac{\pi}{2} - 4\alpha\right) = \cos 4\alpha. \tag{7}$$

Suppose $AD = m$, $AC = n$, $BC = a$.

$\therefore$ $CD = \sqrt{2}m$.

In $\triangle ACD$, according to the Laws of Sines:

$$\left(\sqrt{2}m\right)^2 = m^2 + n^2 - 2mn\cos 2\alpha \Rightarrow \cos 2\alpha = \frac{n^2 - m^2}{2mn}. \tag{8}$$

In isosceles triangle ABC,

$$\sin \alpha = \frac{\frac{1}{2}BC}{AC} = \frac{a}{2n}. \tag{9}$$

$\because$ $\sin 3\alpha = \sin\left(\dfrac{\pi}{2} - 4\alpha\right) = \cos 4\alpha$

$\therefore$ $2(1 - \sin^2 2\alpha)^2 - 1 = 3\sin \alpha - 4\sin^3 \alpha$

$\therefore$ $8\sin^4 \alpha + 4\sin^3 \alpha - 8\sin^2 \alpha - 3\sin \alpha + 1 = 0$

$\therefore$ $(\sin \alpha + 1)(8\sin^3 \alpha - 4\sin^2 \alpha + 1) = 0$

$\therefore$ $-4\sin \alpha \cos 2\alpha - 4\sin^2 \alpha + 1 = 0$.

Therefore by (8) and (9)

$am^2 - an^2 - ma^2 + mn^2 = 0 \Rightarrow (m - a)(am + n^2) = 0 \Rightarrow m = a$.

$\therefore$ $AD = BC$. $\square$

Comment: Research contest problems with triangle can often make the relationship between the quantities in question simple. Convert complex geometric transformation and complex deductive reasoning for operation of trigonometric function. This method is simple and clear thinking. Communication triangle and the geometry relationship, in addition to the direct use of trigonometric function definition and the triangle formula, mainly by means of sine law and cosine law and area formula.

11. In quadrilateral $ABCD$, the diagonal line AC bisects $\angle BAD$, and E is the point on CD. F is intersection of BE and AC. Extend DG to the extended line of CB to G. Prove $\angle GAC = \angle EAC$.

Proof: Suppose $\angle BAC = \angle CAD = \theta, \angle GAC = \alpha$, $\angle EAC = \beta$. According to the Mailer Laws:

$$1 = \frac{BG}{GC} \times \frac{CD}{DE} \times \frac{EF}{FB} = \frac{S_{\triangle ABG}}{S_{\triangle AGC}} \times \frac{S_{\triangle ACD}}{S_{\triangle ADE}} \times \frac{S_{\triangle AEF}}{S_{\triangle AFB}}$$

$$= \frac{AB\sin(\theta - \alpha)}{AC\sin\alpha} \times \frac{AC\sin\theta}{AE\sin(\theta - \beta)} \times \frac{AE\sin\beta}{AB\sin\theta}$$

$$= \frac{\sin(\theta - \alpha)\sin\beta}{\sin\alpha\sin(\theta - \beta)}$$

$$\Rightarrow \sin(\theta - \alpha)\sin\beta = \sin(\theta - \beta)\sin\alpha$$

$\therefore \cos\alpha\sin\beta = \cos\beta\sin\alpha$, $\tan\alpha = \tan\beta$.

$\because y = \tan x$ is monotone increasing in $(0, \pi/2)$

$\therefore \alpha = \beta$

$\therefore \angle GAC = \angle EAC$.   □

Comment: This question uses the Mailer Laws which establish equivalent relation of one side. Solve it by the knowledge of area and trigonometric function. Through this question, the key of application of the Mailer Laws is to combine the condition and select appropriate section line.

12. $\odot O_1$ is tangent to $\odot O_2$ as well as three sides of the triangle ABC, E, F, G, H are points of tangency, P is the intersection of the extended line EG and extended line FH.
Prove: $PA \perp BC$.

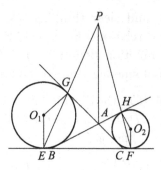

Proof: Extend PA, and D is intersection of BC and PA, connect O_1A, O_1E, O_1G, O_2A, O_2F, O_2H,

Let $\begin{array}{l} \angle EPD = \angle 1, \angle DPF = \angle 2, \angle AGP = \angle 3, \\ \angle AHP = \angle 4, \angle GAO_1 = \angle 5, \angle HAO_2 = \angle 6. \end{array}$

$$\therefore \frac{ED}{DF} = \frac{S_{\triangle PED}}{S_{\triangle PFD}} = \frac{PE \sin \angle 1}{PF \sin \angle 2}$$

in $\triangle PEF$; according to Laws of Sines: $\frac{PE}{PF} = \frac{\sin \angle PFE}{\sin \angle PEF}$

$\because$ CG and CE are the tangent lines of $\odot O$.

$\therefore$ $\angle PEF = \angle CGE = 180° - \angle 3 \Rightarrow \sin \angle PEF = \sin \angle 3.$

Similarly, $\sin \angle PFE = \sin \angle 4.$

$$\therefore \frac{ED}{DF} = \frac{\sin \angle 4 \times \sin \angle 1}{\sin \angle 2 \times \sin \angle 3}$$

in $\triangle PHA$ and $\triangle PGA$, according to Laws of Sines:

$$\frac{\sin \angle 4}{\sin \angle 2} = \frac{PA}{AH}, \frac{\sin \angle 1}{\sin \angle 3} = \frac{AG}{PA} \quad \therefore \frac{ED}{DF} = \frac{PA}{AH} \times \frac{AG}{PA} = \frac{AG}{AH}.$$

$\because$ $\angle 5 = \angle 6$,

$\therefore$ Right-angled $\triangle AGO_1 \sim$ Right-angled $\triangle AHO_2.$

$$\therefore \frac{ED}{DF} = \frac{AG}{AH} = \frac{AO_1}{AO_2}.$$

O_1EFO_2 is right-angled trapezoid, and A is on the line O_1O_2

$\because \frac{ED}{DF} = \frac{AO_1}{AO_2} \Rightarrow AD//O_1E,\ AD \perp BC,\ \therefore\ PA \perp BC.$ □

13. In the non-obtuse triangle ABC, $AB > AC$, $\angle B = 45°$, O and I are the incenter and excenter of triangle ABC, and $\sqrt{2}OI = AB - AC$. Find $\sin A$.

Analysis: $\sqrt{2}OI = AB - AC$, so think of Euler's formula, and then solve it as problem about $\angle A$.

Solution: According to Euler's formula:

$$\left(\frac{c-b}{\sqrt{2}}\right)^2 = OI^2 = R^2 - 2Rr. \tag{10}$$

$$\because\ r = \frac{c+a-b}{2}\tan\frac{B}{2} = \frac{c+a-b}{2}\tan\frac{\pi}{8}$$

$$= \frac{\sqrt{2}-1}{2}(c+a-b). \tag{11}$$

According to (10), (11) and $\frac{a}{\sin A} = \frac{b}{\sin B} = \frac{c}{\sin C} = 2R$:

$$1 - 2(\sin C - \sin B)^2 = 2(\sin A + \sin C - \sin B)(\sqrt{2} - 1).$$

$$\because\ \angle B = \frac{\pi}{4},\ \sin B = \frac{\sqrt{2}}{2},\ \sin C = \sin\left(\frac{3\pi}{4} - A\right)$$

$$= \frac{\sqrt{2}}{2}(\sin A + \cos A),$$

$$\therefore\ 2\sin A\cos A - (2 - \sqrt{2})\sin A - 2\sqrt{2}\cos A + \sqrt{2} - 1 = 0,$$

$$(\sqrt{2}\sin A - 1)(\sqrt{2}\cos A - \sqrt{2} + 1) = 0,$$

$$\therefore\ \sin A = \frac{\sqrt{2}}{2}\ \text{or}\ \cos A = 1 - \frac{\sqrt{2}}{2},$$

$$\therefore\ \sin A = \frac{\sqrt{2}}{2}\ \text{or}\ \sin A = \sqrt{\sqrt{2} - \frac{1}{2}}.$$

14. Shown as in the figure, O and I are the incenter and excenter of triangle ABC. AD is the height of side BC, and I is on the segment OD. Prove radius of circumcircle of triangle ABC equals to radius of escribed circle of side BC.

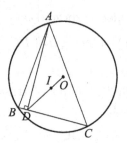

Analysis: Suppose three sides of $\triangle ABC$ are a, b and c. Radius of escribed circle of side BC is r_a, and radius of circumcircle is R. So $S_{\triangle ABC} = \frac{1}{2}bc \sin A = \frac{1}{2}r_a(b+c-a)$. To prove $r_a = R$, only to prove $\frac{r_a}{R} = \frac{2\sin A \sin B \sin C}{\sin B + \sin C - \sin A} = 4\sin\frac{A}{2}\cos\frac{B}{2}\cos\frac{C}{2} = 1$.

Proof: Suppose $AB = c$, $BC = a$, $CA = b$, radius of circumcircle is R, and radius of escribed circle of side BC is r_a. Suppose K is the intersection point of extension lines of AI and circumcircle of triangle ABC, so OK is radius of $\odot O$,

$$\because OK \perp BC$$

$$\therefore OK // AD$$

$$\therefore \frac{AI}{IK} = \frac{AD}{OK} = \frac{C \sin B}{R} = 2\sin B \sin C,$$

$$\because \angle ABI = \angle IBC = \frac{1}{2}\angle B, \angle CBK = \angle CAK = \frac{1}{2}\angle A$$

$$\angle AKB = \angle ACB = \angle C, \angle BAK = \frac{1}{2}\angle A,$$

$$\frac{AI}{IK} = \frac{S_{\triangle ABI}}{S_{\triangle KBI}} = \frac{\frac{1}{2}AB \times BI \sin\frac{B}{2}}{\frac{1}{2}BK \times BI \sin\frac{A+B}{2}} = \frac{AB \sin\frac{B}{2}}{BK \cos\frac{C}{2}}$$

$$= \frac{\sin C}{\sin\frac{A}{2}} \times \frac{\sin\frac{B}{2}}{\cos\frac{C}{2}} = \frac{2\sin\frac{B}{2}\sin\frac{C}{2}}{\sin\frac{A}{2}}$$

$$\therefore 2\sin B \sin C = \frac{2\sin\frac{B}{2}\sin\frac{C}{2}}{\sin\frac{A}{2}},$$

$$\therefore \ 4\sin\frac{A}{2}\cos\frac{B}{2}\cos\frac{C}{2} = 1. \tag{12}$$

$$S_{\triangle ABC} = \frac{1}{2}bc\sin A = \frac{1}{2}r_a(b+c-a),$$

$$r_a = \frac{bc\sin A}{b+c-a} = 2R \times \frac{\sin A\sin B\sin C}{\sin B + \sin C - \sin A}$$

$$= 2R \times \frac{\sin A\sin B\sin C}{2\sin\frac{B+C}{2}\cos\frac{B-C}{2} - 2\sin\frac{B+C}{2}\cos\frac{B+C}{2}}$$

$$= \frac{R\sin A\sin B\sin C}{2\sin\frac{B+C}{2}\sin\frac{B}{2}\sin\frac{C}{2}}$$

$$= 4R\sin\frac{A}{2}\cos\frac{B}{2}\cos\frac{C}{2} = R.$$

$\square$

15. E and F are two points on the side BC of acute triangle, and $\angle BAE = CAF$, $MF \perp AB$, $FN \perp AC$, (M, N are pedals), extend the AE to circumcircle of triangle ABC in D. Prove: area of quadrilateral AMDN equals to area of triangle ABC.

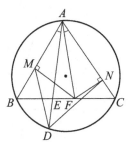

Proof: Suppose $\angle BAE = \alpha$, $\angle CAF = \alpha$, $\angle EAF = \beta$.

So

$$S_{AMDN} = \frac{1}{2}AM \times AD\sin\alpha + \frac{1}{2} \times AD \times AN\sin(\alpha+\beta)$$

$$= \frac{1}{2}AD[AF\cos(\alpha+\beta)\sin\alpha + AF\cos\alpha\sin(\alpha+\beta)]$$

$$= \frac{1}{2}AD \times AF \sin(2\alpha + \beta)$$

$$= \frac{1}{2}AD \times AF \sin \angle BAC$$

$$S_{\triangle ABC} = \frac{1}{2}AB \times AC \sin \angle BAC.$$

Only to prove

$$AD \times AF = AB \times AC. \qquad\qquad (13)$$

$\therefore \ \angle BAD = \angle CAF, \angle BDA = \angle ACF$

$\triangle ABD \sim \triangle AFC.$

Therefore condition (13) holds. $\square$

16. In $\triangle ABC$, $\triangle A = 60°$, through incenter I of triangle to make the parallel line AC, which intersects line AB at F. Take point P at side BC. Prove: $\angle BFP = \frac{1}{2}\angle B$:

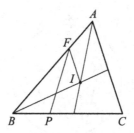

Analysis: Suppose $\triangle BFP = \alpha$, to prove $\alpha = \frac{1}{2}\angle B$.

Only to prove $\alpha = \triangle B - \alpha$

$\because \ \angle B - \alpha$ is an acute angle.

$\therefore$ Only to prove $\sin \alpha = \sin(\angle B - \alpha)$.

Proof: Suppose $BC = 3$, radius of inscribed circle of $\triangle ABC$ is r.

According to Laws of Sines:

$$AB = \frac{3\sin C}{\sin A} = 2\sqrt{3}\sin(\angle B + \angle A) = \sqrt{3}\sin\angle B + 3\cos\angle B.$$

$$\because \; r = 4R\sin\frac{A}{2}\sin\frac{B}{2}\sin\frac{C}{2}$$

(R is radius of circumcircle of $\triangle ABC$)

$$\therefore \; R = \frac{BC}{2\sin A} = \frac{3}{2}\times\frac{1}{\sin A}$$

$$r = \frac{3\sin\frac{B}{2}\sin\frac{C}{2}}{\cos\frac{A}{2}} = 2\sqrt{3}\sin\frac{B}{2}\cos\left(\frac{B}{2} + \frac{A}{2}\right)$$

$$= \frac{3}{2}\sin B + \frac{\sqrt{3}}{2}\cos B - \frac{\sqrt{3}}{2}.$$

$$\because \; \frac{r}{AF} = \sin A, \quad \therefore \; AF = \frac{r}{\sin A} = \sqrt{3}\sin B + \cos B - 1,$$

$$\therefore \; BF = AB - AF = 2\cos B + 1.$$

In $\triangle BFP$, according to Laws of Sines:

$$\frac{\sin(B + \alpha)}{\sin\alpha} = \frac{BF}{BP} = BF = 2\cos B + 1$$

$$\therefore \; \sin\alpha = \sin(\angle B - \alpha). \qquad\qquad \square$$

17. P is the point inside $\triangle ABC$, $\angle APB - \angle ACB = \angle APC - \angle ABC$. D and E are incenters of $\triangle APB$ and $\triangle APC$. Prove: AP, BD and CE intersect at one point.

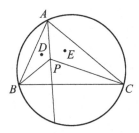

Analysis: To prove AP, BD, CE intersect at one point, the condition is about the relation of angles, so it is difficult to use

CEVA's theorem. Suppose BD and AP are intersect at M, and then M is on the CE. BD and CE are angular bisectors of $\angle ABP$ and $\angle ACP$, so only to prove $\frac{BP}{AB} = \frac{PC}{AC}$. Solve it by condition and cosine law.

Proof: Extend AP, which intersects BC at K and circumcircle of $\triangle ABC$ at F, connect BF and CF

$$\because \angle APC - \angle ABC = \angle AKC + \angle PCK - \angle ABC$$

$$= \angle PCK + \angle KCF = \angle PCF.$$

Similarly, $\angle APB - \angle ACB = \angle PBF$.

$\because \angle PCF = \angle PBF$ and according to Laws of Sines:

$$\frac{PB}{\sin \angle PFB} = \frac{PF}{\sin \angle PBF} = \frac{PF}{\sin \angle PCF} = \frac{PC}{\sin \angle PFC},$$

$$\therefore \frac{PB}{PC} = \frac{\sin \angle PFB}{\sin \angle PFC} = \frac{\sin \angle ACB}{\sin \angle ABC} = \frac{AB}{AC}, \quad \text{so} \quad \frac{PB}{AB} = \frac{PC}{AC}.$$

Suppose BD and AP intersect at M, and BD is bisector of $\angle ABP$, $\therefore \frac{PM}{MA} = \frac{PB}{AB}$.

Suppose CE and AP intersect at N, and CE is bisector of $\angle ACP$, $\therefore \frac{PN}{NA} = \frac{PC}{AC}$.

$$\therefore \frac{PM}{MA} = \frac{PN}{NA}.$$

$\therefore$ M and N are coincidence.

$\therefore$ AP, BD and CE intersect at one point. $\square$

5.2. Area Method

In the plane geometry, the calculation and proof of the area of the triangle and polygon are a basic problem. Using the method of area, especially the cut-fill method, the application of product transformation in solving problems can solve a series of problems related to area, so it is favored by area method.

Triangle area formula: in $\triangle ABC$, a, b and c are opposite sides of A, B and C, and h_a is the height of a. R and r are radius of circumcircle and inscribed circle of $\triangle ABC$, $p = \frac{1}{2}(a + b + c)$. In addition to Sec. 5.2, introduces besides the following formula:

(1) $S_{\triangle ABC} = \sqrt{p(p - a)(p - b)(p - c)}$,

(2) $S_{\triangle ABC} = \frac{1}{2}r(a + b + c) = rp$,

(3) $S_{\triangle ABC} = \frac{abc}{4R}$,

(4) $S_{\triangle ABC} = 2R^2 \sin A \sin B \sin C$,

(5) $S_{\triangle ABC} = \frac{1}{2}r_a(b + c - a)$,

(6) $S_{\triangle ABC} = \frac{1}{2}R^2(\sin 2A + \sin 2B + \sin 2C)$,

(7) $S_{\triangle ABC} = \frac{a^2 \sin B \sin C}{2 \sin(B+C)}$.

In addition to the above formula, the following theorems are often used:

(1) The area of a figure equals the sum of its parts.
(2) Equal areas are two congruent form.
(3) The area of the triangle, the parallel quadrilateral, the trapezoid equal when their base and height are equal (equal base of trapezoid should be understood as sum of two bottom equal).
(4) The area of a similar triangle is equal to the square of the similar ratio.
(5) The ratio of area of a triangle, a trapezoid, or a trapezoid which base (or height) is equal is equal to the ratio of height (or base).
(6) Ratio of isogonal or supplementary triangle area is equal to the ratio of product of both sides of the isogonal included angle or supplementary angle; ratio of equilateral parallelogram area is equal to the ratio of product of both sides of the isogonal included angle.
(7) Common side theorem: lines AB and PQ intersect at M, so $\frac{\triangle PAB}{\triangle QAB} = \frac{PM}{QM}$.

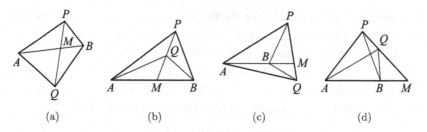

    (a)   (b)   (c)   (d)

(8) Common angle theorem: $\angle ABC$ and $\angle A'B'C'$ is equal or complementary, so $\frac{\triangle ABC}{\triangle A'B'C'} = \frac{AB \cdot BC}{A'B' \cdot B'C'}$.

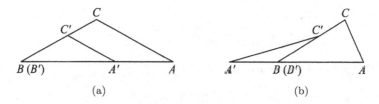

           (a)   (b)

Exercise

1. M is the midpoint of side BC of an arbitrary triangle ABC. Take point E and F on AB and AC. EF and AM intersect at N. Prove $\frac{AM}{AN} = \frac{1}{2}\left(\frac{AB}{AE} + \frac{AC}{AF}\right)$.

Proof: $\because MB = MC$

$$\therefore S\triangle ABC = 2S\triangle ABM = 2S\triangle ACM, \tag{1}$$

$$\because S\triangle AEF = S\triangle AEN + S\triangle AFN. \tag{2}$$

Equation (2) $\div \triangle ABC$, and by (1), we get

$$\frac{S_{\triangle AEF}}{S_{\triangle ABC}} = \frac{S_{\triangle AEN}}{2S_{\triangle ABM}} + \frac{S_{\triangle AFN}}{2S_{\triangle ACM}}. \tag{3}$$

Common side theorem:

$$\frac{AE \times AF}{AB \times AC} = \frac{1}{2}\left(\frac{AE \times AN}{AB \times AM} + \frac{AF \times AN}{AC \times AM}\right)$$

$$= \frac{1}{2}\left(\frac{AE}{AB} + \frac{AF}{AC}\right) \times \frac{AN}{AM}$$

Multiplying both sides with $\frac{AB \times AC \times AM}{AE \times AF \times AN}$, we get

$$\therefore \ \frac{AM}{AN} = \frac{1}{2}\left(\frac{AB}{AE} + \frac{AC}{AF}\right).$$

$\square$

Comment: This question can also be solved by Mailer Laws' theorem.

2. E is the midpoint of side AB of inscribed quadrilateral $ABCD$, $EF \perp AD$, $EH \perp BC$. Prove EG bisect FH

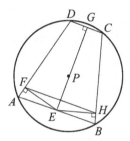

Proof:
Suppose GE and FH intersect at P, $\angle HEP = \alpha, \angle FEP = \beta$, $\angle EPF = \theta$

$\because$ α and $\angle C$ complementary, $\angle A$ and $\angle C$ complementary, $\therefore$ $\alpha = \angle A$,

Similarly, $\beta = \angle B$.

$$\because \ EF = AE \sin A = AE \sin \alpha,$$

$$EH = EB \sin B = EB \sin \beta, AE = EB,$$

$$\therefore \ 1 = \frac{S_{\triangle PEF}}{S_{\triangle PEF}} \times \frac{S_{\triangle PEH}}{S_{\triangle PEH}}$$

$$= \frac{\frac{1}{2}PE \times PF \sin \theta}{\frac{1}{2}PE \times EF \sin \beta} \times \frac{\frac{1}{2}PE \times EH \sin \alpha}{\frac{1}{2}PE \times PH \sin(180° - \theta)}$$

$$= \frac{PE^2 \times PF \times EH \sin\theta \sin\alpha}{PE^2 \times EF \times PH \sin\theta \sin\beta}$$

$$= \frac{PF}{PH} \times \frac{EH \sin\alpha}{EF \sin\beta}$$

$$= \frac{PF}{PH} \times \frac{EB \sin\beta \sin\alpha}{AE \sin\alpha \sin\beta}$$

$$= \frac{PF}{PH}$$

$$\therefore \ PF = PH.$$

$\square$

3. O is center of circumcircle of quadrilateral $ABCD$, and O_1 is center of inscribed circle. AC and BD intersect at E. Prove E, O, O_1 are collinear.

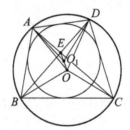

Proof: Suppose OO_1 and BD intersect at E_1, OO_1 and AC intersect at E_2,

so $\dfrac{OE_1}{O_1E_1} = \dfrac{S_{\triangle OBD}}{S_{\triangle O_1 BD}}, \dfrac{OE_2}{O_1E_2} = \dfrac{S_{\triangle OAC}}{S_{\triangle O_1 AC}}.$

If $\dfrac{OE_1}{O_1E_1} = \dfrac{OE_2}{O_1E_2}$, therefore

E, O, O_1 are collinear.

So to prove: $\dfrac{S_{\triangle OBD}}{S_{\triangle O_1 BD}} = \dfrac{S_{\triangle OAC}}{S_{\triangle O_1 AC}}.$ $\hspace{2cm}$ (*)

R is radius of quadrilateral $ABCD$, and r is radius of inscribed circle.

$$\therefore \frac{S_{\triangle OBD}}{S_{\triangle OAC}} = \frac{\frac{1}{2}R^2 \sin \angle BOD}{\frac{1}{2}R^2 \sin \angle AOC} = \frac{\sin 2\angle C}{\sin 2\angle B}$$

$$S_{\triangle O_1 BD} = \frac{1}{2}BO_1 \cdot DO_1 \sin \angle BO_1 D$$

$$= \frac{1}{2} \cdot \frac{r}{\sin \frac{\angle B}{2}} \cdot \frac{r}{\sin \frac{\angle D}{2}} \cdot \sin \left(\frac{\angle B}{2} + \frac{\angle D}{2} + \angle c\right)$$

$$= \frac{1}{2}r^2 \frac{1}{\sin \frac{\angle B}{2} \cdot \cos \frac{\angle B}{2}} \cdot \sin \left(\frac{\pi}{2} + \angle c\right) = \frac{r^2 \cos \angle C}{\sin \angle B};$$

Similarly,

$$S_{\triangle O_1 AC} = \frac{r^2 \cos \angle B}{\sin \angle C},$$

$$\therefore \frac{S_{\triangle O_1 BD}}{S_{\triangle O_1 AC}} = \frac{\sin \angle C \cdot \cos \angle C}{\sin \angle B \cos \angle B} = \frac{\sin 2\angle C}{\sin 2\angle B} = \frac{S_{\triangle OBD}}{S_{\triangle OAC}}$$

$$\therefore \frac{S_{\triangle OBD}}{S_{\triangle O_1 BD}} = \frac{S_{\triangle OAC}}{S_{\triangle O_1 AC}} \quad (*).$$

$\therefore E, O, O_1$ are collinear. $\qquad\square$

Comment: In the first phase of the "secondary mathematics" in 2002, the author gave the proof which is solved by reverse evolution in exchange for proof. The inversion method is not easy to be accepted by students, many readers wrote to ask other new methods, now we give another proof method ("method of identity" + "area method").

Exercise 5

1. In $\triangle ABC$, $B = 2A$. Prove $\frac{a}{b} = \frac{a+b}{a+b+c}$.
2. In $\triangle ABC$, a, b, c are corresponding edges of angles A, B, C. If $a^2 + b^2 = tc^2$, and $\cot C = 2004(\cot A + \cot B)$, find the constant t.
3. In $\triangle ABC$, D is a point on the BC, $BD = p$, $DC = q$. Prove $AD^2 = \frac{b^2 p + c^2 q}{p+q} - pq$.

4. In $\triangle ABC$, corresponding edges of angles A, B, C are in an arithmetic progression.

   (1) Prove $\tan \frac{A}{2} \tan \frac{C}{2} = \frac{1}{3}$.

   (2) Find $5 \cos A - 4 \cos A \cos C + 5 \cos C$.

   (3) If $A - C = \frac{\pi}{2}$, find $\sin A + \sin C$.

5. In $\triangle ABC$, if $B = 2A$, and $B^2 = A \cdot C$, find $\cos A \cos B \cos C$.

6. In $\triangle ABC$, $\angle B = \angle C$, P and Q are respectively on AC and AB, and $AP = PQ = QB = BC$, find $\angle A$.

7. If the area of $\triangle ABC$, $S = a^2 - (b - c)^2$, and $b + c = 8$, find the maximum of S.

8. In $\triangle ABC$, there is an angle of $60°$, and the ratio of the two sides of the angle is 8:5, inscribed circle area is 12π, find the area of $\triangle ABC$.

9. Lengths of inscribed quadrangles are $AB = 2$, $BC = 6$, $CD = DA = 4$, find the area of quadrilateral $ABCD$.

10. Width of a rectangular corridor is 1.5 meters. A car which can promote flexible flat bottom surface is rectangular and width of 1 meter. If you want to smooth over rectangular corridor, the maximum length of the flat car should not to exceed the number of meters?

11. In $\triangle ABC$, a, b, $(b \neq 1)$, c are corresponding edges of angles A, B, C. If $\frac{C}{A}$ and $\frac{\sin B}{\sin A}$ are roots of equation $\log_{\sqrt{b}} x = \log_b(4x - 4)$, judge the shape of $\triangle ABC$.

12. O is a point in $\triangle ABC$, and S_A, S_B and S_C express areas of $\triangle BOC$, $\triangle COA$ and $\triangle AOB$. Prove $S_A \cdot \overrightarrow{OA} + S_B \cdot \overrightarrow{OB}$ and $S_C \cdot \overrightarrow{OC} = \overrightarrow{0}$.

13. In $\triangle ABC$, D, E and F are midpoints of BC, CA and AB. X, Y and Z are foot points of height of BC, CA and AB. ZD and FX intersect at L. ZE and FY intersect at M. DY and XY intersect at N. Prove L, M and N are all on the Euler line of $\triangle ABC$.

14. In acute $\triangle ABC$, $\angle BAC = 60°$, $AB = c$, $AC = b$, and $b > c$, the line through O which is circumcenter of $\triangle ABC$ and AB intersect at X, and the line through M which is the orthocenter

of $\triangle ABC$ and AC intersect at Y. Prove

(1) Perimeter of $\triangle ABC$, $P = b + c$.

(2) $OM = b - c$

15. In $\angle ABC$, AP divides equally $\angle ABC$. AP and BC intersect at P. BQ divides equally $\angle ABC$. BQ and CA intersect at Q. If $\angle BAC = 60°$ and $AB + BP = AQ + QB$, find the interior angles of $\triangle ABC$.

Chapter 6

Trigonometric Substitution and Trigonometric Inequality

6.1. Trigonometric Substitution

1. If $x^2 + y^2 = r^2$, suppose $x = r\cos\theta$, $y = r\sin\theta$.
2. If $x^2 + y^2 \leq r^2$, suppose $x = k\cos\theta$, $y = k\sin\theta$ $(0 \leq k \leq r)$.
3. If $\frac{x^2}{a^2} + \frac{y^2}{b^2} = 1$, suppose $x = a\cos\theta$, $y = b\sin\theta$.
4. If $\frac{x^2}{a^2} - \frac{y^2}{b^2} = 1$, suppose $x = a\sec\theta$, $y = b\tan\theta$.
5. If $\frac{1}{1+x^2}, \frac{x+y}{1-xy}, \frac{x-y}{1+xy}$, suppose $x = \tan\alpha$, $y = \tan\beta$.
6. If $x^2 + y^2 + z^2 = r^2$ $(r > 0)$, suppose
$$x = r\cos\alpha\cos\beta, \quad y = r\cos\alpha\sin\beta, \quad z = r\sin\alpha.$$

Exercise

1. Solve equation
$$\begin{cases} \sqrt{x(1-y)} + \sqrt{y(1-x)} = \dfrac{1}{2}, \\[2mm] \sqrt{xy} + \sqrt{(1-x)(1-y)} = \dfrac{\sqrt{3}}{2}. \end{cases}$$

Analysis: According to the condition $0 \leq x \leq 1, 0 \leq y \leq 1$, we have $x = \sin^2\alpha$, $y = \sin^2\beta$.

Solution: Suppose $x = \sin^2\alpha$, $y = \sin\beta$, $(0 \leq \alpha, \beta \leq \frac{\pi}{2})$.

$$\therefore \begin{cases} \sin\alpha\cos\beta + \sin\beta\cos\alpha = \dfrac{1}{2} \\[2mm] \sin\alpha\sin\beta + \cos\alpha\cos\beta = \dfrac{\sqrt{3}}{2} \end{cases} \Rightarrow \begin{cases} \sin(\alpha+\beta) = \dfrac{1}{2}, \\[2mm] \cos(\alpha-\beta) = \dfrac{\sqrt{3}}{2}. \end{cases}$$

$$\because 0 \leq \alpha + \beta \leq \pi, \quad -\frac{\pi}{2} \leq \alpha - \beta \leq \frac{\pi}{2},$$

$$\therefore \begin{cases} \alpha + \beta = \dfrac{\pi}{6} \quad \text{or} \quad \dfrac{5}{6}\pi \\[2mm] \alpha - \beta = \dfrac{\pi}{6} \quad \text{or} \quad -\dfrac{\pi}{6} \end{cases}$$

$$\Rightarrow \begin{cases} \alpha = \dfrac{\pi}{6} \\[2mm] \beta = 0 \end{cases} \text{or} \quad \begin{cases} \alpha = \dfrac{\pi}{2} \\[2mm] \beta = \dfrac{\pi}{3} \end{cases} \text{or} \quad \begin{cases} \alpha = 0 \\[2mm] \beta = \dfrac{\pi}{6} \end{cases} \text{or} \quad \begin{cases} \alpha = \dfrac{\pi}{3} \\[2mm] \beta = \dfrac{\pi}{2} \end{cases}$$

$$\therefore \begin{cases} x = \dfrac{1}{4} \\[2mm] y = 0 \end{cases} \text{or} \quad \begin{cases} x = 1 \\[2mm] y = \dfrac{3}{4} \end{cases} \text{or} \quad \begin{cases} x = 0 \\[2mm] y = \dfrac{1}{4} \end{cases} \text{or} \quad \begin{cases} x = \dfrac{3}{4}, \\[2mm] y = 1. \end{cases}$$

Comment: The role of triangle substitution is to remove the radical from the equations, make it easier.

2. If $x, y \in (-2, 2)$ and $xy = -1$, find the minimum of $\mu = \dfrac{4}{4-x^2} + \dfrac{9}{9-y^2}$.

Analysis: This function is related to the problem of multiple values, and the structure is presented $4 - x^2$ and $9 - y^2$, which can be considered for the triangular element method.

Solution: Let $x = 2\cos\alpha, y = 3\cos\beta, \alpha, \beta \in (0, \pi)$

$$\therefore \cos\alpha \cdot \cos\beta = -\frac{1}{6}$$

$$\therefore \mu = \frac{1}{\sin^2\alpha} + \frac{1}{\sin^2\beta} = 2 + \cot^2\alpha + \cot^2\beta$$

$$\geq 2 + 2|\cot\alpha\cot\beta| = 2 + 2\left|\frac{\cos\alpha\cos\beta}{\sin\alpha\sin\beta}\right|$$

$$= 2 + \frac{1}{3} \cdot \frac{1}{|\sin\alpha\sin\beta|}$$

$$\because |\sin\alpha\sin\beta - \cos\alpha\cos\beta| = |\cos(\alpha+\beta)| \leq 1$$

$$\therefore \ |\sin\alpha\sin\beta + \frac{1}{6}| \le 1$$

$$\therefore \ |\sin\alpha\sin\beta| \le \frac{5}{6}$$

$$\therefore \ \mu \ge 2 + \frac{1}{3}\cdot\frac{6}{5} = \frac{12}{5}.$$

If and only if the equality was $\cot^2\alpha = \cot^2\beta$ and $|\sin\alpha\sin\beta| = \frac{5}{6}$, therefore

$$\begin{cases} x = \dfrac{\sqrt{6}}{3} \\ y = -\dfrac{\sqrt{6}}{2} \end{cases} \quad \text{or} \quad \begin{cases} x = -\dfrac{\sqrt{6}}{3} \\ y = \dfrac{\sqrt{6}}{2} \end{cases} \Rightarrow \mu_{\min} = \frac{12}{5}.$$

Comment: The purpose of the introduction of the triangular element is to simplify the structure of μ, so that the triangle operation and the use of the properties are obtained.

3. If sequence $\{a_n\}$ meets $a_1 = \frac{\sqrt{2}}{2}, a_{n+1} = a_1\cdot\sqrt{1 - \sqrt{1 - a_n^2}}$, and if sequence $\{b_n\}$ meets $b_1 = 1, b_{n+1} = \frac{\sqrt{1+b_n^2}-1}{b_n}$. Prove $2^{n+1}\cdot a_n < \pi < 2^{n+1}\cdot b_n$.

Analysis: Use trigonometric substitution to simplify recursive formula a_{n+1} and b_{n+1}, and further apply sequence knowledge to solve the problem.

Proof:

$$\because \ 0 < a_n < 1, \therefore \ a_n = \sin\alpha_n \ \left(0 < \alpha_n \le \frac{\pi}{2}\right)$$

$$\therefore \ a_{n+1} = \sin\alpha_{n+1} = \frac{\sqrt{2}}{2}\cdot\sqrt{1 - \sqrt{1 - \sin^2\alpha_n}}$$

$$= \frac{\sqrt{2}}{2}\cdot\sqrt{1 - \cos\alpha_n} = \sin\frac{\alpha_n}{2}$$

$$\because a_1 = \frac{\sqrt{2}}{2}, \quad \therefore \alpha_1 = \frac{\pi}{4}, \text{ Common ratio } q = \frac{1}{2}$$

$$\therefore \alpha_n = \frac{\pi}{4}\left(-\frac{1}{2}\right)^{n+1} = \frac{\pi}{2^{n+1}}, \quad \therefore \alpha_n = \sin\frac{\pi}{2^{n+1}}.$$

Let

$$b_n = \tan\beta_n \left(0 < \beta_n < \frac{\pi}{2}\right), \quad \therefore b_{n+1} = \tan\beta_{n+1}$$

$$\because b_{n+1} = \frac{\sqrt{1+b_n^2}-1}{b_n}$$

$$\therefore \tan\beta_{n+1} = \frac{\sqrt{1+\tan^2\beta_n}-1}{\tan\beta_n} = \frac{\sec\beta_n - 1}{\tan\beta_n}$$

$$= \frac{1-\cos\beta_n}{\sin\beta_n} = \tan\frac{\beta_n}{2}$$

$$\therefore \beta_{n+1} = \frac{\beta_n}{2}$$

$$\because b_1 = 1 \quad \therefore \beta_1 = \frac{\pi}{4}, \text{ common ratio } q = \frac{1}{2}$$

$$\therefore \beta_n = \frac{\pi}{4}\left(\frac{1}{2}\right)^{n-1} = \frac{\pi}{2^{n+1}}, \quad \therefore b_n = \tan\frac{\pi}{2^{n+1}}$$

$$\because \sin\frac{\pi}{2^{n+1}} < \frac{\pi}{2^{n+1}} < \tan\frac{\pi}{2^{n+1}}, \quad \therefore a_n < \frac{\pi}{2^{n+1}} < b_n$$

$$\therefore 2^{n+1} \cdot a_n < \pi < 2^{n+1} \cdot b_n.$$

$\square$

Comment: Recursive sequence containing radicals, so using the triangle substitution method would be more effective.

4. Suppose $f_1(x) = x^2 - 2$, for all $n \geq 2$, $f_n(x) = f_1[f_{n-1}(x)]$. Prove: roots of equation $f_n(x) = x$ are not equal.

Analysis: $f_n(x) = f_1[f_{n-1}(x)] - 2 \Rightarrow \cos 2\theta = 2\cos^2\theta - 1$ and then use triangle substitution.

Proof: If $|x| > 2 \Rightarrow f_n(x) > f_{n-1}(x) > \cdots > f_1(x) = x^2 - 2 > x$, contradict to $f_n(x) = x$, $\therefore |x| \leq 2$.

Let $x = 2\cos\theta \ (0 < \theta < \pi)$

$$\therefore \ \frac{f_1(x)}{2} = 2\cos^2\theta - 1 = \cos 2\theta \Rightarrow f_1(x) = 2\cos 2\theta$$

$$\frac{f_2(x)}{2} = 2(\cos 2\theta)^2 - 1 = \cos 4\theta \Rightarrow f_2(x) = 2\cos 4\theta$$

$$\vdots$$

Suppose $f_n(x) = 2\cos 2^n\theta$.

(1) When $n = 1$, $f_1(x) = 2\cos 2\theta$.

(2) Suppose $n = k(k \in N^*)$, $f_k(x) = 2\cos 2^k\theta$.

$\therefore \ f_{k+1}(x) = f_1(f_k(x)) = 4\cos^2 2^k\theta - 2$

$= 2(2\cos^2 2^k\theta - 1) = 2\cos 2^{k+1}\theta;$

$\therefore$ when $n = k + 1$, $f_n(x) = 2\cos 2^n\theta$.

$\therefore \ f_n(x) = x \Rightarrow 2\cos 2^n\theta - 2\cos\theta = 0$

$$\Rightarrow 2\sin\frac{2^n + 1}{2}\theta \cdot \cos\frac{2^n - 1}{2}\theta = 0$$

$\therefore \ \theta_k = \dfrac{2k\pi}{2^n - 1} \ (k = 1, 2, 3, \ldots, 2^{n-1})$ or

$$\theta_k = \frac{2k\pi}{2^n + 1} \ (k = 0, 1, 2, \ldots, 2^{n-1} - 1)$$

$\therefore \ x_k = 2\cos\theta_k$

$\because \ \theta_k \in (0, \pi)$, $\therefore \ \cos x$ monotonic decline in$(0, \pi)$.

$\therefore$ These 2^n are not equal to each other.

So there are no other roots in the original equation.   □

Comment: This problem is the use of triangular substitution to find the root of equation.

5. Suppose $f(x, y) = \dfrac{ax^2 + xy + y^2}{x^2 + y^2}$ meet $\max_{x^2+y^2\neq 0} f(x, y) - \min_{x^2+y^2\neq 0} f(x, y) = 2$, find a.

Analysis: Starting from the structure of $x^2 + y^2$, and solve it with triangular substitution.

Solution:

$$\because f(kx, ky) = f(x, y);$$

$$\therefore \max_{x^2+y^2\neq 0} f(x, y) = \max_{x^2+y^2=1} f(x, y)$$

$$= \max_{0\leq\varphi\leq 2\pi}(a\cos^2\varphi + \cos\varphi\sin\varphi + \sin^2\varphi)$$

$$= \max_{0\leq\phi\leq 2\pi}\left[\sqrt{\left(\frac{a-1}{2}\right)^2 + \frac{1}{4}}\cdot\sin(2\varphi + a) + \frac{a+1}{2}\right]$$

$$= \sqrt{\left(\frac{a-1}{2}\right)^2 + \frac{1}{4}} + \frac{a+1}{2}.$$

Similarly

$$\min_{x^2+y^2\neq 0} f(x, y) = \min_{x^2+y^2=1} f(x, y)$$

$$= \min_{0\leq\varphi\leq 2\pi}(a\cos^2\varphi + \cos\varphi\sin\varphi + \sin^2\varphi)$$

$$= -\sqrt{\left(\frac{a-1}{2}\right)^2 + \frac{1}{4}} + \frac{a+1}{2}$$

$$\therefore 2\sqrt{\left(\frac{a-1}{2}\right)^2 + \frac{1}{4}} = 2, \quad \therefore a = 1 \pm \sqrt{3}.$$

Comment: When we meet the most valuable problems of multivariate function, we use triangle substitution to solve the difficulties of multiple and fraction.

6. Suppose $a, b \in R^+, |x| < \frac{1}{2}$, find the maximum of $f(x) = \sqrt[3]{\frac{4a^2}{(1+2x)^2} + \frac{4b^2}{(1-2x)^2}}$.

Analysis: Because $(1 + 2x) + (1 - 2x)$ is constant, think of $a \cdot \frac{1+2x}{2a} + b \cdot \frac{1-2x}{2b} = 1$, and then use the triangle substitution.

Solution: Suppose

$$\frac{2a}{1+2x} = m \cdot \frac{2b}{1-2x} = n \ (m, n > 0), \ \therefore \ \frac{a}{m} + \frac{b}{n} = 1.$$

∴ We can suppose

$$\begin{cases} \dfrac{a}{m} = \cos^2 \alpha \\[2mm] \dfrac{b}{n} = \sin^2 \alpha \end{cases} \left(0 < \alpha < \frac{\pi}{2} \right),$$

$$\therefore \ f^3(x) = m^2 + n^2 = \frac{a^2}{\cos^4 \alpha} + \frac{b^2}{\sin^4 \alpha}$$

$$= a^2 + b^2 + (a^2 \tan^4 \alpha + b^2 \cot^2 \alpha + b^2 \cot^2 \alpha)$$

$$+ (b^2 \cot^4 \alpha + a^2 \tan^2 \alpha + a^2 \tan^2 \alpha)$$

$$\geq a^2 + b^2 + 3\sqrt[3]{a^2 + b^4} + 3\sqrt[3]{a^4 + b^2} = (\sqrt[3]{a^2} + \sqrt[3]{b^2})^3$$

$$\therefore \ f(x) \geq \sqrt[3]{a^2} + \sqrt[3]{b^2}$$

if and only if

$$\begin{cases} a^2 \tan^4 \alpha = b^2 \cot^2 \alpha, \\ b^2 \cot^4 \alpha = a^2 \tan^2 \alpha. \end{cases}$$

$$\therefore \ \text{When} \ \tan \alpha = \sqrt[3]{\frac{b}{a}}, \ f(x)_{\min} = \sqrt[3]{a^2} + \sqrt[3]{b^2}.$$

Comment: Substitution relationship is implicit in this problem. The key is to find the implicit relationship and reveal the key to solving the problem.

7. Suppose $a, b, c \in R^+$, and $abc + a + c = b$, find the maximum of $p = \frac{2}{a^2+1} - \frac{2}{b^2+1} + \frac{3}{c^2+1}$.

Analysis: Change $abc + a + c = b$ into $b = \frac{a+c}{1-ac}$; suppose $a = \tan \alpha$, $c = \tan \gamma$. Then solve it.

Solution: According to the condition $a + c = (1 - ac)b$,

$$\because a > 0, \ c > 0, \ b > 0$$

$$\therefore 1 - ac \neq 0 \Rightarrow b = \frac{a+c}{1-ac}.$$

Let $a = \tan\alpha, b = \tan\beta, c = \tan\gamma$

$$(\alpha, \beta, \gamma \text{ are all acute angles})$$

$$\therefore \tan\beta = \frac{\tan\alpha + \tan\gamma}{1 - \tan\alpha\tan\gamma} = \tan(\alpha + \gamma)$$

$$\therefore \beta = \alpha + \gamma$$

$$\therefore p = \frac{2}{1 + \tan^2\alpha} - \frac{2}{1 + \tan^2\beta} + \frac{3}{1 + \tan^2\gamma}$$

$$= 2\cos^2\alpha - 2\cos^2(\alpha + \gamma) + 3\cos^2\gamma$$

$$= (\cos 2\alpha + 1) - [\cos(2\alpha + 2\gamma) + 1] + 3\cos^2\gamma$$

$$= 2\sin\gamma\sin(2\alpha + \gamma) + 3\cos^2\gamma$$

$$\leq 2\sin\gamma + 3(1 - \sin^2\gamma) = \frac{10}{3} - 3\left(\sin\gamma - \frac{1}{3}\right)^2 \leq \frac{10}{3}$$

if and only if $\alpha + \beta = \frac{\pi}{2}, \sin\gamma = \frac{1}{3}$, when $a = \frac{\sqrt{2}}{2}, b = \sqrt{2}, c = \frac{\sqrt{2}}{4}$, which implies $p_{max} = \frac{10}{3}$.

Comment: When we meet the most valuable problem of multivariate function, boundedness of trigonometric function and elimination are the main function when solving problem with triangle substitution.

8. Real numbers x, y meet $3x^2 - xy + 3y^2 = 20$, find the maximum of $8x^2 + 23y^2$.

Solution: Suppose

$$8x^2 + 23y^2 = s \Rightarrow \begin{cases} x = \sqrt{\dfrac{s}{8}}\cos\theta, \\[2mm] y = \sqrt{\dfrac{s}{23}}\sin\theta, \end{cases}$$

$$\theta \in [0, 2\pi], \quad xy = \frac{s}{4\sqrt{46}} \sin 2\theta,$$

$$\therefore \frac{3s}{8} \cos^2 \theta + \frac{3s}{23} \sin^2 \theta - \frac{s}{4\sqrt{46}} \sin 2\theta = 20,$$

$$\frac{20}{s} = \frac{3}{8} \cdot \frac{1 + \cos 2\theta}{2} + \frac{3}{23} \cdot \frac{1 - \cos 2\theta}{2} - \frac{1}{4\sqrt{46}} \sin 2\theta$$

$$= \frac{93}{16 \times 23} + \frac{45}{16 \times 23} \cos 2\theta - \frac{1}{4\sqrt{46}} \sin 2\theta$$

$$\geq \frac{93}{16 \times 23} - \sqrt{\left(\frac{45}{16 \times 23}\right)^2 + \left(\frac{1}{4\sqrt{46}}\right)^2} = \frac{1}{8}$$

$$\therefore \ s \leq 160.$$

Suppose $y = tx$, so

$$\begin{cases} 3x^2 - tx^2 + 3t^2x^2 = 20 \\ 8x^2 + 23t^2x^2 = 160 \end{cases} \Rightarrow t = 4.$$

$\therefore$ When

$$\begin{cases} x = \dfrac{2}{47}\sqrt{235} \\ y = \dfrac{8}{47}\sqrt{235} \end{cases} \quad \text{or} \quad \begin{cases} x = -\dfrac{2}{47}\sqrt{235} \\ y = -\dfrac{8}{47}\sqrt{235} \end{cases}$$

the maximum of $8x^2 + 23y^2$ is 160.

9. If $x, y > 0$, $x^2 + (y-2)^2 = 1$, find the range of $u = \frac{3x^2 + 2\sqrt{3}xy + 5y^2}{x^2 + y^2}$.

Solution: Suppose $x^2 + y^2 = r^2 (r > 0)$, $x = r \cos \theta$, $y = r \sin \theta$. Substitute $x^2 + (y - 2)^2 = 1$;

$$\therefore \ r^2 - 4r \sin \theta + 3 = 0, \quad 4r \sin \theta = r^2 + 3 \geq 2\sqrt{3}r, \quad \sin \theta \geq \frac{\sqrt{3}}{2},$$

$$\because \ x = r \cos \theta \geq 0, \ \therefore \ \cos \theta \geq 0 \ \left(\frac{\pi}{3} \leq \vartheta \leq \frac{\pi}{2}\right),$$

$$\therefore u = 3\cos^2\theta + 2\sqrt{3}\sin\theta\cos\theta + 5\sin^2\theta = 2\sin\left(2\theta - \frac{\pi}{6}\right) + 4,$$

$$\frac{\pi}{3} \le \vartheta \le \frac{\pi}{2} \Rightarrow \frac{\pi}{2} \le 2\theta - \frac{\pi}{6} \le \frac{5\pi}{6} \Rightarrow u \in [5,6].$$

10. If $x, y, z \in R^+$, $x^2 + y^2 + z^2 = 1$, find the minimum of $s = \frac{yz}{x} + \frac{zx}{y} + \frac{xy}{z}$.

Solution:

$$\because x, y, z \in R^+, x^2 + y^2 + z^2 = 1.$$

Suppose

$$x = \cos\alpha\cos\beta, \ y = \cos\alpha\sin\beta, \ z = \sin\alpha, \ \text{and} \ \alpha, \beta \in \left(0, \frac{\pi}{2}\right);$$

$$s = \frac{\sin\alpha\sin\beta}{\cos\beta} + \frac{\sin\alpha\cos\beta}{\sin\beta} + \frac{\cos^2\alpha\sin\beta\cos\beta}{\sin\alpha}$$

$$= \frac{2}{\sin 2\beta}\sin\alpha + \frac{1 - \sin^2\alpha}{2\sin\alpha}\sin 2\beta$$

$$= \left(\frac{2}{\sin 2\beta} - \frac{\sin 2\beta}{2}\right)\sin\alpha + \frac{\sin 2\beta}{2\sin\alpha}$$

$$= \frac{4 - \sin^2 2\beta}{2\sin 2\beta} \times \sin\alpha + \frac{\sin 2\beta}{2\sin\alpha}$$

$$\ge 2\sqrt{\frac{4 - \sin^2 2\beta}{2\sin 2\beta}\sin\alpha \times \frac{\sin 2\beta}{\sin\alpha}}$$

$$= \sqrt{4 - \sin^2 2\beta} \ge \sqrt{3}$$

if and only if

$$\begin{cases} \sin 2\beta = 1, \\ \sin\alpha = \dfrac{\sqrt{3}}{3}, \\ \cos\alpha = \dfrac{\sqrt{6}}{3}. \end{cases}$$

$$\therefore \text{When } x = y = z = \frac{\sqrt{3}}{3}, \text{ the equality holds.}$$

11. If $x, y, z \in R, xy + yz + zx = 1$, find the maximum of $xyz(x + y + z)$.

Solution: Suppose $xy = \cos^2 \alpha \cos^2 \beta$, $yz = \cos^2 \alpha \sin^2 \beta$, $zx = \sin^2 \alpha$;

$$(xyz)^2 = \cos^4 \alpha \sin^2 \alpha \cos^2 \beta \sin^2 \beta,$$

$$x^2 yz = \cos^2 \alpha \cos^2 \beta \sin^2 \alpha,$$

$$xy^2 z = \cos^4 \alpha \cos^2 \beta \sin^2 \beta,$$

$$xyz^2 = \cos^2 \alpha \sin^2 \beta \sin^2 \alpha,$$

$$xyz(x + y + z) = \cos^2 \alpha \sin^2 \alpha + \cos^4 \alpha \cos^2 \beta \sin^2 \beta$$

$$= \frac{1}{4} \sin^2 2\alpha + \frac{(1 + \cos 2\alpha)^2}{16} \sin^2 2\beta$$

$$\leq \frac{1}{4} \sin^2 2\alpha + \frac{1}{16}(1 + 2\cos 2\alpha + \cos^2 2\alpha)$$

$$= \frac{1}{3} - \frac{3}{16}\left(\cos 2\alpha - \frac{1}{3}\right)^2 \leq \frac{1}{3}$$

if and only if

$$\begin{cases} \cos 2\alpha = 1, \\ \sin^2 2\beta = 1. \end{cases}$$

$\therefore$ When $x = y = z = \dfrac{\sqrt{3}}{3}$, the equality holds.

6.2. Trigonometric Inequality

The inequality of trigonometric functions is called triangle inequality. The following inequalities are common in $\triangle ABC$:

$$\sin A + \sin B + \sin C \leq \frac{3\sqrt{3}}{2};$$

$$\sin A \sin B \sin C \leq \frac{3\sqrt{3}}{8};$$

$$\cos 2A + \cos 2B + \cos 2C \geq -\frac{3}{2};$$

$$\cos^2 A + \cos^2 B + \cos^2 C \geq \frac{3}{4};$$

$$\cos \frac{A}{2} + \cos \frac{B}{2} + \cos \frac{C}{2} \leq \frac{3\sqrt{3}}{8};$$

$$\cos A \cos B \cos C \leq \frac{1}{8}.$$

Exercise

1. In $\triangle ABC$, prove:

(1) $\dfrac{1}{b+c-a} + \dfrac{1}{c+a-b} + \dfrac{1}{a+b-c} \geq \dfrac{9}{a+b+c}$;

(2) $abc \geq (b+c-a)(c+a-b)(a+b-c)$;   (*)

(3) $a^2b(a-b) + b^2c(b-c) + c^2a(c-a) \geq 0$

Proof:

(1) Replacement equation (*)

$$\frac{1}{b+c-a} + \frac{1}{c+a-b} + \frac{1}{a+b-c} \geq \frac{9}{a+b+c}$$

$$\Leftrightarrow \frac{1}{x} + \frac{1}{y} + \frac{1}{z} \geq \frac{9}{x+y+z}$$

$$\Leftrightarrow (x+y+z)\left(\frac{1}{x} + \frac{1}{y} + \frac{1}{z}\right) \geq 9$$

$$\because x, y, z > 0,$$

$$\therefore x+y+z \geq 3^3\sqrt{xyz}, \quad \frac{1}{x} + \frac{1}{y} + \frac{1}{z} \geq 3^3\sqrt{\frac{1}{xyz}}$$

$$\therefore (x+y+z)\left(\frac{1}{x} + \frac{1}{y} + \frac{1}{z}\right) \geq 9$$

$$\therefore \frac{1}{b+c-a} + \frac{1}{c+a-b} + \frac{1}{a+b-c} \geq \frac{9}{a+b+c}.$$

(2) Replacement equation (*)

$$abc = (y+z)(z+x)(x+y)$$

$$\geq 2\sqrt{yz} \cdot 2\sqrt{zx} \cdot 2\sqrt{xy} = 2x \cdot 2y \cdot 2z$$

$$= (b+c-a)(c+a-b)(a+b-c).$$

(3) Replacement equation (*)

$$(y+z)^2(z+x)(y-x) + (z+x)^2(x+y)(z-y)$$

$$+(x+y)^2(y+z)(x-z) \geq 0.$$

Expansion and simplification:

$$xy^3 + yz^3 + zx^3 - xyz(x+y+z) \geq 0$$

$$xyz\left(\frac{y^2}{z} + \frac{z^2}{x} + \frac{x^2}{y} - x - y - z\right) \geq 0 \qquad (1)$$

$$\because x, y, z > 0, \therefore (1) \Rightarrow \frac{x^2}{y} + \frac{y^2}{z} + \frac{z^2}{x} \geq x+y+z.$$

Add the following inequalities

$$\frac{x^2}{y} + y \geq 2x, \quad \frac{y^2}{z} + z \geq 2y, \quad \frac{z^2}{x} + x \geq 2x$$

$$\therefore a^2b(a-b) + b^2c(b-c) + c^2a(c-a) \geq 0.$$

$\square$

2. In $\triangle ABC$, prove:

(1) $p^3 \geq 27(p-a)(p-b)(p-c)$;

(2) $\dfrac{a}{p-a} + \dfrac{b}{p-b} + \dfrac{c}{p-c} \geq 6$; \hfill (*)

(3) $\dfrac{a^2}{(p-b)(p-c)} + \dfrac{b^2}{(p-c)(p-a)} + \dfrac{c^2}{(p-a)(p-b)} \geq 12.$

Proof:

(1) Replacement equation (*)

$$p^3 = (x + y + z)^3 \geq (3 \cdot \sqrt[3]{xyz})^3$$
$$= 27xyz = 27(p - a)(p - b)(p - c).$$

(2) Replacement equation (*)

$$p - a = x, p - b = y, p - c = z$$

$$\therefore \frac{a}{p-a} + \frac{b}{p-b} + \frac{c}{p-c} \geq \frac{y+z}{x} + \frac{z+x}{y} + \frac{x+y}{z}$$

$$= \left(\frac{y}{x} + \frac{x}{y}\right) + \left(\frac{z}{x} + \frac{x}{z}\right) + \left(\frac{y}{z} + \frac{z}{y}\right) \geq 6.$$

(3) Replacement equation (*)

$$\frac{a^2}{(p-b)(p-c)} + \frac{b^2}{(p-c)(p-a)} + \frac{c^2}{(p-a)(p-b)}$$

$$= \frac{(y+z)^2}{yz} + \frac{(z+x)^2}{zx} + \frac{(x+y)^2}{xy}$$

$$\geq \frac{(2\sqrt{yz})^2}{yz} + \frac{(2\sqrt{zx})^2}{zx} + \frac{(2\sqrt{xy})^2}{xy} = 4 + 4 + 4 = 12. \qquad \square$$

3. In $\triangle ABC$, prove:

(1) $\tan \dfrac{A}{2} \tan \dfrac{B}{2} \tan \dfrac{C}{2} \leq \dfrac{\sqrt{3}}{9}$;

(2) $\cot \dfrac{A}{2} + \cot \dfrac{B}{2} + \cot \dfrac{C}{2} \geq 3\sqrt{3}$;

$$(*)$$

(3) $\dfrac{1}{\sin^2 A} + \dfrac{1}{\sin^2 B} + \dfrac{1}{\sin^2 C} \geq 4$;

(4) $(1 - \cos A)(1 - \cos B)(1 - \cos C) \leq \dfrac{1}{8}$.

Proof:

(1) Replacement equation (*)

$$\tan\frac{A}{2}\tan\frac{B}{2}\tan\frac{C}{2}$$

$$=\sqrt{\frac{yz}{x(x+y+z)}}\cdot\sqrt{\frac{zx}{y(x+y+z)}}\cdot\sqrt{\frac{xy}{z(x+y+z)}}$$

$$=\frac{(xyz)^{\frac{1}{2}}}{(x+y+z)^{\frac{3}{2}}}\leq\frac{(xyz)^{\frac{1}{2}}}{\left[3(xyz)^{\frac{1}{3}}\right]^{\frac{3}{2}}}=\frac{(xyz)^{\frac{1}{2}}}{3\sqrt{3}(xyz)^{\frac{1}{2}}}=\frac{\sqrt{3}}{9}.$$

(2) Replacement equation (*)

$$\cot\frac{A}{2}+\cot\frac{B}{2}+\cot\frac{C}{2}$$

$$=\sqrt{\frac{x(x+y+z)}{yz}}+\sqrt{\frac{y(x+y+z)}{zx}}+\sqrt{\frac{z(x+y+z)}{xy}}$$

$$=(x+y+z)\cdot\sqrt{\frac{x+y+z}{xyz}}=\frac{(x+y+z)^{\frac{3}{2}}}{(xyz)^{\frac{1}{2}}}$$

$$\geq\frac{\left[3(xyz)^{\frac{1}{3}}\right]^{\frac{3}{2}}}{(xyz)^{\frac{1}{2}}}=\frac{3\sqrt{3}(xyz)^{\frac{1}{2}}}{(xyz)^{\frac{1}{2}}}=3\sqrt{3}.$$

(3) Replacement equation (*)

$$\frac{1}{\sin^2 A}=\frac{(z+x)^2(x+y)^2}{4(x+y+z)xyz}\geq\frac{(2\sqrt{zx})^2\cdot(2\sqrt{xy})^2}{4(x+y+z)xyz}$$

$$=\frac{4x}{x+y+z}.$$

Similarly,

$$\frac{1}{\sin^2 B} \geq \frac{4y}{x+y+z}, \frac{1}{\sin^2 C} \geq \frac{4z}{x+y+z}$$

$$\frac{1}{\sin^2 A} + \frac{1}{\sin^2 B} + \frac{1}{\sin^2 C}$$

$$\geq \frac{4x}{x+y+z} + \frac{4y}{x+y+z} + \frac{4z}{x+y+z}$$

$$= \frac{4(x+y+z)}{x+y+z} = 4.$$

(4) Replacement equation (*):

$$1 - \cos A = 1 - \left[1 - \frac{2yz}{(z+x)(x+y)}\right] = \frac{2yz}{(z+x)(x+y)}$$

$$\leq \frac{2yz}{2\sqrt{zx} \cdot 2\sqrt{xy}} = \frac{\sqrt{yz}}{2x}$$

Similarly,

$$1 - \cos B \leq \frac{\sqrt{zx}}{2y}, \quad 1 - \cos C \leq \frac{\sqrt{xy}}{2z}.$$

$$(1 - \cos A)(1 - \cos B)(1 - \cos C) \leq \frac{\sqrt{yz}}{2x} \cdot \frac{\sqrt{zx}}{2y} \cdot \frac{\sqrt{xy}}{2z}$$

$$= \frac{xyz}{8xyz} = \frac{1}{8}.$$

□

4. In $\triangle ABC$, prove:

(1) $R \geq 2r$;

(2) $\dfrac{1}{\sin A} + \dfrac{1}{\sin B} + \dfrac{1}{\sin C} \geq 2\sqrt{3};$　　　　　(*)

(3) $\dfrac{r_a}{h_a} + \dfrac{r_b}{h_b} + \dfrac{r_c}{h_c} \geq 3.$

Proof:

(1) Replacement equation (*):

$$R = \frac{(x+y)(y+z)(z+x)}{4 \cdot \sqrt{xyz(x+y+z)}}, r = \sqrt{\frac{xyz}{x+y+z}}$$

$$\therefore R \geq 2r \Leftrightarrow (x+y)(y+z)(z+x) \geq 8xyz$$

$$\because y+z \geq 2\sqrt{yz}, \ z+x \geq \sqrt{xz}, \ x+y = 2\sqrt{xy}$$

$$\therefore (y+z)(z+x)(x+y) \geq 8xyz$$

$$\therefore R \geq 2r.$$

(2) According to the Law of Sine: $\frac{a}{\sin A} = \frac{b}{\sin B} = \frac{c}{\sin C} = 2R$,

$$\therefore \frac{\sqrt{3}}{3}R\left(\frac{1}{a}+\frac{1}{b}+\frac{1}{c}\right) \geq 1$$

$$\because \frac{1}{3}R^2\left(\frac{1}{a}+\frac{1}{b}+\frac{1}{c}\right)^2$$

$$= \frac{1}{3}R^2\left[\frac{1}{a^2}+\frac{1}{b^2}+\frac{1}{c^2}+2\left(\frac{1}{ab}+\frac{1}{bc}+\frac{1}{ca}\right)\right]$$

$$\geq \frac{1}{3}R^2 \cdot 3\left(\frac{1}{ab}+\frac{1}{bc}+\frac{1}{ca}\right)$$

$$= R^2 \cdot \frac{a+b+c}{abc} = R^2 \cdot \frac{1}{2\frac{\Delta}{p} \cdot \frac{abc}{4\Delta}} = R^2 \cdot \frac{1}{2rR} = \frac{R}{2r},$$

$$\because R \geq 2r \quad \therefore \frac{1}{3}R^2\left(\frac{1}{a}+\frac{1}{b}+\frac{1}{c}\right)^2 \geq 1,$$

$$\therefore \frac{\sqrt{3}}{3}R\left(\frac{1}{a}+\frac{1}{b}+\frac{1}{c}\right) \geq 1$$

$$\therefore \frac{1}{\sin A}+\frac{1}{\sin B}+\frac{1}{\sin C} \geq 2\sqrt{3}.$$

(3) Replacement equation (*):

$$\therefore \frac{r_a}{h_a} = \frac{\sqrt{(x+y+z)xyz}}{x} \Big/ \frac{2\sqrt{(x+y+z)xyz}}{y+z} = \frac{y+z}{2x}$$

Similarly, $\dfrac{r_b}{h_b} = \dfrac{x+z}{2y}, \dfrac{r_c}{h_c} = \dfrac{y+x}{2z}$

$$\therefore \frac{r_a}{h_a} + \frac{r_b}{h_b} + \frac{r_c}{h_c} = \frac{y+z}{2x} + \frac{z+x}{2y} + \frac{x+y}{2z}$$

$$= \frac{1}{2}\left[\left(\frac{y}{x}+\frac{x}{y}\right) + \left(\frac{y}{z}+\frac{z}{y}\right) + \left(\frac{z}{x}+\frac{x}{z}\right)\right]$$

$$\geq \frac{1}{2}(2+2+2) = 3.$$

$\square$

Comment: (1)$R \geq 2r$ is called Euler formula, if and only if triangle is equilateral triangle, and inequality holds.

$$(2) \because \triangle = \frac{1}{2}ab\sin C = \frac{1}{2}ac\sin B = \frac{1}{2}bc\sin A,$$

$$\therefore \frac{1}{\sin C} = \frac{ab}{2\triangle}, \frac{1}{\sin B} = \frac{ac}{2\triangle}, \frac{1}{\sin A} = \frac{bc}{2\triangle}$$

$$\therefore ab + bc + ac \geq 4\sqrt{3}\triangle.$$

This is the strengthening form of the Burke inequality $(a^2 + b^2 + c^2 \geq 4\sqrt{3}\triangle)$ for the third IMO test question $(a^2 + b^2 + c^2 \geq ab + bc + ca)$.

5. In $\triangle ABC$, if $a + b + c = 1$, prove: $a^2 + b^2 + c^2 + 4abc < \frac{1}{2}$.

Proof:

Method 1: Replacement equation (*):

$$\because a+b+c=1, \therefore x+y+z = \frac{1}{2},$$

$$x^2 + y^2 + z^2 + 2(xy + yz + zx) = \frac{1}{4}$$

$$\therefore a^2 + b^2 + c^2 + 4abc = (y+z)^2 + (z+x)^2$$

$$+ (x+y)^2 + 4(y+z)(z+x)(x+y)$$

$$= 2(x^2 + y^2 + z^2) + 2(xy + yz + zx)$$

$$+ 4(x^2 y + x^2 z + y^2 x + y^2 z + z^2 x + z^2 y + 2xyz)$$

$$= \frac{1}{4} + x^2 + y^2 + z^2 + 4[xy(x+y+z)$$

$$+ yz(x+y+z) + zx(x+y+z) - xyz]$$

$$= \frac{1}{4} + x^2 + y^2 + z^2 + 2(xy + yz + zx) - 4xyz$$

$$= \frac{1}{2} - 4xyz < \frac{1}{2}$$

Method 2: Suppose $a = \sin^2 \alpha \cos^2 \beta, b = \cos^2 \alpha \cos^2 \beta, c = \sin^2 \beta, \beta \in (0, \frac{\pi}{2})$.

$\because$ a, b, c are three sides of the triangle

$\therefore$ $c < \frac{1}{2}, c > |a - b|, \therefore \beta \in \left(0, \frac{\pi}{4}\right), \sin^2 \beta > |\cos 2\alpha \cdot \cos^2 \beta|$

$\because$ $1 = (a + b + c)^2 = a^2 + b^2 + c^2 + 2(ab + bc + ca)$

$\therefore$ $a^2 + b^2 + c^2 + 4abc = 1 - 2(ab + bc + ca - 2abc)$

$\because$ $ab + bc + ca - 2abc = c(a + b) + ab(1 - 2c)$

$$= \sin^2 \beta \cos^2 \beta + \sin^2 \alpha \cos^2 \alpha \cos^4 \beta \cos 2\beta$$

$$= \frac{1}{4}[1 - \cos^2 2\beta + (1 - \cos^2 2\alpha) \cos^4 \beta \cos 2\beta]$$

$$= \frac{1}{4} + \frac{1}{4} \cos 2\beta(\cos^4 \beta - \cos^2 2\alpha \cos^4 \beta - \cos 2\beta)$$

$$> \frac{1}{4} + \frac{1}{4} \cos 2\beta(\cos^4 \beta - \sin^4 \beta - \cos 2\beta) = \frac{1}{4}$$

$\therefore$ $a^2 + b^2 + c^2 + 4abc < 1 - \frac{1}{4} = \frac{1}{2}.$

Method 3: Suppose $a = \sin^2 \alpha \cos^2 \beta, b = \cos^2 \alpha \cos^2 \beta, c = \sin^2 \beta$.

$$a^2 + b^2 + c^2 + 4abc$$

$$= \sin^4 \alpha \cos^4 \beta + \cos^4 \alpha \cos^4 \beta + \sin^4 \beta$$

$$\quad + 4 \sin^2 \alpha \cos^2 \alpha \cos^4 \beta \sin \beta$$

$$= (1 - 2 \sin^2 \alpha \cos^2 \alpha) \cos^4 \beta + \sin^4 \beta$$

$$\quad + 4 \sin^2 \alpha \cos^2 \alpha \cos^4 \beta \sin^2 \beta$$

$$= (\cos^4 \beta + \sin^4 \beta) + 2 \sin^2 \alpha \cos^2 \alpha \cos^4 \beta (2 \sin^2 \beta - 1)$$

$$= (1 - 2 \cos^2 \beta \sin^2 \beta) + \frac{1}{2} \sin^2 2\alpha \cos^4 \beta (2 \sin^2 \beta - 1)$$

$$= 1 - \frac{1}{2} \sin^2 2\beta - \frac{1}{2} \cos^2 2\beta \cos^4 \beta \sin 2\alpha$$

$$= 1 - \frac{1}{2}(1 - \cos^2 2\beta) - \frac{1}{2} \cos 2\beta \cos^4 \beta (1 - \cos^2 2\alpha)$$

$$= 1 - \frac{1}{2} + \frac{1}{2} \cos^2 2\beta - \frac{1}{2} \cos 2\beta \cos^4 \beta$$

$$\quad + \frac{1}{2} \cos 2\beta \cos^4 \beta \cos^2 2\alpha$$

$$< \frac{1}{2} + \frac{1}{2} \cos^2 2\beta - \frac{1}{2} \cos 2\beta \cos^4 \beta + \frac{1}{2} \cos 2\beta \sin^4 \beta$$

$$= \frac{1}{2} + \frac{1}{2} \cos^2 2\beta - \frac{1}{2} \cos 2\beta (\cos^4 \beta - \sin^4 \beta)$$

$$= \frac{1}{2} + \frac{1}{2} \cos^2 2\beta - \frac{1}{2} \cos^2 2\beta = \frac{1}{2}$$

Method 4: $\because a + b > c, b + c > a, c + a > b,$

$$\therefore 2a < a + b + c = 1, a < \frac{1}{2}$$

$$\therefore b < \frac{1}{2}, \quad c < \frac{1}{2}$$

$$\therefore (2a-1)(2b-1)(2c-1) < 0,$$

$$8abc - 4ab - 4ac + 2a - 4bc + 2b + 2c - 1 < 0$$

$$\therefore 4abc - 2ab - 2ac - 2bc + (a+b+c) < \frac{1}{2}$$

$$\therefore 4abc - 2ab - 2ac - 2bc + (a+b+c)^2 < \frac{1}{2},$$

$$a^2 + b^2 + c^2 + 4abc < \frac{1}{2}.$$

$\square$

6. In $\triangle ABC$, prove:

$$\tan\frac{A}{2} + \tan\frac{B}{2} + \tan\frac{C}{2} \le \frac{9R^2}{4\Delta}$$

(Δ is the area of $\triangle ABC$). $\qquad (*)$

Proof: Replacement equation $(*)$

$$\tan\frac{A}{2} + \tan\frac{B}{2} + \tan\frac{C}{2} \le \frac{9R^2}{4\Delta} \Leftrightarrow \frac{yz + zx + xy}{\Delta}$$

$$\therefore \le \frac{9R^2}{4\Delta} \Leftrightarrow yz + zx + xy \le \frac{9}{4} \cdot \left[\frac{(x+y)(y+z)(z+x)}{4 \cdot \sqrt{xyz(x+y+z)}}\right]^2$$

$$\Leftrightarrow \left(\frac{1}{x} + \frac{1}{y} + \frac{1}{z}\right)(x+y+z)$$

$$\le \frac{9}{64}\left[\left(1+\frac{y}{x}\right)\left(1+\frac{z}{y}\right)\left(1+\frac{x}{z}\right)\right]^2$$

Suppose $\frac{y}{x} = m, \frac{z}{y} = n, \frac{x}{z} = t, \therefore mnt = 1,$

and $\left(\frac{1}{x} + \frac{1}{y} + \frac{1}{z}\right)(x+y+z)$

$$\le \frac{9}{64}\left[\left(1+\frac{y}{x}\right)\left(1+\frac{z}{y}\right)\left(1+\frac{x}{z}\right)\right]^2.$$

$$\Leftrightarrow 3 + (m+n+t) + (mn+nt+tm)$$

$$\leq \frac{9}{64}[2 + (m + n + t) + (mn + nt + tm)]^2.$$

Suppose $(m + n + t) + (mn + nt + tm) = q$,

$$\therefore\ q \geq 6\sqrt[6]{\mathrm{mnt}(mn)(\mathrm{nt})(tm)} = 6,$$

and $3 + (m + n + t) + (mn + nt + tm)$

$$\leq \frac{9}{64}[2 + (m + n + t) + (mn + nt + tm)]^2$$

$$\Leftrightarrow 3 + q \leq \frac{9}{64}(2 + q)^2 \Leftrightarrow \left(3q - \frac{14}{3}\right)^2 \geq \frac{1600}{9}.$$

$\because\ q \geq 6$, $\therefore$ Above inequality clearly holds.

□

Comment: This question can also be solved as follows:

$$\tan\frac{A}{2} + \tan\frac{B}{2} + \tan\frac{C}{2}$$

$$\leq \frac{9R^2}{4\Delta} \Leftrightarrow 64xyz(xy + yz + zx)(x + y + z)$$

$$\leq 9(y + z)^2(z + x)^2(x + y)^2$$

$\because\ 64xyz(xy + yz + zx)(x + y + z)$

$$= 64xyz(x^2y + xy^2 + y^2z + yz^2 + z^2x + z^2x + 3xyz)$$

$$= 64xyz[(x + y)(y + z)(z + x) + xyz],$$

$$(x + y)(y + z)(z + x) \geq 8xyz$$

$$\therefore\ xyz \leq \frac{1}{8}(x + y)(y + z)(z + x)$$

$$\therefore\ 64xyz(xy + yz + zx)(x + y + z)$$

$$\leq 64 \cdot \frac{1}{8}(x + y)(y + z)(z + x)$$

$$\times \left[(x+y)(y+z)(z+x) + \frac{1}{8}(x+y)(y+z)(z+x) \right]$$

$$= 9(y+z)^2(z+x)^2(x+y)^2$$

$\therefore$ The original inequality is proved.

7. From a point in triangle ABC to three sides for vertical line OM, ON, OP, prove: $OA + OB + OC \geq 2(OM + ON + OP)$ (Aidesi Modeer Theorem).

Prove: Suppose OM, ON, OP respectively x, y, z. By A, M, O, P four points of the circle and the cosine law $MP = \sqrt{x^2 + z^2 + 2xz \cos A}$,

and the sine law $OA = \dfrac{MP}{\sin A} = \dfrac{\sqrt{x^2 + z^2 + 2xz \cos A}}{\sin A}$,

substitute $\cos A = -\cos(B+C)$

$$= \sin B \sin C - \cos B \cos C$$

into above type,

$$OA = \frac{\sqrt{(x\sin B + z\sin C)^2 + (x\cos B - z\cos C)^2}}{\sin A}$$

$$\geq \frac{x\sin B + z\sin C}{\sin A}$$

Similarly,

$$OB \geq \frac{x\sin A + y\sin C}{\sin B}, \quad OC \geq \frac{y\sin B + z\sin A}{\sin C}$$

$\therefore$ $OA + OB + OC$

$$\geq x\left(\frac{\sin B}{\sin A} + \frac{\sin A}{\sin B}\right) + y\left(\frac{\sin C}{\sin B} + \frac{\sin B}{\sin C}\right) + z\left(\frac{\sin C}{\sin A} + \frac{\sin A}{\sin C}\right)$$

$$\geq 2(x+y+z).$$

8. In non-right-angled triangle ABC, prove:

$$\tan^2 A + \tan^2 B + \tan^2 C$$

$$\geq \tan A \tan B + \tan B \tan C + \tan C \tan A.$$

Proof: Suppose $\tan A \geq \tan B \geq \tan C$.

By the sort inequality

$$\tan^2 A + \tan^2 B + \tan^2 C$$

$$\geq \tan A \tan B + \tan B \tan C + \tan C \tan A.$$

Comment: This problem also can be proved by Cauchy inequality,

$\because$ $\tan^2 A, \tan^2 B, \tan^2 C$ are all real number

$\therefore$ $(\tan A \tan B + \tan B \tan C + \tan C \tan A)^2$

$\qquad \leq (\tan^2 A + \tan^2 B + \tan^2 C) \; (\tan^2 B + \tan^2 C + \tan^2 A)$

$\qquad = (\tan^2 A + \tan^2 B + \tan^2 C)^2$

$\therefore$ $\tan^2 A + \tan^2 B + \tan^2 C$

$\qquad \geq |\tan A \tan B + \tan B \tan C + \tan C \tan A|$

$\qquad \geq \tan A \tan B + \tan B \tan C + \tan C \tan A.$ $\square$

9. (1) In $\triangle ABC$, prove: $\frac{2\pi}{3} p \leq aA + bB + cC < \pi p$ ($p = a + b + c$ is the perimeter of $\triangle ABC$).

(2) In $\triangle ABC$, prove: $\frac{\sin A}{h_a} + \frac{\sin B}{h_b} + \frac{\sin C}{h_c} \geq \frac{\sqrt{3}}{R}$ (R is the radius of circumcircle of $\triangle ABC$).

Proof:

(1) Note $M = aA + bB + cC$ (M is ordered sum),

$\qquad \therefore$ $M \geq aB + bC + cA$, $M \geq aC + bA + cB$.

By above three formulas
$$3M \geq a(A+B+C)+b(A+B+C)+c(A+B+C) = \pi(a+b+c).$$

$$\therefore M \geq \frac{2\pi}{3}p$$

$$\because a < p, b < p, c < p, \therefore M < pA + pB + pC = \pi P$$

$$\therefore \frac{2\pi}{3}p \leq M \leq \pi p.$$

(2)   Note $M = \dfrac{\sin A}{h_a} + \dfrac{\sin B}{h_b} + \dfrac{\sin C}{h_c}$.

By the Chebyshev inequality,

$$M \geq \frac{1}{3}\left(\frac{1}{h_a} + \frac{1}{h_b} + \frac{1}{h_c}\right)(\sin A + \sin B + \sin C).$$

$\square$

By inequality $\frac{1}{h_a} + \frac{1}{h_b} + \frac{1}{h_c} \geq \frac{2}{\sqrt{3}}(\frac{1}{a} + \frac{1}{b} + \frac{1}{c})$ and $\sin A = \frac{a}{2R}$ (sine law), we have

$$M \geq \frac{1}{3} \cdot \frac{2}{\sqrt{3}}\left(\frac{1}{a} + \frac{1}{b} + \frac{1}{c}\right)\left(\frac{a}{2R} + \frac{b}{2R} + \frac{c}{2R}\right)$$

$$= \frac{1}{3\sqrt{3}R}\left(\frac{1}{a} + \frac{1}{b} + \frac{1}{c}\right)(a+b+c) \geq \frac{1}{3\sqrt{3}R} \cdot 3^2 = \frac{\sqrt{3}}{R}.$$

Comment: In any $\triangle ABC$, if we have the three sides $a \leq b \leq c$, then the three angles will be $A \leq B \leq C$.
Moreover, we have the following:

trigonometric functions:
$\sin A \leq \sin B \leq \sin C$, $\cos A \leq \cos B \leq \cos C$;
three heights: $h_a \geq h_b \geq h_c$, $\frac{1}{h_a} \leq \frac{1}{h_b} \leq \frac{1}{h_c}$;
three midlines: $m_a \geq m_b \geq m_c$, $\frac{1}{m_a} \leq \frac{1}{m_b} \leq \frac{1}{m_c}$;
three angular bisectors: $t_a \geq t_b \geq t_c$, $\frac{1}{t_a} \leq \frac{1}{t_b} \leq \frac{1}{t_c}$.

The order of the above-mentioned equations makes it as a good basis for the application of the inequality of the common laws (sine law, cosine law, area formula) and the use of common inequality:

$$\sin A + \sin B + \sin C \le \frac{3\sqrt{3}}{2}, \quad \cos A + \cos B + \cos C \le \frac{3}{2},$$

$$a^2 + b^2 + c^2 \ge 4\sqrt{3}\Delta \quad (\Delta \text{ is area})$$

$$h_a + h_b + h_c \le \frac{\sqrt{3}}{2}(a+b+c) \quad \text{and}$$

$$\frac{1}{h_a} + \frac{1}{h_b} + \frac{1}{h_c} \ge \frac{2}{\sqrt{3}}\left(\frac{1}{a} + \frac{1}{b} + \frac{1}{c}\right)$$

$$t_a + t_b + t_c \le \frac{\sqrt{3}}{2}(a+b+c) \quad \text{and}$$

$$\frac{1}{t_a} + \frac{1}{t_b} + \frac{1}{t_c} \ge \frac{2}{\sqrt{3}}\left(\frac{1}{a} + \frac{1}{b} + \frac{1}{c}\right)$$

$$m_a + m_b + m_c > \frac{3}{4}(a+b+c) \quad \text{and}$$

$$m_a + m_b + m_c \le \frac{\sqrt{3}}{2}(a+b+c).$$

In addition to the above example, you can also get the following series of results:

(1) $a \sin A + b \sin B$
$\quad + c \sin C \ge \frac{\sqrt{3}}{2}p$;

(2) $a \cos A + b.\cos B$
$\quad + c \cos C \le \frac{1}{2}p$;

(3) $A \sin A + B \sin B + C \sin C \ge \pi$;

(4) $Ah_a + Bh_b + Ch_c \le \frac{3}{6}\pi p$;

(5) $\dfrac{a}{h_a} + \dfrac{b}{h_b} + \dfrac{c}{h_c} \ge 2\sqrt{3}$;

(6) $\dfrac{A}{t_a} + \dfrac{B}{t_b} + \dfrac{C}{t_c} \ge \dfrac{2\sqrt{3}\pi}{p}$;

(7) $at_a + bt_b + ct_c \le \dfrac{\sqrt{3}}{6}p^2$;

(8) $\dfrac{a}{t_a} + \dfrac{b}{t_b} + \dfrac{c}{t_c} \ge 2\sqrt{3}$;

(9) $t_a \sin A + t_b \sin B + t_c \sin C \le \frac{3}{4}p$;

(10) $\dfrac{m_a}{a} + \dfrac{m_b}{b} + \dfrac{m_c}{c} \ge \dfrac{3\sqrt{3}}{2}$;

(11) $am_a + bm_b + cm_c \le \dfrac{1}{3}p^2$;

(12) $Am_a + Bm_b + Cm_c \le \dfrac{\pi}{3}p$;

(13) $m_a \sin A + m_b \sin B$
$\quad + m_c \sin C \le \dfrac{\sqrt{3}}{2}p$.

10. If $x, y, z \in R, 0 < x < y < z < \frac{\pi}{2}$, prove:

$$\frac{\pi}{2} + 2\sin x \cos y + 2\sin y \cos z > \sin 2x + \sin 2y + \sin 2z.$$

Proof:

$$\frac{\pi}{4} + \sin x \cos y + \sin y \cos z > \sin x \cos x + \sin y \cos y + \sin z \cos z$$

$$\Leftrightarrow \frac{\pi}{4} > \sin x(\cos x - \cos y) + \sin y(\cos y - \cos z) + \sin z \cos z. \quad (*)$$

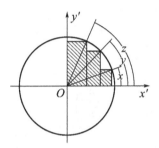

As shown in figure, make a circle of units, so $\frac{\pi}{4}$ is area of unit circle in the first quadrant. The area of the shadow is the right of the equation (*). □

Comment: This problem can be generalized to a finite rectangle: $0 < x_i < \frac{\pi}{2}$ $(i = 1, 2, 3, \ldots, n)$
so $\frac{\pi}{2} + 2\sin x_1 \cos x_2 + 2\sin x_2 \cos x_3 + \cdots + 2\sin x_{n-2} \cos x_{n-1} + 2\sin x_{n-1} \cos x_n > \sin 2x_1 + \sin 2x_2 + \cdots + 2\sin x_n.$

11. Prove
$$2\sin^2\left(\frac{\pi}{4} - \frac{\sqrt{2}}{2}\right) \leq \cos(\sin x) - \sin(\cos x) \leq 2\sin^2\left(\frac{\pi}{4} + \frac{\sqrt{2}}{2}\right).$$

Proof:

$$\cos(\sin x) - \sin(\cos x) = \cos(\sin x) - \cos\left(\frac{\pi}{2} - \cos x\right)$$

$$= 2\sin\left(\frac{\pi}{4} - \frac{\cos x + \sin x}{2}\right) \cdot \sin\left(\frac{\pi}{4} - \frac{\cos x - \sin x}{2}\right)$$

$$\because \ -\sqrt{2} \le \cos x \pm \sin x \le \sqrt{2}$$

$$\therefore \ \frac{\pi}{4} - \frac{\sqrt{2}}{2} \le \frac{\pi}{4} - \frac{\cos x \pm \sin x}{2} \le \frac{\pi}{4} + \frac{\sqrt{2}}{2}$$

$$\because \ 0 < \frac{\pi}{4} - \frac{\sqrt{2}}{2} < \frac{\pi}{4} + \frac{\sqrt{2}}{2} < \frac{\pi}{2},$$

Sine function is increasing function in $\left[0, \dfrac{\pi}{2}\right]$

$$\therefore \ 2\sin^2\left(\frac{\pi}{4} - \frac{\sqrt{2}}{2}\right) \le \cos(\sin x) - \sin(\cos x)$$

$$\le 2\sin^2\left(\frac{\pi}{4} + \frac{\sqrt{2}}{2}\right).$$

$\square$

12. In $\triangle ABC$, prove:

$$\sin\frac{A}{2}\cos\frac{B}{2}\cos\frac{C}{2} + \sin\frac{B}{2}\cos\frac{A}{2}\cos\frac{C}{2} + \sin\frac{C}{2}\cos\frac{A}{2}\cos\frac{B}{2} \le \frac{9}{8}.$$

Proof:

$$\sin\frac{A}{2}\cos\frac{B}{2}\cos\frac{C}{2} = \frac{1}{2}\left[\sin\frac{A+B}{2} + \sin\frac{A-B}{2}\right]\cos\frac{C}{2}$$

$$= \frac{1}{2}\left[\cos\frac{C}{2}\cos\frac{C}{2} + \sin\frac{A-B}{2}\sin\frac{A+B}{2}\right]$$

$$= \frac{1}{4}[1 + \cos C + \cos B - \cos A] \tag{3}$$

Similarly,

$$\sin\frac{B}{2}\cos\frac{A}{2}\cos\frac{C}{2} = \frac{1}{4}[1 + \cos A + \cos C - \cos B] \qquad (4)$$

$$\sin\frac{C}{2}\cos\frac{A}{2}\cos\frac{B}{2} = \frac{1}{4}[1 + \cos A + \cos B - \cos C] \qquad (5)$$

Adding $(3) - (5)$, we get

$$\sin\frac{A}{2}\cos\frac{B}{2}\cos\frac{C}{2} + \sin\frac{B}{2}\cos\frac{A}{2}\cos\frac{C}{2} + \sin\frac{C}{2}\cos\frac{A}{2}\cos\frac{B}{2}$$

$$= \frac{3}{4} + \frac{1}{4}(\cos A + \cos B + \cos C).$$

But when $A = B = C$, $(\cos A + \cos B + \cos C)_{max} = \dfrac{3}{2}$.

$\because$ $y = \cos x$ is convex function in $\left[0, \dfrac{\pi}{2}\right]$.

$\because$ Jean inequality: If $f(x)$ is convex function in (a, b),

$$f\left(\frac{\sum_{i=1}^{n} x_i}{n}\right) \geq \frac{1}{n}\sum_{i=1}^{n} f(x_i),$$

$x_1, x_2, \ldots, x_n \in (a, b)$, if and only if $x_1 = x_2 = \cdots = x_n$.

Therefore, the inequality holds.

$$\because \quad \cos\frac{A + B + C}{3} \geq \frac{1}{3}\sum \cos A,$$

$$\because \quad (\cos A + \cos B + \cos C)_{max} = \frac{3}{2}$$

$$\therefore \quad \sin\frac{A}{2}\cos\frac{B}{2}\cos\frac{C}{2} + \sin\frac{B}{2}\cos\frac{A}{2}\cos\frac{C}{2}$$

$$+ \sin\frac{C}{2}\cos\frac{A}{2}\cos\frac{B}{2} \leq \frac{9}{8}. \qquad \square$$

13. If $\sin^2 A + \sin^2 B + \sin^2 C = 1$ (A, B, C are acute angles), prove $\frac{\pi}{2} < A + B + C \leq 3\arcsin\frac{\sqrt{3}}{3}$.

Proof:

$$\sin^2 A = 1 - \sin^2 B + \sin^2 C = \cos^2 B - \sin^2 C$$

$$= \cos(B + C) \cdot \cos(B - C)$$

$\because$ B, C are acute angles, $B - C \in \left(-\dfrac{\pi}{2}, \dfrac{\pi}{2}\right)$,

$\therefore$ $\cos(B - C) > 0, \cos(B + C) \cdot \cos(B - C) = \sin^2 A > 0,$

$\therefore$ $\cos(B + C) > 0,$

$\therefore$ $0 < B + C < \dfrac{\pi}{2}, A + B + C < \pi,$

$\because$ $0 \leq |B - C| < B + C < \dfrac{\pi}{2},$

$\therefore$ $\cos(|B - C|) > \cos(B + C), \therefore \cos(B - C) > \cos(B + C),$

$\therefore$ $\sin^2 A = \cos(B + C) \cos(B - C) > \cos^2(B + C)$

$$= \sin^2 \left[\dfrac{\pi}{2} - (B + C)\right],$$

$\because$ $y = \sin^2 x$ is increasing function in $\left[0, \dfrac{\pi}{2}\right]$

$\therefore$ $A > \dfrac{\pi}{2} - (B + C) \Rightarrow A + B + C > \dfrac{\pi}{2},$

$\because$ $y = \cos x$ is convex function in $\left[0, \dfrac{\pi}{2}\right]$

$\therefore$ $A + B, B + C, C + A$ are all acute angles

$\therefore$ $\cos \dfrac{(A + B) + (B + C) + (C + A)}{3}$

$$\geq \dfrac{\cos(A + B) + \cos(B + C) + \cos(C + A)}{3}$$

$\therefore$ $3 \cos 2 \left(\dfrac{A + B + C}{3}\right)$

$$\geq \cos(A + B) + \cos(B + C) + \cos(C + A)$$

$$\geq \cos(A + B) \cos(A - B) + \cos(B + C) \cos(B - C)$$

$$+\cos(C+A)\cos(C-A)$$

$$=\cos 2A+\cos 2B+\cos 2C$$

$$\therefore\ 3\left[1-2\sin^2\frac{A+B+C}{3}\right]$$

$$\geq 1-2\sin^2 A+1-2\sin^2 B+1-2\sin^2 C$$

$$\sin\frac{A+B+C}{3}\leq\frac{\sqrt{3}}{3}$$

$$\therefore\ A+B+C\leq 3\arcsin\frac{\sqrt{3}}{3}$$

$$\therefore\ \frac{\pi}{2}<A+B+C\leq 3\arcsin\frac{\sqrt{3}}{3}.$$

□

Comment:

(1) If $y=f(x)$ is convex up function in (a,b), $x_1,x_2,\ldots,x_n\in(a,b)$,

$$\therefore\ f\left(\frac{x_1+x_2+\cdots+x_n}{n}\right)\geq\frac{f(x_1)+f(x_2)+\cdots+f(x_n)}{n}.$$

If and only if $x_1=x_2=\cdots=x_n$, the inequality holds in above type.

(2) If $y=f(x)$ is convex down function in (a,b), $x_1,x_2,\ldots,x_n\in(a,b)$

$$f\left(\frac{x_1+x_2+\cdots+x_n}{n}\right)\leq\frac{f(x_1)+f(x_2)+\cdots+f(x_n)}{n}.$$

If and only if $x_1=x_2=\cdots=x_n$, the inequality holds in above type.

14. If $\angle A$, $\angle B$, $\angle C$, are three interior angles in acute $\triangle ABC$, prove

$$\sin A+\sin B+\sin C+\tan A+\tan B+\tan C>2\pi.$$

Proof 1: According to universal formula:

$$\sin A + \tan A = \frac{2\tan\frac{A}{2}}{1+\tan^2\frac{A}{2}} + \frac{2\tan\frac{A}{2}}{1-\tan^2\frac{A}{2}}$$

$$= 2\tan\frac{A}{2}\left(\frac{1}{1+\tan^2\frac{A}{2}} + \frac{1}{1-\tan^2\frac{A}{2}}\right) = \frac{4\tan\frac{A}{2}}{1-\tan^4\frac{A}{2}};$$

$\therefore$ $\angle A$ is an acute angle, $0 < \tan\dfrac{A}{2} < 1$,

$\therefore$ $0 < 1 - \tan^4\dfrac{A}{2} < 1$, $\dfrac{4\tan\frac{A}{2}}{1-\tan^4\frac{A}{2}} > 4\tan\dfrac{A}{2}$;

$\because$ $\dfrac{\angle A}{2}$ is an acute angle, $\tan\dfrac{A}{2} > \dfrac{\angle A}{2}$,

$\therefore$ $\sin A + \tan A > 2\angle A$.

$\therefore$ Similarly, $\sin B + \tan B > 2\angle B$, $\sin C + \tan C > 2\angle C$

$\therefore$ $\sin A + \sin B + \sin C + \tan A + \tan B + \tan C$

$\qquad > 2(\angle A + \angle B + \angle C) = 2\pi$.

$\square$

Proof 2: Suppose $\angle A$'s end side OM and unit circle intersect at M, and make $MN \perp Ox$ through M. Make tangent line QP through Q and line OM intersect at P. Known by trigonometric function:

$$\sin A = NM, \ \tan A = QP, \ \angle A = M\overset{\frown}{Q}.$$

Prove $\sin A + \tan A > 2\angle A$ by calculating the area. Make tangent line of circle through M and line PQ intersect at T.

$\qquad \because \angle PMT = 90°, \ \therefore \ PT > MT = QT.$

$\qquad \therefore S_{\triangle PMT} > S_{\triangle MTP} >$ segment of a circle$_{MQ}$

$\qquad \therefore S_{\triangle OMQ} + S_{\triangle OPQ} > 2S_{\text{sector}} \cdot OMQ$

$\qquad \therefore NM + QP > 2\,M\overset{\frown}{Q}, \sin A + \tan A > 2\angle A.$

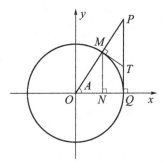

15. In acute $\triangle ABC$, prove $\sin A + \sin B + \sin C + \tan A + \tan B + \tan C \geq \frac{9\sqrt{3}}{2}$.

Solution: $0 < A, B, C < \frac{\pi}{2}$, $\therefore$ $0 < \frac{A}{2}, \frac{B}{2}, \frac{C}{2} < \frac{\pi}{4}$, $\therefore$ $\tan \frac{A}{2}, \tan \frac{B}{2}, \tan \frac{C}{2} \in (0, 1)$.
Let

$$\tan \frac{A}{2} = t, \ \tan \frac{B}{2} = r, \ \tan \frac{C}{2} = s,$$

so $\sin A + \tan A = \frac{2t}{1+t^2} + \frac{2t}{1-t^2} = \frac{4t}{1-t^4} = 4t(1 + t^4 + t^8 + \cdots) = 4(t + t^5 + t^9 \cdots)$.
Similarly, $\sin B + \tan B = 4(r + r^5 + r^9 + \cdots)$, $\sin C + \tan C = 4(s + s^5 + s^9 + \cdots)$.
So the left side of the inequality is equal to:

$$4[(t + r + s) + (t^5 + r^5 + s^5) + (t^9 + r^9 + s^9) + \cdots]$$

$$\geq 4\left[(t + r + s) + \frac{(t + r + s)^5}{3^4} + \frac{(t + r + s)^9}{3^8} + \cdots\right]$$

$$= \frac{4(t + r + s)}{1 - \left(\frac{t+r+s}{3}\right)^4} \tag{6}$$

$$\therefore \ f(x) = \frac{4x}{1 - \left(\frac{x}{3}\right)^4} \ \text{is increasing function in } (0, 3),$$

$$\because \ \tan\frac{A}{2} + \tan\frac{B}{2} + \tan\frac{C}{2} \geq \sqrt{3}$$

$$\therefore \ \text{equality (6)} \leq \frac{4\sqrt{3}}{1 - (\frac{\sqrt{3}}{3})^4} = \frac{9\sqrt{3}}{2}.$$

$\therefore$ The inequality holds.

16. If A, B, C are three interior angles in $\triangle ABC$, find the maximum of $W = \tan\frac{A}{2}\tan\frac{B}{2}\tan\frac{C}{2} + \cot\frac{A}{2}\cot\frac{A}{2}\cot\frac{C}{2}$.

Solution: $\because$ A, B, C are three interior angles in $\triangle ABC$

$$\therefore \ \tan\frac{A}{2}, \tan\frac{B}{2}, \tan\frac{C}{2}, \cot\frac{A}{2}, \cot\frac{A}{2}, \cot\frac{C}{2} > 0.$$

Let

$$x = \tan\frac{A}{2}, y = \tan\frac{B}{2}, z = \tan\frac{C}{2}, \ \text{so } xy + yz + zx = 1.$$

$$\therefore \ xyz = \sqrt{xy \cdot yz \cdot zx} \leq \sqrt{\left(\frac{xy + yz + zx}{3}\right)^3}$$

$$= \sqrt{\frac{1}{27}} = \frac{\sqrt{3}}{9} < 1$$

When $0 < a \leq b < 1$, $a + \dfrac{1}{a} \geq b + \dfrac{1}{b}$

$$\therefore \ xyz + \frac{1}{xyz} \geq \frac{\sqrt{3}}{9} + \frac{9}{\sqrt{3}} = \frac{28}{9}\sqrt{3}.$$

$$\therefore \ \text{When } A = B = C, \ W_{\min} = \frac{28}{9}\sqrt{3}.$$

Exercise 6:

1. Find the range of $y = \sqrt{4x - 1} + \sqrt{2 - x}$.
2. Find the range of $y = x + 3 + \sqrt{-3x^2 + 6x + 12}$.

3. In $\triangle ABC$, $\tan A = \frac{1}{4}$, $\tan B = \frac{3}{5}$. If the maximum side of $\triangle ABC$ is $\sqrt{17}$, find the minimum side.

4. If sequence $\{a_n\}$ meet $a_1 = 1$ and $a_{n+1} = \frac{a_n + 2 - \sqrt{3}}{1 - a_n(2 - \sqrt{3})}$, find a_{2005}.

5. If $a^2 + b^2 = 1$, find the minimum of $(a^2 + \frac{1}{a^2})(b^2 + \frac{1}{b^2})$.

6. Find the maximum and minimum of $\sqrt{5 - x^2} + \sqrt{3}x$.

7. Solve inequality $\frac{x}{\sqrt{x^2+1}} + \frac{1-x^2}{1+x^2} > 0$.

8. Find the maximum and minimum of $y = \frac{1 + x - 2x^2 + x^3 + x^4}{1 + 2x^2 + x^4}$.

9. If x, y, z are not all zero, find the maximal of $u = \frac{xy + 2yz + 2zx}{x^2 + y^2 + z^2}$.

10. Prove: $0 \le xy + yz + zx - 2xyz \le \frac{7}{27}$ (x, y, z are non-negative real number, and $x + y + z = 1$).

11. Any given 13 different real numbers. Prove: there are at least two of them (suppose x and y) meet $\frac{x-y}{1+xy} \le 2 - \sqrt{3}$.

12. If the sequence $\{a_n\}$ is defined recursive method, $a_0 = \frac{1}{3}$, $a_n = \sqrt{\frac{1 + a_{n-1}}{2}}$, and $n = 1, 2, 3, \ldots$, prove: $\{a_n\}$ is monotonous.

13. Known x, y are positive integers, and $x - y = 1$, $A = (\sqrt{x} - \frac{1}{\sqrt{x}})(\sqrt{y} + \frac{1}{\sqrt{y}}) \cdot \frac{1}{x}$, prove: $0 < A < 1$.

14. The sequence $\{a_n\}$ meets $a_1 = \sqrt{2}$, $a_{n+1} = \sqrt{\frac{2a_n}{1 + a_n}}$, and the sequence $\{b_n\}$ meets: $b_1 = \sqrt{2}$, $b_{n+1} = \sqrt{2} \cdot \sqrt{b_n^2 + b_n\sqrt{b_n^2 - 1}}$, prove: $\frac{1}{a_n^2} + \frac{1}{b_n^2} = 1$.

15. The area of a planar convex quadrilateral is $32\,\text{cm}^2$, where sum of length of a pair of edges and a diagonal is 16 cm. Try to find all possible length of the other diagonal.

16. Solve:

$$\begin{cases} \left(1 + \frac{12}{3x+y}\right)\sqrt{x} = 2, \\[2mm] \left(1 - \frac{12}{3x+y}\right)\sqrt{y} = 6. \end{cases}$$

17. If a, b, c are real numbers and $abc + a + c = b$, find the maximum of $p = \frac{2}{a^2+1} - \frac{2}{b^2+1} + \frac{3}{c^2+1}$.

18. In $\triangle ABC$, find integer part of

$$S = \sqrt{3\tan\frac{A}{2}\tan\frac{B}{2}+1} + \sqrt{3\tan\frac{B}{2}\tan\frac{C}{2}+1}$$

$$+ \sqrt{3\tan\frac{C}{2}\tan\frac{A}{2}+1}.$$

19. If α, β, γ are three interior angles that correspond to vertex A, B, C in $\triangle ABC$, find one point D in AB that CD is mean terms of proportion of AD and BD, and its necessary and sufficient condition is $\sin\alpha\sin\beta\sin\gamma \leq \sin^2\frac{\gamma}{2}$.

20. In $\triangle ABC$, find the maximum of $f(A,B,C) = \frac{\sin A}{\sqrt{1-\sin B\sin C}} + \frac{\sin B}{\sqrt{1-\sin C\sin A}} + \frac{\sin C}{\sqrt{1-\sin A\sin B}}$.

21. If $x, y, z, a, b, c \in R^+$ and $cy + bz = a, az + cx = b, bx + ay = c$, find the minimum of $f(x, y, z) = \frac{x^2}{1+x} + \frac{y^2}{1+y} + \frac{z^2}{1+z}$.

22. If $a, b \in R^+, n \in N^*, N \geq 1$, find the minimum of

$$y = \frac{a}{\sqrt{(1+x)^n}} + \frac{b}{\sqrt{(1-x)^n}}, \quad x \in (-1, 1).$$

23. If $x_1, x_2, \ldots, x_n \in \left(-\frac{\pi}{2}, \frac{\pi}{2}\right)$, $n \in N^*, N \geq 2$, $x_1 + x_2 + \cdots + x_n = \frac{n\pi}{6}$, find the minimum of

$$y = \sum_{i=1}^{n} \frac{6x_i \sin x_i - \pi \sin x_i}{3 + 2\sin x_i}.$$

24. BC, CA, AB tangent inscribed circle of $\triangle ABC$ at D, E and F, $DG \perp EF$ at G. Connect and extend BG and CG which intersect AC and AB at H and I. If $HE = 3$, $IF = 4$, $BC = 21$, find the area of $\triangle ABC$.

25. If radius of inscribed circle of the triangle is 1, find the minimum area of the triangle.

26. In acute $\triangle ABC$, prove:

$$\frac{\cos(B-C)}{\cos A} + \frac{\cos(C-A)}{\cos B} + \frac{\cos(A-B)}{\cos C} \geq 6.$$

27. If $\alpha, \beta \in (0, \frac{\pi}{2}), \sin^2 \alpha + \sin^2 \beta + \sin^2 \gamma = 1$, prove:

$$\frac{\sin^3 \alpha}{\sin \beta} + \frac{\sin^3 \beta}{\sin \gamma} + \frac{\sin^3 \gamma}{\sin \alpha} \geq 1.$$

28. In the right-angled triangle $\triangle ABC$, $AC = BC$, $\triangle ACB = 90°$, D, E in AB, the circle through C, D and E intersects AC at P, and intersects BC at Q. Prove the necessary and sufficient condition of $AP + BQ = PQ$ is $\triangle DCE = 45°$.

29. In $\triangle ABC$, h_a, h_b, h_c are heights of $\triangle ABC$, R, r, p are respectively the radius of circumcircle, the radius of inscribed circle and semi-perimeter of $\triangle ABC$. Prove:

$$(1) \sum \frac{\cos A}{h_a} = \frac{1}{R}; \quad (2) \sum \frac{\sin A}{h_a} = \frac{p^2 4Rr - r^2}{2pRr}.$$

30. If I is the center of inscribed circle of equilateral $\triangle ABC$ and $r = 2000$, if P is one arbitrary point in $\odot I$, and d_1, d_2, d_3 are distances from P to side BC, CA, AB, prove: $\sqrt{d_1}, \sqrt{d_2}, \sqrt{d_3}$ can be three sides of one triangle.

31. If $\theta_1, \theta_2, \theta_3 \in (0, \frac{\pi}{2})$ and $\tan \theta_1 \tan \theta_2 \tan \theta_3 = 2\sqrt{2}$, prove: $\cos \theta_1 + \cos \theta_2 + \cos \theta_3 < 2$.

32. In $\triangle ABC$, prove:

$$(1) \; 8 \sin \frac{C}{2} \sin \frac{B}{2} \sin \frac{C}{2} \leq \cos \frac{A-B}{2} \cos \frac{B-C}{2} \cos \frac{C-A}{2};$$

$$(2) \; 8 \cos A \cos B \cos C \leq \cos^2 \frac{A-B}{2} \cos^2 \frac{B-C}{2} \cos^2 \frac{C-A}{2}.$$

33. If h_a, h_b, h_c are heights in non-obtuse $\triangle ABC$, prove: $\frac{h_a}{a} + \frac{h_b}{b} + \frac{h_c}{c} \geq \frac{5}{2}$.

34. If α, β and γ are three interior angles in arbitrary triangle, find the minimum of

$$u = \frac{\left(\frac{1}{\beta} + \frac{1}{\gamma}\right) \sin \alpha + \left(\frac{1}{\gamma} + \frac{1}{\alpha}\right) \sin \beta + \left(\frac{1}{\alpha} + \frac{1}{\beta}\right) \sin \gamma}{\frac{\sin \alpha}{\alpha} + \frac{\sin \beta}{\beta} + \frac{\sin \gamma}{\gamma}}.$$

35. If $0 < \alpha < \beta < \gamma < \frac{\pi}{2}$, $\sin^3 \alpha + \sin^3 \beta + \sin^3 \gamma = 1$,
   prove: $\tan^2 \alpha + \tan^2 \beta + \tan^2 \gamma \geq \frac{3}{\sqrt[3]{9}-1}$.

36. If $\theta \in (0, \frac{\pi}{2})$, find minimum positive a that meets the following two conditions:

(i) $\dfrac{\sqrt{a}}{\cos \theta} + \dfrac{\sqrt{a}}{\sin \theta} > 1$;

(ii)   There is $x \in [1 - \dfrac{\sqrt{a}}{\sin \theta}, \dfrac{\sqrt{a}}{\sin \theta}]$ which makes

$$[(1 - x) \sin \theta - \sqrt{a - x^2 \cos^2 \theta}]^2$$

$$+ [x \cos \theta - \sqrt{a - (1 - x)^2 \sin^2 \theta}]^2 \leq a.$$

37. If $n \in N^*$ and $0 < nx < \frac{\pi}{4}$, prove: $\frac{\sin nx}{\sin x} \geq \frac{\sqrt{3}}{3}(2n - 1)^{\frac{3}{4}}$.

38. In $\triangle ABC$, $\angle C \geq 60°$, prove: $(a + b)(\frac{1}{a} + \frac{1}{b} + \frac{1}{c}) \geq 4 + \frac{1}{\sin \frac{C}{2}}$.

39. In parallelogram $ABCD$, $\triangle ABD$ is an acute triangle, $AB = a$, $AD = 1$, $\angle BAD = \alpha$. Prove: KA, KB, KC, KD (A, B, C, D are the center, the radius is 1) can cover parallelogram $ABCD$, and its necessary and sufficient condition is $a \leq \cos \alpha + \sqrt{3} \sin \alpha$.

PART II
Complex Number

Chapter 7

Concept of Complex Number

Imaginary Unit, Complex Number, Modulus of Complex Number

1. $i^2 = -1$, the number i is called imaginary unit; suppose $a, b \in R$, any number of the form $z = a + bi$ is called a complex number, a is the real part of z, and b is the imaginary part of z. We write $a = \text{Re } z$, $b = \text{Im } z$. To complex number $z = a + bi$, when $b = 0$, z is the real number; when $b \neq 0$, z is called an imaginary number; when $a = 0$ and $b \neq 0$, z is called a purely imaginary number. A collection of all complex numbers is called complex set, and we write it C.

2. In the plane rectangular coordinate system, $Z(a, b)$ expresses complex number $a + bi$. The distance from Z to origin is called modulus of complex number. The modulus of $z = a + bi$ $(a, b \in R)$ is written $|z|$ or $|a + bi|$, and $|z| = |a + bi| = \sqrt{a^2 + b^2}$.

3. To learn the concept of complex number, pay attention to the following:

   (1) When we write the algebraic form of complex number $a + bi$, do not leave out $a, b \in R$. If not, a, b do not necessarily express the real and imaginary parts of the complex number, and in this case, the definition of equality of complex numbers and formula for modulus of complex numbers cannot be used. Do not think that a complex number must be an imaginary number, and do not even think a complex number is certainly not a real number. Should master the following relations: complex number $a + bi$ $(a, b \in R)$,

$$\begin{cases} \text{real number} & \text{(when } b = 0\text{)}, \\ \text{imaginary number} & \text{(when } b \neq 0\text{)}, \end{cases}$$

$$\begin{cases} \text{purely imaginary number} \\ \text{(when } a = 0, b \neq 0), \\ \text{imaginary number which is not purely} \\ \text{(when } ab \neq 0). \end{cases}$$

(2) If two complex numbers are not real numbers, the size of them cannot be compared. That is two imaginary numbers cannot be compared the size and a real number and an imaginary number cannot be compare the size. Only two real numbers can be compared the size.

(3) In complex plane, the origin is not a point on the imaginary axis (y-axis), which is a point on the real axis (x-axis).

(4) After the concept of number is extended to the complex number, real number of computing properties, concepts and relations is not necessarily applicable in complex set, such as the property of the inequality, the definition of absolute value and even power of the number is not negative and so on.

Exercise

1. Complex number $z = \frac{m^2 - m - 12}{m + 4} + (m^2 - 10m + 24)i$ $(m \in R)$, so what is the real number m when z is a real number? The imaginary number? Purely imaginary number? Zero?

Solution:

$$\text{Re } z = \frac{m^2 - m - 12}{m + 4} = \frac{(m + 3)(m - 4)}{(m + 4)};$$

$$\text{Im } z = m^2 - 10m + 24 = (m - 4)(m - 6).$$

(1) When $\text{Im } z = m^2 - 10m + 24 = (m - 4)(m - 6) = 0$, namely $m = 4$ or 6, z is a real number.

(2) When $\text{Im } z = m^2 - 10m + 24 = (m - 4)(m - 6) \neq 0$, namely $m \neq 4$, $m \neq 6$, and $m \neq -4$, z is an imaginary number.

(3) When

$$\begin{cases} \text{Re}\, z = 0 \\ \text{Im}\, z \neq 0 \end{cases} \Rightarrow \begin{cases} (m+3)(m-4) = 0 \\ (m-4)(m-6) \neq 0 \end{cases} \Rightarrow m = -3,$$

then z is a purely imaginary number.

(4) When

$$\begin{cases} \text{Re}\, z = 0 \\ \text{Im}\, z = 0 \end{cases} \Rightarrow \begin{cases} (m+3)(m-4) = 0 \\ (m-4)(m-6) = 0 \end{cases} \Rightarrow m = 4,$$

then $z = 0$.

2. Complex number $z = (a^2-4)+(\frac{1}{a}-a)i$ $(a \in R, a \neq 0)$. If the point corresponding to z in complex plane is in the second quadrant, find the range of a.

Solution:

$$\begin{cases} a^2 - 4 < 0 \\ \dfrac{1}{a} - a > 0 \end{cases} \Rightarrow a \in (-2, -1) \cup (0, 1).$$

3. (1) Find the modulus of $z = 4a - 3ai$ $(a < 0)$.
   (2) If $\left|z + \frac{1}{z}\right| = 1$, find $|z|_{\max} - |z|_{\min}$.
   (3) If $|z| = 1$, $\omega = z^3 - 3z - 2$, find $|\omega|_{\max} + |\omega|_{\min}$.
   (4) If for all $\theta \in R$, the modulus of $z = (a + \cos\theta) + (2a - \sin\theta)i$ is no more than 2, find the range of a.

Solution:

(1) $|z| = \sqrt{16a^2 + 9a^2} = 5|a|$.

(2) $\left||z| - \dfrac{1}{|z|}\right| \leq \left|z + \dfrac{1}{z}\right| = 1 \leq |z| + \dfrac{1}{|z|} \Rightarrow \begin{cases} |z| - \dfrac{1}{|z|} \leq 1 \\ |z| - \dfrac{1}{|z|} \geq -1 \end{cases}$

$$\Rightarrow \frac{\sqrt{5}-1}{2} \leq |z| \leq \frac{\sqrt{5}+1}{2}.$$

$$\therefore |z|_{\max} - |z|_{\min} = \frac{\sqrt{5}+1}{2} - \frac{\sqrt{5}-1}{2} = 1.$$

(3) $\because |z^3 - 3z - 2| = |z+2|^2|z-2|,$

let $z = x + yi \, (x, y \in R) \Rightarrow x^2 + y^2 = 1 \, (-1 \leq x \leq 1)$

$\therefore |z+1|^2 = 2 + 2x, \quad |z-2| = \sqrt{5 - 4x}$

$$\therefore |z^3 - 3z - 2| \leq \sqrt{\left[\frac{(2+2x) + (2+2x) + (5-4x)}{3}\right]^3}$$

$$= 3\sqrt{3}$$

if and only if $2x + 2 = 5 - 4x \Rightarrow x = \frac{1}{2}, y = \pm\frac{\sqrt{3}}{2}$, namely $z = \frac{1}{2} \pm \frac{\sqrt{3}}{2}i$, the equality holds. When $x = -1$, $|w|_{\min} = 0$, $\therefore |w|_{\max} + |w|_{\min} = 3\sqrt{3}$.

(4) $|z| \leq 2 \Leftrightarrow (a + \cos\theta)^2 + (2a - \sin\theta)^2 \leq 4$

$$\Leftrightarrow 2a(\cos\theta - 2\sin\theta) \leq 3 - 5a^2$$

$$\Leftrightarrow -2\sqrt{5}a\sin(\theta - \varphi) \leq 3 - 5a^2 \quad \left(\varphi = \arcsin\frac{1}{\sqrt{5}}\right)$$

(for all real number θ)

$$\Rightarrow 2\sqrt{5}|a| \leq 3 - 5a^2 \Rightarrow |a| \leq \frac{\sqrt{5}}{5},$$

$$\therefore \text{ the range of } a \text{ is } \left[-\frac{\sqrt{5}}{5}, \frac{\sqrt{5}}{5}\right].$$

4. If $|z| = 1$ and $z^{2001} + z = 1$, find the complex number z.

Solution: $z^{2001} = 1 - z$, two sides take modulus, $|z^{2001}| = |1 - z|$, $|z|^{2001} = |1 - z|$, take $|z| = 1$ into the equation. $\therefore |z - 1| = 1$.

$\therefore$ Solutions to the equation should be satisfied

$$\begin{cases} |z| = 1, \\ |z - 1| = 1. \end{cases}$$

The plural of the corresponding two circle is $z_1 = \frac{1}{2} + \frac{\sqrt{3}}{2}i$, $z_2 = \frac{1}{2} - \frac{\sqrt{3}}{2}i$.

After test, they are not the solution of the original equation, so there is no solution to the original equation.

Complex Conjugates

1. **Complex conjugates: complex numbers** $a + bi$ and $a - bi$ are called complex conjugates. If $z = a + bi$, we write its conjugate as $\bar{z} = a - bi \, (a, b \in R)$.

2. **Property of complex conjugates**

   (i) Operations property:

   (1) $\overline{z_1 + z_2} = \overline{z_1} + \overline{z_2}$;

   (2) $\overline{z_1 - z_2} = \overline{z_1} - \overline{z_2}$;

   (3) $\overline{z_1 z_2} = \overline{z_1} \cdot \overline{z_2}$;

   (4) $\overline{\left(\frac{z_1}{z_2}\right)} = \frac{\overline{z_1}}{\overline{z_2}} (z_2 \equiv 0)$.

   Properties (1) and (3) can be generalized to the case of n:
   $\overline{z_1 + z_2 + \cdots + z_n} = \overline{z_1} + \overline{z_2} + \cdots + |\overline{z_n}$; $\overline{z_1 z_2 \ldots z_n} = \overline{z_1} \cdot \overline{z_2} \cdot \ldots \cdot z_n$.

   (ii) Important conclusion:

   $$|z_1 \cdot z_2| = |z_1| \cdot |z_2|;$$

   $$\left|\frac{z_1}{z_2}\right| = \frac{|z_1|}{|z_2|} \quad (z_2 \neq 0);$$

   $$|z|^2 = |\bar{z}|^2 = z \cdot \bar{z};$$

   $$||z_1| - |z_2|| \leq |z_1 \pm z_2| \leq |z_1| + |z_2|,$$

   if and only if vectors $\overrightarrow{OZ_1}, \overrightarrow{OZ_2}$ corresponding to z_1, z_2 have the same direction, inequality holds in right; if and only if vectors $\overrightarrow{OZ_1}, \overrightarrow{OZ_2}$ corresponding to z_1, z_2 have the opposite direction, inequality holds in left.

$$z \in R \Leftrightarrow z - \bar{z} = 0;$$

z is an imaginary number if and only if $z + \bar{z} = 0, z \neq 0$.

Exercise

5. If imaginary number z meets $|2z + 15| = \sqrt{3}|\bar{z} + 10|$,

(1) find $|z|$;

(2) if $\dfrac{z}{a} + \dfrac{a}{z} \in R$, find real number a.

Solution: (1) Method 1: Suppose $z = x + yi$, $x, y \in R$.

$$\because |2z + 15| = \sqrt{3}|\bar{z} + 10|$$

$$\therefore |2(x + yi) + 15| = \sqrt{3}|x - yi + 10|$$

$$\therefore |(2x + 15) + 2yi| = \sqrt{3}|(x + 10) - yi|$$

$$\therefore (2x + 15)^2 + 4y^2 = 3[(x + 10)^2 + y^2]$$

$$\therefore x^2 + y^2 = 75, \ \therefore |z| = \sqrt{x^2 + y^2} = 5\sqrt{3}.$$

Method 2:

$$\because |2z + 15| = \sqrt{3}|\bar{z} + 10|$$

$$\therefore |2z + 15|^2 = 3|\bar{z} + 10|^2$$

$$\therefore (2z + 15)(2\bar{z} + 15) = 3(\bar{z} + 10)(z + 10),$$

$$4|z|^2 + 30z + 30\bar{z} + 225 = 3|z|^2 + 30z + 30\bar{z} + 300,$$

$$\therefore |z|^2 = 75, \quad \therefore |z| = 5\sqrt{3}.$$

(2) Method 1: Suppose $z = x + yi$, $x, y \in R$,

$$\frac{z}{a} + \frac{a}{z} = \frac{x + yi}{a} + \frac{a}{x + yi} = \frac{x}{a} + \frac{y}{a}i + \frac{a(x - yi)}{x^2 + y^2}$$

$$= \left(\frac{x}{a} + \frac{ax}{x^2 + y^2} \right) + y \left(\frac{1}{a} - \frac{a}{x^2 + y^2} \right) i$$

$$\because \quad \frac{z}{a} + \frac{a}{z} \in R, \ y\left(\frac{1}{a} - \frac{a}{x^2 + y^2}\right) = 0,$$

$\because \quad z$ is imaginary number,

$$\therefore \quad y \neq 0, \ \therefore \ \frac{1}{a} - \frac{a}{x^2 + y^2} = 0, \ \therefore \ a^2 = x^2 + y^2 = 75,$$

$$\therefore \quad a = \pm 5\sqrt{3}.$$

Method 2:

$$\because \quad \frac{z}{a} + \frac{a}{z} \text{ is a real number,}$$

$$\because \quad \overline{\frac{z}{a} + \frac{a}{z}} = \frac{z}{a} + \frac{a}{z}, \ \therefore \ \frac{\bar{z}}{a} + \frac{a}{\bar{z}} = \frac{z}{a} + \frac{a}{z},$$

$$\therefore \quad \frac{z - \bar{z}}{a} + \frac{a(\bar{z} - z)}{z\bar{z}} = 0, \ \therefore \ (z - \bar{z})\left(\frac{1}{a} - \frac{a}{|z|^2}\right) = 0.$$

$\because \quad z$ is imaginary number,

$$\therefore \quad z - \bar{z} \neq 0, \ \therefore \ \frac{1}{a} - \frac{a}{|z|^2} = 0, \ \therefore \ a^2 = |z|^2 = 75,$$

$$\therefore \quad a = \pm 5\sqrt{3}.$$

Comment: The complex number has algebraic form $a + bi$ $(a, b \in R)$ with whole form z. Using the property of modulus of complex numbers and complex conjugates, sometimes it can be directly analyzed by whole form z. In addition, pay attention to the distinction between $|z|^2 = z\bar{z}$ in complex number field and $|a|^2 = a^2$ in real number field.

5. Suppose $z_1, z_2 \in C$, prove: $|z_1 + z_2|^2 + |z_1 - z_2|^2 = 2(|z_1|^2 + |z_2|^2)$.

Proof:

$$\because \quad |z_1 + z_2|^2 = (z_1 + z_2)(\overline{z_1 + z_2}) = (z_1 + z_2)(\bar{z}_1 + \bar{z}_2)$$

$$= z_1 \cdot \bar{z}_1 + z_2 \cdot \bar{z}_2 + z_1 \cdot \bar{z}_2 + \bar{z}_1 \cdot z_2$$

$$= |z_1|^2 + |z_2|^2 + (z_1 \cdot \bar{z}_2 + \bar{z}_1 \cdot z_2)$$

$$|z_1 - z_2|^2 = (z_1 - z_2)(\overline{z_1 - z_2})$$
$$= (z_1 - z_2)(\bar{z}_1 - \bar{z}_2)$$
$$= z_1 \cdot \bar{z}_1 + z_2 \cdot \bar{z}_2 - z_1 \cdot \bar{z}_2 - \bar{z}_1 \cdot z_2$$
$$= |z_1|^2 + |z_2|^2 - (z_1 \cdot \bar{z}_2 + \bar{z}_1 \cdot z_2)$$
$$\therefore \ \text{left} = [|z_1|^2 + |z_2|^2 + (z_1 \cdot \bar{z}_2 + \bar{z}_1 \cdot z_2)]$$
$$+ [|z_1|^2 + |z_2|^2 - (z_1 \cdot \bar{z}_2 + \bar{z}_1 \cdot z_2)]$$
$$= 2(|z_1|^2 + |z_2|^2) = \text{right}.$$

$\square$

7. If complex number $z_1 \neq z_2$, $|z_1| = \sqrt{2}$, find $\left| \frac{z_1 - \bar{z}_2}{2 - z_1 z_2} \right|$.

Solution 1:

$$\because \ |z_1| = \sqrt{2}, \ \therefore \ z_1 \cdot z_1 = 2.$$

$$\left| \frac{z_1 - \bar{z}_2}{2 - z_1 \cdot z_2} \right|^2 = \left(\frac{z_1 - \bar{z}_2}{2 - z_1 \cdot z_2} \right) \overline{\left(\frac{z_1 - z_2}{2 - z_1 \cdot z_2} \right)}$$

$$= \frac{(z_1 - \bar{z}_2)(\bar{z}_1 - z_2)}{(2 - z_1 \cdot z_2)(2 - \bar{z}_1 \cdot \bar{z}_2)}$$

$$= \frac{z_1 \cdot \bar{z}_1 + z_2 \cdot \bar{z}_2 - z_1 \cdot z_2 - \bar{z}_1 \cdot \bar{z}_2}{4 + z_1 \cdot \bar{z}_1 \cdot z_2 \cdot \bar{z}_2 - 2(z_1 \cdot z_2 + \bar{z}_1 \cdot \bar{z}_2)}$$

$$= \frac{2 + |z_2|^2 - z_1 \cdot z_2 - \bar{z}_1 \cdot \bar{z}_2}{2(2 + |z_2|^2 - z_1 \cdot z_2 - \bar{z}_1 \cdot \bar{z}_2)} = \frac{1}{2}$$

$$\therefore \ \left| \frac{z_1 - \bar{z}_2}{2 - z_1 \cdot z_2} \right| = \frac{\sqrt{2}}{2}$$

Solution 2:

$$\because \ |z_1| = \sqrt{2}, \ \therefore \ z_1 \cdot \bar{z}_1 = 2,$$

$$\left| \frac{z_1 - \bar{z}_2}{2 - z_1 \cdot z_2} \right|$$

$$= \left| \frac{z_1 - \bar{z}_2}{z_1 \cdot \bar{z}_1 - z_1 \cdot z_2} \right| = \left| \frac{z_1 - \bar{z}_2}{z_1(\bar{z}_1 - z_2)} \right|$$

$$= \frac{|z_1 - \bar{z}_2|}{|z_1||\bar{z}_1 - z_2|} = \frac{|\bar{z}_1 - z_2|}{|z_1||\bar{z}_1 - z_2|}$$

$$= \frac{1}{|z_1|} = \frac{\sqrt{2}}{2}.$$

8. If $|z_1| = |z_2| = |z_3| = r$, find

$$\left| \frac{\frac{1}{z_1} + \frac{1}{z_2} + \frac{1}{z_3}}{z_1 + z_2 + z_3} \right|.$$

Solution:

$$\because \ |z_1| = |z_2| = |z_3| = r(r \neq 0)$$

$$\therefore \ z_1 \cdot \bar{z}_1 = z_2 \cdot \bar{z}_2 = z_3 \cdot \bar{z}_3 = r^2$$

$$\frac{1}{z_1} = \frac{\bar{z}_1}{r^2}, \quad \frac{1}{z_2} = \frac{\bar{z}_2}{r^2}, \quad \frac{1}{z_3} = \frac{\bar{z}_3}{r},$$

$$\therefore \ \text{The original type} = \left| \frac{\frac{1}{r^2}\bar{z}_1 + \frac{1}{r^2}\bar{z}_2 + \frac{1}{r^2}\bar{z}_3}{z_1 + z_2 + z_3} \right|$$

$$= \frac{\mathrm{I}}{r^2} \frac{|\overline{z_1 + z_2 + z_3}|}{|z_1 + z_2 + z_3|} = \frac{1}{r^2}.$$

9. If $|z| = 1$, find range of $|z^2 - z + 1|$.

Solution:

$$\because \ |z| = 1 \ \therefore \ z \cdot \bar{z} = 1,$$

$$\therefore \ |z^2 - z + 1| = |z^2 - z + z \cdot \bar{z}|$$

$$|z| \cdot |z - 1 + \bar{z}| = |z + \bar{z} - 1|.$$

Suppose

$$z = x + yi \ (x, y \in R),$$
$$\therefore \ x^2 + y^2 = 1,$$
$$\because \ |x| \leq 1$$
$$\therefore \ |z^2 - z + 1| = |2x - 1| \leq |2x| + 1$$
$$= 2|x| + 1 \leq 3$$

and when $x = -1$, $|2x - 1| = 3$

$$\because \ |2x - 1| \geq 0 \quad \text{and when } x = \frac{1}{2},$$
$$|2x - 1| = 0.$$
$$\therefore \ \text{The range of } |z^2 - z + 1| \text{ is } [0, 3].$$

10. Suppose z is imaginary number, $\omega = z + \frac{1}{z}$ is real number, and $-1 < \omega < 2$.

(1) Find $|z|$ and range of real part of z.
(2) Prove: $u = \frac{1-z}{1+z}$ is a purely imaginary number.
(3) Find the minimum of $\omega - u^2$.

Analysis: $\omega = z + \frac{1}{z}$ is a real number, so it can be analyzed by definition of complex numbers and property of **complex conjugates**.

Solution: (1) Suppose $z = a + bi, \ a, b \in R,$

$$\omega = z + \frac{1}{z} = a + bi + \frac{1}{a + bi} = a + bi + \frac{a - bi}{a^2 + b^2}$$
$$= \left(a + \frac{a}{a^2 + b^2} \right) + \left(b - \frac{b}{a^2 + b^2} \right) i$$

$\because \ \omega$ is a real number

$\therefore \ b - \dfrac{b}{a^2 + b^2} = 0$, z is an imaginary number

$\therefore \ b \neq 0$

$$\therefore\ a^2 + b^2 = 1, \therefore\ |z| = \sqrt{a^2 + b^2} = 1$$

$$\therefore\ \omega = 2a,$$

$$\because\ -1 < \omega < 2 \therefore\ -\frac{1}{2} < a < 1.$$

(2) **Method 1:** Suppose $z = a + bi, a, b \in R$.

$$\therefore\ u = \frac{1-z}{1+z} = \frac{1-a-bi}{1+a+bi} = \frac{(1-a-bi)(1+a-bi)}{(1+a)^2 + b^2}$$

$$= \frac{[(1-bi)-a][(1-bi)+a]}{a^2 + b^2 + 1 + 2a}$$

$$= \frac{(1-bi)^2 - a^2}{2a+2} = \frac{1-(a^2+b^2)-2bi}{2(a+1)} = \frac{-b}{a+1}i, b \neq 0,$$

$\therefore\ u$ is a purely imaginary number.

Method 2:

$$u + \bar{u} = \frac{1-z}{1+z} + \overline{\left(\frac{1-z}{1+z}\right)} = \frac{(1-z)(1+\bar{z}) + (1+z)(1-\bar{z})}{(1+z)(1+\bar{z})}$$

$$= \frac{1+\bar{z}-z-z\bar{z}+1-\bar{z}+z-z\bar{z}}{(1+z)(1+\bar{z})} = \frac{2-2|z|^2}{(1+z)(1+\bar{z})} = 0$$

$\because\ u \neq 0, \therefore\ u$ is a purely imaginary number.

(3) By (1) and (2),

$$\omega - u^2 = 2a - \left(\frac{-b}{a+1}i\right)^2 = 2a + \frac{b^2}{(a+1)^2}$$

$$= 2a + \frac{1-a^2}{(a+1)^2} = 2a + \frac{2}{a+1} - 1$$

$$= 2(a+1) + \frac{2}{a+1} - 3 \geq 2\sqrt{2(a+1)\left(\frac{2}{a+1}\right)} - 3 = 1.$$

If and only if $2(a+1) = \frac{2}{a+1}$, namely when $a = 0$, the equality holds.

$\therefore$ The minimum of u is 1.

Comment: Method 2 to question (2) uses the property of **complex conjugates** and the overall application of complex number z; pay attention to equality conditions when using the basic inequality in question (3).

Equality of Complex Numbers

Two complex numbers are equal when their real parts are equal and their imaginary parts are equal.

$$a + bi = c + di \Leftrightarrow a = c \quad \text{and} \quad b = d.$$

If z_1 and z_2 are not real numbers, their size cannot be compared.

According to the definition of complex concepts and complex equal, the complex problem is transformed into the problem of real, namely "complex transform to real". This is a commonly used method to solve complex problems, which is the mathematics reduction thinking method in complex applications.

Exercise

11. If $2z + |z| = 2 + 6i$, find z.

Solution: Suppose $z = x + yi$ $(x, y \in R)$.

$$\therefore 2(x + yi) + \sqrt{x^2 + y^2} = 2 + 6i$$

$$\therefore \begin{cases} 2x + \sqrt{x^2 + y^2} = 2 \\ 2y = 6 \end{cases} \Rightarrow \begin{cases} x = \dfrac{4 + \sqrt{31}}{3} \\ y = 3 \end{cases}$$

$$\text{(rejection) or} \begin{cases} x = \dfrac{4 - \sqrt{31}}{3}, \\ y = 3. \end{cases}$$

When $z = \dfrac{4 + \sqrt{31}}{3} + 3i$,

$$2\left(\frac{4+\sqrt{31}}{3}+3i\right)+\sqrt{\frac{16+31+2\sqrt{31}}{9}}+9$$

$$=\frac{8+2\sqrt{31}}{2}+\frac{2\sqrt{32+2\sqrt{31}}}{3}+6i$$

$$=\frac{8+2\sqrt{31}}{3}+\frac{2\sqrt{(\sqrt{31}+1)^2}}{3}+6i$$

$$=\frac{8+2\sqrt{31}}{3}+\frac{2(\sqrt{31}+1)}{3}+6i$$

$$=\frac{8}{3}+\frac{4\sqrt{31}}{3}+\frac{2}{3}6i\neq 6i+2,\ \text{so rejection.}$$

$$\therefore\ \begin{cases} x=\dfrac{4-\sqrt{31}}{3}, \\ y=3. \end{cases}$$

12. (1) Solve $|z-(1-2i)|^2+z-(3+4i)=|z+(3-i)|^2+(2-3i)$, where $z\in C$.

(2) If $|z+(3+4i)|=|(z+(1+2i))|=|z-(1-2i)|$, find z.

Solution: (1) Suppose $z=x+yi\ (x,y\in R)$,

$$|(x+yi)-(1-2i)|^2+(x+yi)-(3+4i)$$

$$=|(x+yi)+(3-i)|^2+(2-3i),$$

$$|(x-1)+(y+2)i|^2+(x-3)+(y-4)i$$

$$=|(x+3)+(y-1)i|^2+(2-3i),$$

$$[(x-1)^2+(y+2)^2+(x-3)]+(y-4)i$$

$$=[(x+3)^2+(y-1)^2+2]-3i.$$

$$\therefore \begin{cases} (x-1)^2 + (y+2)^2 + x - 3 = (x+3)^2 - (y-1)^2 + 2, \\ y - 4 = -3. \end{cases}$$

$$\therefore x = -\frac{4}{7}, \ y = 1, \ \therefore z = -\frac{4}{7} + i$$

(2) Suppose $z = x + yi(x, y \in R)$,

$$|(x+yi) + (3+4i)| = |(x+yi) + (1+2i)|$$
$$= |(x+yi) - (1-2i)|,$$
$$|(x+3) + (y+4)i| = |(x+1) + (y+2)i|$$
$$= |(x-1) + (y+2)i|.$$

$$\therefore \sqrt{(x+3)^2 + (y+4)^2} = \sqrt{(x+1)^2 + (y+2)^2}$$
$$= \sqrt{(x-1)^2 + (y+2)^2}.$$

$$\therefore \begin{cases} (x+3)^2 + (y+4)^2 = (x+1)^2 + (y+2)^2, \\ (x+1)^2 + (y+2)^2 = (x-1)^2 + (y+2)^2. \end{cases}$$

$$\therefore \begin{cases} x = 0, \\ y = -5. \end{cases}$$

$$\therefore z = -5i.$$

13. If $(3+2i)z + (5i-4)\bar{z} = -11 - 7i$, find z.

Analysis: Suppose $z = a + bi$, $a, b \in R$, or solving complex conjugates on both sides of the equation is also obtained.

Solution 1: Suppose $z = a + bi$, $a, b \in R$.

$$\therefore (3+2i)(a+bi) + (5i-4)(a-bi) = -11 - 7i$$
$$\therefore -a + 3b + (7a + 7b)i = -11 - 7i$$
$$\therefore \begin{cases} -a + 3b = -11 \\ 7a + 7b = -7 \end{cases}$$
$$\therefore a = 2, b = -3, \text{ namely } z = 2 - 3i. \tag{1}$$

Solution 2:

$$\because (3+2i)z + (5i-4)\bar{z} = -11-7i,$$

$$\therefore \overline{(3+2i)z + (5i-4)\bar{z}} = \overline{-11-7i},$$

$$\therefore (3-2i)\bar{z} + (-4-5i)z = -11+7i, \qquad (2)$$

$$\because \text{By (1) and (2)},$$

$$z = 2 - 3i.$$

Comment: Method 1 uses complex transform to real from equality of complex numbers, which is the basic method; method 2 takes the complex conjugates of both sides, structure equations, which is global analysis method.

14. Suppose two complex sets $M = \{z|z = a + i(1-a^2),\ a \in R\}$, $N = \{z|z = \sin\theta + i(m - \frac{\sqrt{3}}{2}\sin 2\theta),\ m \in R, \theta \in [0, \frac{\pi}{2}]\}$. If $M \cap N \neq \Phi$, find the range of m.

Analysis:

$$\because M \cap N \neq \Phi, \therefore \begin{cases} z = a + i(1-a^2) \\ z = \sin\theta + i\left(m - \dfrac{\sqrt{3}}{2}\sin 2\theta\right) \end{cases}$$

has solution; eliminate z, separate variable m after finishing, and find the range of m.

Solution: $\because M \cap N \neq \Phi$

$$\therefore z_0 = a + i(1-a^2) = \sin\theta + i\left(m - \frac{\sqrt{3}}{2}\sin 2\theta\right)$$

$$\therefore \begin{cases} a = \sin\theta \\ 1-a^2 = m - \frac{\sqrt{3}}{2}\sin 2\theta \end{cases} \Rightarrow 1 - \sin^2\theta$$

$$= \cos^2\theta = m - \frac{\sqrt{3}}{2}\sin 2\theta$$

$$m = \frac{\sqrt{3}}{2} \sin 2\theta + \frac{1}{2} \cos 2\theta + \frac{1}{2} = \sin\left(2\theta + \frac{\pi}{6}\right) + \frac{1}{2},$$

$$\therefore \ m \in \left[0, \frac{3}{2}\right].$$

Comment: To find range of parameters in the equation, separate variable is commonly used to find the range.

Complex Plane: The complex plane is a geometric representation of the complex numbers established by the real axis and the orthogonal imaginary axis. It can be thought of as a modified Cartesian plane, with the real part of a complex number represented by a displacement along the x-axis, and the imaginary part by a displacement along the y-axis. The complex set C, set of all the points z in the complex plane and vectors $\overrightarrow{OZ}$ are one-to-one correspondence.

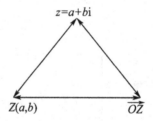

Complex problem has four tools (algebraic tools, tool geometry and vector tools and triangular tool), especially effective number shape union thinking. To the problem solving, bring the flexibility and intuitive.

15. If $z \cdot \bar{z} + (3 + 4i)z + (3 - 4i)\bar{z} = 11$, find the maximum and minimum of $|z|$.

Solution: $|z|_{\max} = 11$, $|z|_{\min} = 1$.
Suppose

$$z = x + yi(x, y \in R).$$

$$\therefore \ z \cdot \bar{z} + (3 + 4i)z + (3 - 4i)\bar{z} = 11$$

$$\therefore \ (x+3)^2 + (y-4)^2 = 36$$

$$\therefore \ |z|_{\max} = 6 + 5 = 11, \ |z|_{\min} = 6 - 5 = 1$$

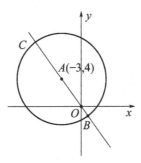

16. If elements $a_i(i = 1, 2, \ldots, 8)$ are non-zero real numbers which belong to set $A = \{a_1, a_2, \ldots, a_8\}$, there is at least one element which is non-negative real numbers in set $B = \{a_1 a_3 + a_2 a_4, a_1 a_5 + a_2 a_6, a_1 a_7 + a_2 a_8, a_3 a_5 + a_4 a_6, a_3 a_7 + a_4 a_8, a_5 a_7 + a_6 a_8\}$.

Analysis: In six given numbers, $a_1, a_2, \ldots, a_6$ come in pairs. They can be constructed into four complex numbers $z_1 = a_1 + a_2 i$, $z_2 = a_3 + a_4 i$, $z_3 = a_5 + a_6 i$, $z_4 = a_7 + a_8 i$, six number of modules corresponding to each two complex numbers differential modes appear.

Solution: Suppose $z_1 = a_1 + a_2 i$, $z_2 = a_3 + a_4 i$, $z_3 = a_5 + a_6 i$, $z_4 = a_7 + a_8 i$.

$$\therefore \ |z_1 - z_2|^2 = |(a_1 - a_3) + (a_2 - a_4)i|^2$$

$$= (a_1 - a_3)^2 + (a_2 - a_4)^2$$

$$= a_1^2 + a_2^2 + a_3^2 + a_4^2 - 2(a_1 a_3 + a_2 a_4)$$

$$= |z_1|^2 + |z_2|^2 - 2(a_1a_3 + a_2a_4)$$
$$\therefore \ |z_1|^2 + |z_2|^2 - |z_1 - z_2|^2 = 2(a_1a_3 + a_2a_4), \qquad (3)$$

Similarly,

$$|z_1|^2 + |z_3|^2 - |z_1 - z_3|^2 = 2(a_1a_5 + a_2a_6), \qquad (4)$$

$$|z_1|^2 + |z_4|^2 - |z_1 - z_4|^2 = 2(a_1a_7 + a_2a_8), \qquad (5)$$

$$|z_2|^2 + |z_3|^2 - |z_2 - z_3|^2 = 2(a_3a_5 + a_4a_6), \qquad (6)$$

$$|z_2|^2 + |z_4|^2 - |z_2 - z_4|^2 = 2(a_3a_7 + a_4a_8), \qquad (7)$$

$$|z_3|^2 + |z_4|^2 - |z_3 - z_4|^2 = 2(a_5a_7 + a_6a_8). \qquad (8)$$

To prove that there is at least one non-negative real number in right from (3) to (8) according to geometric meaning of complex number $(z_1, z_2, z_3, z_4$ corresponding four vectors $\overrightarrow{OZ_1}, \overrightarrow{OZ_2}, \overrightarrow{OZ_3}, \overrightarrow{OZ_4})$ and cosine laws:

$$\cos \angle Z_1 O Z_2 = \frac{|\overrightarrow{OZ_1}|^2 + |\overrightarrow{OZ_2}|^2 - |\overrightarrow{Z_1 Z_2}|^2}{2|\overrightarrow{OZ_1}||\overrightarrow{OZ_2}|}$$

$$= \frac{|z_1|^2 + |z_2|^2 - |z_1 - z_2|^2}{2|z_1||z_2|},$$

$$\cos \angle Z_3 O Z_4 = \frac{|\overrightarrow{OZ_3}|^2 + |\overrightarrow{OZ_4}|^2 - |\overrightarrow{Z_3 Z_4}|^2}{2|\overrightarrow{OZ_3}||\overrightarrow{OZ_4}|}$$

$$= \frac{|z_3|^2 + |z_4|^2 - |z_3 - z_4|^2}{2|z_3||z_4|}, \text{ and so on.}$$

Only to prove that there is at least the minimum angle between two vectors less than or equal to $\frac{\pi}{2}$. This is clearly established.

17. Suppose $z_1, z_2, \ldots, z_n \in C$ and $|z_1| + |z_2| + \cdots + |z_n| = 1$, prove: there are several complex numbers where sum of their modulus is not less than $\frac{1}{6}$ in these n complex numbers.

Solution:
Suppose $z_j = a_j + bj$, $j = 1, 2, 3, \ldots, n$.

$$\because |z_1| + |z_2| + \cdots + |z_n| = 1,$$

$$\therefore 1 \leq \sum_{j=1}^{n} |a_j| + \sum_{j=1}^{n} |b_j|$$

$$= \sum_{a_j < 0} |a_j| + \sum_{a_j \geq 0} |a_j| + \sum_{b_j < 0} |b_j| + \sum_{b_j \geq 0} |b_j|.$$

From the drawer principle of the right side of the four in at least one paragraph $\geq \frac{1}{4}$.
Suppose

$$\left| \sum_{b_j \geq 0} b_j \right| \geq \frac{1}{4}, \therefore \left| \sum_{b_j \geq 0} z_j \right| \geq \left| \sum_{b_j \geq 0} b_j \right| = \sum_{b_j \geq 0} |b_j| \frac{1}{4} > \frac{1}{6}.$$

Comment: The complex number inequality has been used, suppose $z = a + bi\,(a, b \in R)$, so $|z| \leq |a| + |b|$.
We can also think about the problem from another point of view: use line $y = x$, $y = -x$ to divide the plane into four ranges.
$\because |z_1| + |z_2| + \cdots + |z_n| = 1$, $\therefore$ in these ranges, sum of modulus of all complex numbers are no more than $\frac{1}{4}$ at least one range. For simplification, we can suppose this region to be a range of the positive x-axis ($\because$ take the modulus, $\therefore$ we can rotate it if it is not this range).
Suppose $z_{k_t} = a_t + ib_t$, $t = 1, 2, 3, \ldots, m$ $(1 \leq m \leq n, m \in N^*)$, so $a_t > 0, \sum_{t=1}^{m} |z_{k_t}| \geq \frac{1}{4}$;

$$\sum_{t=1}^{m} |z_{k_t}| = \sqrt{\left(\sum_{t=1}^{m} a_t \right)^2 + \left(\sum_{t=1}^{m} b_t \right)^2}$$

$$\geq \left| \sum_{t=1}^{m} a_t \right| = \sum_{t=1}^{m} |a_t|$$

$$\geq \frac{1}{\sqrt{2}} \sum_{t=1}^{m} \sqrt{a_t^2 + b_t^2} = \frac{1}{\sqrt{2}} \sum_{t=1}^{m} |z_{k_t}|$$

$$\geq \frac{1}{4\sqrt{2}} > \frac{1}{6}.$$

This problem is a definitive solution.

18. Suppose complex z_1, z_2 meet:

(1) $z_1 = \dfrac{\cos x^2}{\sin y^2} + i \cdot \dfrac{\cos y^2}{\sin x^2}$;

(2) $z_2 = x + yi$ $(x, y \in R)$;

(3) for all $x, y \in \left[-\sqrt{\dfrac{\pi}{2}}, \sqrt{\dfrac{\pi}{2}}\right]$, $|z_1| \equiv \sqrt{2}$,

find the maximum and minimum of $|z_1 - z_2|$.

Solution: As in the figure, according to (3), arbitrarily $x^2, y^2 \in \left[0, \frac{\pi}{2}\right]$, it has

$$\left(\frac{\cos x^2}{\sin y^2}\right)^2 + \left(\frac{\cos y^2}{\sin x^2}\right)^2 = 2. \tag{9}$$

Following prove $x^2 + y^2 = \frac{\pi}{2}$, if not:

① if $x^2 + y^2 < \dfrac{\pi}{2}$, then $0 \leq x^2 < \dfrac{\pi}{2} - y^2 \leq \dfrac{\pi}{2}$.

Inverse monotonic property of sine, cosine function in $\left[0, \frac{\pi}{2}\right]$, it has

$$\cos x^2 > \sin y^2, \cos y^2 > \sin x^2 \Rightarrow \left(\frac{\cos x^2}{\sin y^2}\right)^2 + \left(\frac{\cos y^2}{\sin x^2}\right)^2 > 2,$$

which contradict with (9);

② if $x^2 + y^2 > \dfrac{\pi}{2}$, then $\dfrac{\pi}{2} \geq x^2 > \dfrac{\pi}{2} - y^2 \geq 0$.

Inverse monotonic property of sine, cosine function in $\left[0, \frac{\pi}{2}\right]$, it has

$$\cos x^2 < \sin y^2, \cos y^2 < \sin x^2 \Rightarrow \left(\frac{\cos x^2}{\sin y^2}\right)^2 + \left(\frac{\cos y^2}{\sin x^2}\right)^2 < 2,$$

which also contradict with $(*)$

$\therefore$ Arbitrarily $x^2, y^2 \in \left[0, \frac{\pi}{2}\right]$, it has $x^2 + y^2 = \frac{\pi}{2}$.

$\therefore$ The locus of z_1 only shows the point $(1,1)$. The locus of z_2 has the origin as the center and $\sqrt{\frac{\pi}{2}}$ as the radius of circle.

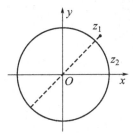

According to the figure:

$$\text{when } z_2 = \frac{\sqrt{\pi}}{2}(1 + i), \ |z_1 - z_2|_{\min} = \sqrt{2} - \sqrt{\frac{\pi}{2}};$$

$$\text{when } z_2 = \frac{\sqrt{\pi}}{2}(1 - i), \ |z_1 - z_2|_{\max} = \sqrt{2} + \sqrt{\frac{\pi}{2}}.$$

Comment: When finding the locus of z_2, it is also used to guess the idea of the card. When proving guess $x^2 + y^2 = \frac{\pi}{2}$, use proof by contradiction to rule out $x^2 + y^2 > \frac{\pi}{2}$ and $x^2 + y^2 < \frac{\pi}{2}$.

19. If $|z - z_1| = \lambda|z - z_2|$ (λ is positive constant number) and $z_1 \neq z_2$, discuss the locus of z in complex plane.

Analysis: Using $|z|^2 = z \cdot \bar{z}$ and $|z - z_1|^2 = \lambda^2|z - z_2|^2$, find locus of z in complex plane.

Solution: (1) When $\lambda = 1$, the locus is perpendicular bisector of segment $Z_1 Z_2$.

(2) When $\lambda \neq 1$, the equation is equivalent to $(z - z_1)(\bar{z} - \bar{z}_1) = \lambda^2(z - z_2)(\bar{z} - \bar{z}_2)$.

Expand:

$$|z|^2 - \frac{z_1 - \lambda^2 z_2}{1 - \lambda^2}\bar{z} - \frac{\bar{z}_1 - \lambda^2 \bar{z}_2}{1 - \lambda^2}z = \frac{\lambda^2 |z_2|^2 - |z_1|^2}{1 - \lambda^2}$$

$$\therefore \ |z|^2 - \frac{z_1 - \lambda^2 z_2}{1 - \lambda^2}\bar{z} - \frac{\overline{z_1} - \lambda^2 \overline{z_2}}{1 - \lambda^2}z + \left|\frac{z_1 - \lambda^2 z_2}{1 - \lambda^2}\right|^2$$

$$= \frac{\lambda^2 |z_2|^2 - |z_1|^2}{1 - \lambda^2} + \left|\frac{z_1 - \lambda^2 z_2}{1 - \lambda^2}\right|^2$$

$$\therefore \ \left|z - \frac{z_1 - \lambda^2 z_2}{1 - \lambda^2}\right|^2 = \left|\frac{z_1 - \lambda^2 z_2}{1 - \lambda^2}\right|^2 - \frac{|z_1|^2 - \lambda^2|z_2|^2}{1 - \lambda^2}.$$

According to inequality of modulus and reverse Cauchy inequality,

$$\left|\frac{z_1 - \lambda^2 z_2}{1 - \lambda^2}\right|^2 \geq \left|\frac{|z_1|^2 - \lambda^2|z_2|^2}{1 - \lambda^2}\right| \geq \frac{|z_1|^2 - \lambda^2|z_2|^2}{1 - \lambda^2}$$

$$\therefore \ \left|z - \frac{z_1 - \lambda^2 z_2}{1 - \lambda^2}\right| = \sqrt{\left|\frac{z_1 - \lambda^2 z_2}{1 - \lambda^2}\right|^2 - \frac{|z_1|^2 - \lambda^2|z_2|^2}{1 - \lambda^2}}.$$

$\therefore$ When $\lambda \neq 1$, the locus of z is a circle with center

$$z_0 = \frac{z_1 - \lambda^2 z_2}{1 - \lambda^2} \text{ and radius}$$

$$r = \sqrt{\left|\frac{z_1 - \lambda^2 z_2}{1 - \lambda^2}\right|^2 - \frac{|z_1|^2 - \lambda^2|z_2|^2}{1 - \lambda^2}}.$$

Comment: When $\lambda \neq 1$, the circle in plane geometry is called Apollonius circle about $\overrightarrow{Z_1 Z_2}$.

Exercise 7

1. If $|z| = 1$ and $z^2 + 2z + \frac{1}{z} < 0$, find z.
2. If $|z - 4| = |z - 4i|$ and $z + \frac{14 - z}{z - 1} \in R$, find z.

3. $z = \log(x^2 - 9x + 21) + i(\log \frac{x}{10} + \log(x - 3))$ $(x \in R)$, what is x
   (1) when z is a real number; (2) z is an imaginary number; (3) z
   is a purely imaginary number.

4. If $z_1 = \cos \alpha + i \sin \alpha$, $z_2 = \cos \beta + i \sin \beta$ and $\bar{z}_1 + \bar{z}_2 = \frac{1}{2} - \frac{1}{4}i$,
   find $\cos(\alpha - \beta)$.

5. O is the origin in complex plane, and Z_1 and Z_2 are two points
   in complex plane, which meet the following:

   (i) arguments corresponding complex number Z_1 and Z_2 are,
       respectively, fixed value θ and $-\theta$;
   (ii) suppose area of $\triangle OZ_1Z_2$ is S,

   find minimum of modulus of focus z of $\triangle OZ_1Z_2$.

6. Two points A and B correspond to -3 and z in complex plane,
   and $|z| = 1$. P is a trisection points of segment AB which close
   to point A.

   (1) Find the locus of P.
   (2) $\overrightarrow{AP}$ corresponds to z', find the locus of z'.

7. If A, B and C are three non-collinear points corresponding to
   complex numbers $Z_0 = ai$, $Z_1 = \frac{1}{2} + bi$, $Z_2 = 1 + ci$ $(a, b, c \in R)$,
   prove: curve $Z = Z_0 \cos^4 t + 2Z_1 \cos^2 t \sin^2 t + Z_2 \sin^4 t$ $(t \in R)$
   and the median line that parallel to line AC in $\triangle ABC$ have only
   one intersection point, and then find the point.

8. If $|z| = 1$ and $u = z^4 - z^3 - 3z^2 i - z + 1$, find the maximum of $|u|$
   and complex number z when maximum of $|u|$ is obtained.

Chapter 8

Operations with Complex Number

8.1. Operations with Complex Number

1. Addition and subtraction:

$$(a + bi) \pm (c + di) = (a \pm c) + (b \pm d)i \ (a, b, c, d \in R).$$

Geometrical meaning of addition and subtraction operation:
Suppose $z_1 = a + bi, z_2 = c + di \ (a, b, c, d \in R)$.
Addition: Suppose points Z_1 and Z_2 correspond to $z_1 = a + bi, z_2 = c + di \ (a, b, c, d \in R)$ in complex plane

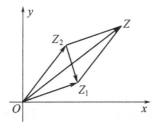

(1) Make points Z_1 and Z_2 correspond to $z_1 = a + bi, z_2 = c + di \ (a, b, c, d \in R)$ in complex plane, and connect OZ_1, OZ_2; where vectors $\overrightarrow{OZ_1}, \overrightarrow{OZ_2}$.

(2) Make a parallelogram OZ_1ZZ_2 where $\overrightarrow{OZ_1}, \overrightarrow{OZ_2}$ are adjacent edges.

(3) Diagonal vector $\overrightarrow{OZ}$ is the geometrical meaning of $z_1 + z_2$.

(4) Diagonal vector $\overrightarrow{Z_2Z_1}$ is the geometrical meaning of $z_1 - z_2$.

2. Multiplication: $(a + bi)(c + di) = (ac - bd) + (bc + ad)i$ $(a, b, c, d \in R)$.

Product of n complex number z is written z^n. When $z \neq 0$, $z^0 = 1$; when $z \neq 0$, $n \in R$, $z^{-n} = \frac{1}{z^n}$. The arithmetic of real number power of integers is still applicable to the complex number; when $m, n \in Z$ and $z \neq 0$, $z^m \cdot z^n = z^{m+n}$, $(z^n)^m = z^{mn}$, $(z_1 z_2)^n = z_1^n z_2^n$. Operation with i: $i^{4n} = 1$, $i^{4n+1} = i$, $i^{4n+2} = -1$, $i^{4n+3} = -i$ $(n \in z)$,

$(a \pm ai)^2 = \pm 2a^2 i$, $\omega^3 = 1$, $1 + \omega + \omega^2 = 0$, where $\omega = -\frac{1}{2} + \frac{\sqrt{3}}{2}i$.

3. **Division:** $\frac{a+bi}{c+di} = \frac{ac+bd}{c^2+d^2} + \frac{bc-ad}{c^2+d^2}i$ $(a, b, c, d \in R)$.

Operation skill: $\dfrac{1+i}{1-i} = i$, $\dfrac{1+i}{1-i} = -i$, $\dfrac{a+bi}{b-ai} = i$.

4. **The addition and multiplication of the complex number satisfy commutative law and associative law, and satisfy distribution law of the multiplication.**

Suppose $z_1 = a_1 + b_1 i$, $z_2 = a_2 + b_2 i$, $z_3 = a_3 + b_3 i$ $(a_1, a_2, a_3, b_1, b_2, b_3 \in R)$.

Commutative law: $z_1 + z_2 = z_2 + z_1$;

Associative law $(z_1 + z_2) + z_3 = z_1 + (z_2 + z_3)$.

Exercise

1. Calculate: (1) $(3 + 2i) + (1 - i) - (2 - 3i)$; (2) $i^{2009} + i^{-2009}$; (3) $\frac{(1+i)^{312}}{(1-i)^{311}}$.

Analysis: Calculate i^4 in question (2), and first calculate $(1 \pm i)^2$ in question (3).

Solution:

(1) $(3+2i)+(1-i)-(2-3i) = (3+1-2)+(2-1+3)i = 2+4i$.

(2) $i^{2009} + i^{-2009} = (i^4)^{502}i + \dfrac{1}{(i^4)^{502}i} = i + \dfrac{1}{i} = 0$.

(3) $\dfrac{(1+i)^{312}}{(1-i)^{311}} = \dfrac{(1+i)^{312}(1-i)}{(1-i)^{312}} = \left[\dfrac{(1+i)^2}{(1-i)^2}\right]^{156}(1-i) = \left(\dfrac{2i}{-2i}\right)^{156}(1-i) = 1-i$.

Comment: $(a \pm ai)^2 = \pm 2a^2 i$ $(a \in R)$. The results of the above often make the operation of the complex number relatively simple.

2. Calculate: $i^{29} + i^{30} + i^{31} + i^{32} + \cdots + i^{250}$.

Solution:

$\because$ $i^{4k+1} = i, i^{4k+2} = -1, i^{4k+3} - i, i^{4k} = 1$,

$\therefore$ $i + i^2 + i^3 + i^4 = 0$

$\therefore$ $i^{29} + i^{30} + i^{31} + i^{32} + \cdots + i^{250} = -1 + i$.

3. If $k \in N$, calculate: $(i^k + i^{k+2} + i^{k+3})^{48}$.

Solution: According to the periodicity of i:

$$i^k + i^{k+2} + i^{k+3} = -i^{k+1},$$

$$\therefore (i^k + i^{k+2} + i^{k+3})^{48} = (-i^{k+1})^{48} = (i^4)^{2k+12} = 1.$$

4. If $z = 1 + i$ and $\frac{z^2 + az + b}{z^2 - z + 1} = 1 - i$, find a, b.

Solution:

$$\frac{z^2 + az + b}{z^2 - z + 1} = \frac{(1+i)^2 + a(1+i) + b}{(1+i)^2 - (1+i) + 1} = (a+2) - (a+b)i = 1 - i \Rightarrow$$
$$\begin{cases} a = -1, \\ b = 2. \end{cases}$$

5. Calculate: (1) $\frac{(2+2i)^4}{(1-\sqrt{3}i)^5}$; (2) $\left[\frac{\sqrt{3}+i}{-1+\sqrt{3}i}\right]^{2010}$.

Solution:

(1) $\frac{(2+2i)^4}{(1-\sqrt{3}i)^5} = \frac{2^4(1+i)^4}{2^5(\frac{-1+\sqrt{3}i}{2})^5} = -1 + \sqrt{3}i$.

(2) $\left[\dfrac{\sqrt{3}+i}{-1+\sqrt{3}i}\right]^{2010} = \dfrac{(-2i)^{2010}(-\frac{1}{2}+\frac{\sqrt{3}}{2}i)^{2010}}{2^{2010}(-\frac{1}{2}-\frac{\sqrt{3}}{2}i)^{2010}}$

$= -1 \times \left[\dfrac{(-\frac{1}{2}+\frac{\sqrt{3}}{2}i)^3}{(-\frac{1}{2}+\frac{\sqrt{3}}{2}i)^3}\right]^{670} = -1$.

Comment: Use $\omega^3 = 1, 1 + \omega + \omega^2 = 0$, where $\omega = -\frac{1}{2} + \frac{\sqrt{3}}{2}i$.

Distance Formula Between Two Points in the Plane and Express Common Curves with Complex Numbers

1. If $Z_1(a, b)$ and $Z_2(c, d)$ correspond to $z_1 = a + bi$ and $z_2 = c + di$ in complex plane, then $|z_1 - z_2| = |\overrightarrow{Z_1 Z_2}| = |(a - c) + (b - d)i| = \sqrt{(a - c)^2 + (b - d)^2}$.

2. Express common curves with complex numbers:

    (1) The center of the circle is origin O and r is the radius: $|z| = r$.

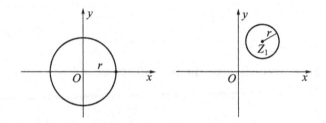

$z_1 = a + bi$ $(a, b \in R)$ is the center of the circle and r is the radius: $|z - z_1| = r$.

Circle region (excluding boundary), where $z_1 = a + bi$ $(a, b \in R)$ is the center of the circle and r is the radius: $|z - z_1| < r$.

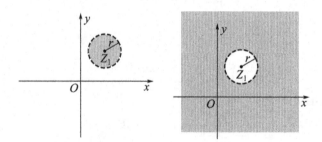

Circle out area (excluding boundary), where $z_1 = a + bi$ $(a, b \in R)$ is the center of the circle and r is the radius: $|z - z_1| > r$.

Ring region (excluding boundary), where $z_1 = a + bi$ $(a, b \in R)$ is the center of the circle and r is the radius, and $z_2 = c + di$ $(c, d \in R)$ is the center of the circle and R is the radius: $r < |z - z_1| < R$.

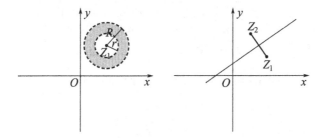

(2) If points Z_1 and Z_2 correspond to $z_1 = a + bi$ and $z_2 = c + di$ $(a, b, c, d \in R)$ in complex plane, midperpendicular of segment of $Z_1 Z_2$: $|z - z_1| = |z - z_2|$.

(3) Elliptic equation:

$$|z - z_1| + |z - z_2| = 2a(|F_1 F_2| = |z_1 - z_2| < 2a), \qquad (1)$$

where z_1, z_2 are focus and major axis is $2a$. When $|F_1 F_2| = |z_1 - z_2| = 2a$, equation (1) expresses segment $Z_1 Z_2$; when $|F_1 F_2| = |z_1 - z_2| > 2a$, no locus.

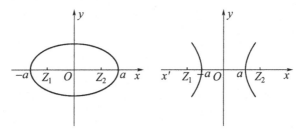

(4) Hyperbolic equation:

$$|z - z_1| - |z - z_2| = \pm 2a(|F_1 F_2| = |z_1 - z_2| > 2a), \qquad (2)$$

where z_1, z_2 are focus and real axis is $2a$. When $|F_1 F_2| = |z_1 - z_2| = 2a$, equation (2) expresses ray $Z_1 x'$ and $Z_2 x$; when $|F_1 F_2| = |z_1 - z_2| < 2a$, no locus.

Exercise

6. If $z_1 \in \{z | |z - i| = |z + 1|\}$, $z_2 \in \{z | |z - 2| = 1\}$, find the range of $|z_1 - z_2|$.

Solution: $|z - i| = |z + 1|$ is the midperpendicular of segment $AB(A(0, 1), B(-1, 0))$. $|z - 2| = 1$ is the circle where the center is $(2, 0)$ and 1 is the radius.

The distance from center of the circle to the straight line:

$$d = \frac{|2 + 0|}{\sqrt{1^2 + (-1)^2}} = \sqrt{2}$$

$\therefore \ |z_1 - z_2|_{\min} = |DE| = |CE| - |CD| = \sqrt{2} - 1$, no maximum.
$\therefore \ |z_1 - z_2| \le \sqrt{2} - 1$.

7. If $|z| = 3$, find the minimum and maximum of $|z - 1 + \sqrt{3}i|$.

Solution:

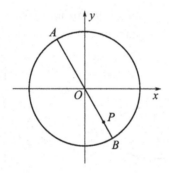

$|z - 1 + \sqrt{3}i|_{\max} = |AP| = 3 + |OP| = 5$,

$|z - 1| + \sqrt{3}i|_{\min} = |PB| = 3 - |OP| = 3 - 2 = 1$.

8. If $|z + 3 - 4i| = 2$, find the minimum and maximum of $|z|$.

Solution: $\because \ |z + 3 - 4i| = 2$

$\therefore \ |z|_{\max} = |OB| = |OA| + |OB| = 5 + 2 = 7$,

$|z|_{\min} = |OC| = |OA| - |AC| = 5 - 2 = 3$.

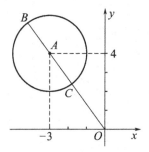

9. If $||z - i| - 2| + |z - i| - 2 = 0$, find S (the area of the graph) corresponding to the point set in the complex plane.

Solution:

$$\because \; ||z - i| - 2| + |z - i| - 2 = 0$$

$$\therefore \; ||z - i| - 2| = 2 - |z - i|$$

$$\therefore \; ||z - i| - 2| \geq 0, \;\; \therefore \; |z - i| \leq 2 \therefore \; S = 4\pi.$$

10. The complex plane region K is composed of the point Z corresponding to the complex number z. If real parts and imaginary parts of $\frac{z}{40}$ and $\frac{40}{z^2}$ all belong to [0, 1], find the area of K.

Solution:

Suppose

$$z = x + yi \; (x, y \in R), \; \therefore \; 0 \leq x \leq 40, 0 \leq y \leq 40 \qquad (1)$$

$$\because \; \frac{40}{\bar{z}} = \frac{40x}{x^2 + y^2} + \frac{40y}{x^2 + y^2}i$$

$$\therefore \; \begin{cases} 0 \leq \dfrac{40x}{x^2 + y^2} \leq 1 \\[2mm] 0 \leq \dfrac{40y}{x^2 + y^2} \leq 1 \end{cases} \Leftrightarrow \begin{cases} (x - 20)^2 + y^2 \geq 20^2, & (2) \\[2mm] x^2 + (y - 20)^2 \geq 20^2. & (3) \end{cases}$$

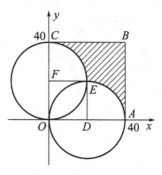

According to (1)–(3), the area of K is given by

$$S_{OABC} - S_{ODEF} - S_{\text{sector}-F-EC} - S_{\text{sector}-D-AE}$$

$$= 1600 - 400 - \frac{1}{2}\pi \times 20^2 = 1200 - 200\pi.$$

11. $S = \{z||z-1| \le 3, z \in C\}$, $T = \{z|z = \frac{w+2}{3}i + t, w \in S, t \in R\}$.

(1) If $S \cap T = \Phi$, find the range of t.

(2) If $S \cup T = S$, find the range of t.

Analysis: Analysis from the geometric significance of modulus of complex numbers.

Solution: Suppose $z \in T$, so $z = \frac{w+2}{3}i + t, w \in S, t \in R$.

$$\therefore w = \frac{3z - 3t}{i} - 2,$$

$$\because w \in S, \therefore |w - 1| \le 3,$$

$$\therefore \left| \frac{3z - 3t}{i} - 2 - 1 \right| \le 3$$

$$\therefore |z - (t + i)| \le 1.$$

So set T expresses the circle and its internal where the center is $(t, 1)$ and 1 is the radius. Set S expresses the circle and its internal where the center is $(1, 0)$ and 3 is the radius. See the below figure.

(1) If $S \cap T = \Phi$, two circles are out:

$$\therefore \sqrt{(t-1)^2 + 1} > 4, \therefore (t-1)^2 > 15, \therefore |t - 1| > \sqrt{15}$$

$$\therefore t < 1 - \sqrt{15} \text{ or } t > 1 + \sqrt{15}.$$

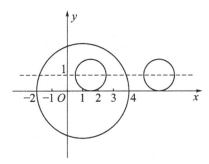

(2) If $S \cup T = S$, two circles contain or inscribe.

$$\therefore \sqrt{(t-1)^2 + 1} \le 4, \therefore (t-1)^2 \le 3, \therefore |t-1| \le \sqrt{3}$$

$$\therefore 1 - \sqrt{3} \le t \le 1 + \sqrt{3}.$$

Comment: About the problem of the complex number Z, if it can be determined that the graphical background of Z, it is very effective to take the composition method to solve, intuitively!

12. If imaginary number z meets $|\bar{z} - 3| = |\bar{z} - 3i|$ and $z - 1 + \frac{9}{z-1} = m \ (m \in R)$, find z.

Solution:

Suppose $z = x + yi \ (x, y \in R)$,

$$\because |\bar{z} - 3| = |\bar{z} - 3i| \quad \therefore (x-3)^2 + y^2 = x^2 + (y+3)^2$$

$$\therefore y = -x, \quad \therefore z = x - xi$$

$$\because z - 1 + \frac{9}{z-1} = m \in R,$$

$$\therefore z - 1 + \frac{9}{z-1} = \overline{z - 1 + \frac{9}{z-1}} = \bar{z} - 1 + \frac{9}{\bar{z} - 1}$$

$$\therefore \ z - \bar{z} + \frac{9}{z-1} - \frac{9}{\bar{z}-1} = 0$$

$$\therefore \ (z - \bar{z}) \left[1 - \frac{9}{(z-1)(\bar{z}-1)} \right] = 0.$$

$\because \ z$ is imaginary number

$$\therefore \ z \neq \bar{z} \text{ and } \frac{9}{(z-1)(\bar{z}-1)} = 1$$

$$\therefore \ |z-1|^2 = 9.$$

$\because \ z = x - xi, \ \therefore \ (x-1)^2 + x^2 = 9$

$$\therefore \ x^2 - x - 4 = 0$$

$$\therefore \ x = \frac{1 \pm \sqrt{17}}{2},$$

$$\therefore \ z = \frac{1 + \sqrt{17}}{2}(1 - i) \text{ or } z = \frac{1 - \sqrt{17}}{2}(1 - i).$$

13. If $11z^{10} + 10iz^9 + 10iz - 11 = 0$, find $|z|$.

Solution: According to the conditions $z^9 = \frac{11-10iz}{11z+10i}$,

suppose $z = a + bi \ (a, b \in \mathbf{R})$,

then $|z|^9 = \left| \dfrac{11 - 10iz}{11z + 10i} \right| = \sqrt{\dfrac{11^2 + 220b + 10^2(a^2 + b^2)}{11^2(a^2 + b^2) + 220b + 10^2}}.$

Suppose $f(a, b) = 11^2 + 220b + 10^2(a^2 + b^2)$,
$g(a, b) = 11^2(a^2 + b^2) + 220b + 10^2$.

If $a^2 + b^2 > 1$, then $g(a, b) > f(a, b), |z|^9 < 1$,

$\therefore \ |z| > 1, a^2 + b^2 < 1$, contradiction with subject.

If $a^2 + b^2 < 1$, then $g(a, b) < f(a, b), |z|^9 > 1$.

$\therefore \ |z| > 1, a^2 + b^2 > 1$, contradiction with subject.

$\therefore \ a^2 + b^2 = 1, \ \therefore \ |z| = 1.$

Comment: First guess $a^2 + b^2 = 1$, and then prove $a^2 + b^2 = 1$.

Exercise 8

1. If $(1+i)z_1 = -1 + 5i$, $z_2 = a - 2 - i$, where i is imaginary unit, $a \in R$, and $|z_1 - \overline{z_2}| < |z_1|$, find the range of a.

2. If three vertices of a square are $A(1,2), B(-2,1), C(-1,-2)$ in complex plane, find the complex number that corresponds to its fourth vertices D.

3. A, B and C are three points that correspond to complex numbers z_1, z_2, z_3. If $\frac{z_2 - z_1}{z_3 - z_1} = 1 + \frac{4}{3}i$, find the ratio of three sides in $\triangle ABC$.

4. Two points Z_1, Z_2 correspond to two non-zero complex numbers z_1, z_2, which meet:
   ① $z_2 = z_1 \cdot ai$ $(a > 0)$, ② midpoint of segment Z_1, Z_2 corresponding to $3 + 4i$;
   Find the maximum area of $\triangle Z_1 O Z_2$ (O is origin) and its z_1, z_2.

5. If $|z + 2 + 3i|^2 + \|z - 2 - 3i\|^2 = 40$, find $|z|$.

6. If A and B are two points correspond to the roots of equation $x^2 - 2x + 2 = 0$ in complex plane, and the complex number that corresponds to C meets $(1+i)^2(1+z) = -6$, find the maximum angle of $\triangle ABC$.

7. If $z_1 = 2 - \sqrt{3}a + ai$, $z_2 = \sqrt{3}b - 1 + (\sqrt{3} - b)i$ and their modulus are equal, and argument of $\overline{z_1} \cdot z_2$ is $\frac{\pi}{2}$, find a and b.

8. If non-zero complex numbers x and y meet $x^2 + xy + y^2 = 0$, find $(\frac{x}{x+y})^{2010} + (\frac{y}{x+y})^{2010}$.

9. If real numbers x and y meet $z_1 = x + \sqrt{3} + yi$, $z_2 = x - \sqrt{3} + yi$, $|z_1| + |z_2| = 4$, find the minimum and maximum of $f(x,y) = |2x - 4y - 9|$.

10. The equation of the curve C in the complex plane is $|z + \sqrt{a^2 - b^2}| + |z - \sqrt{a^2 - b^2}| = 2a$ $(a > b > 0)$, z_1, z_2, z_3 are three points in C, point Z_1 corresponding to complex bi and $\overrightarrow{Z_1 Z_2}i = \overrightarrow{Z_1 Z_3}$, find the number of $\triangle Z_1 Z_2 Z_3$.

11. P_1 and P_2 are the two point sets of the complex plane:

   $P_1 = \{z | z\overline{z} + 3i(\overline{z} - z) + m = 0, z \in C, m \in R, m < 9\}$ (m is constant),

   $P_2 = \{\omega | \omega = 2iz, z \in P_1\}$.

   (1) When m changes in what range, $P_1 \cap P_2 \neq \phi$.

(2) If $m = 5$, $z_1 \in P_1$, $z_2 \in P_2$, find the range of the minimum and maximum of $|z_1 - z_2|$.

12. If $|z| = 5$ and corresponding point $(3 + 4i)z$ in the complex plane is on the angular bisector of the second and fourth quadrant, $|\sqrt{2}z - m| = 5\sqrt{2}$, $(m \in R)$, find z and m.

13. If x is an imaginary number and $x + \frac{1}{x}$ is the real root of equation $y^2 - ay + a + 1 = 0$, find the range of a.

14. If $|z| = 1$, find the maximum of $u = |(z - a)^2(z + \beta)|$ $(\alpha, \beta > 0)$.

15. $A = \{x | |z - c| + |z + c| = 2a, z \in C, a > c > 0\}$. If a and c take all positive real numbers and $z + i \in A$, try to make the image corresponds to set A in the complex plane.

Chapter 9

Trigonometric Form of a Complex Number

Trigonometric Form of a Complex Number

1. Argument of Complex Number

Arg z is the angle θ from the positive real axis to the vector $\overrightarrow{OZ}$ representing $z = a + bi$, and arg $\in [0, 2\pi)$.

2. Trigonometric Form of Complex Number

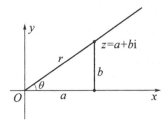

As shown in the figure,

$$\begin{cases} a = r\cos\theta, \\ b = r\sin\theta, \end{cases}$$

and any complex number $z = a + bi$ can be expressed as $z = r(\cos\theta + i\sin\theta)$; $z = r(\cos\theta + i\sin\theta)$ is called trigonometric form of complex number. Moreover,

$$\begin{cases} \cos\theta = \dfrac{a}{r}, \\ \sin\theta = \dfrac{b}{r}, \\ r = \sqrt{a^2 + b^2}. \end{cases}$$

Pay attention to: (1) Every non-zero complex number has only modulus and argument, and can be uniquely determined by it, so sufficient necessary condition that two non-zero complex

227

numbers are equal is that their modulus and arguments are equal.

(2) Trigonometric form of complex number: ① $r \geq 0$; ② sine and cosine are the same; ③ $\cos\theta, \sin\theta$ are connected by the plus; ④ real part is $r\cos\theta$, imaginary part is $r\sin\theta$.

Exercise

1. If $z = \frac{1}{2} + \frac{\sqrt{3}}{2}i$, $w = \frac{\sqrt{2}}{2} + \frac{\sqrt{2}}{2}i$, find trigonometric form of complex number of $zw + zw^3$.

   Solution: $zw + zw^3 = zw(1+w^2) = \left(\frac{1}{2} + \frac{\sqrt{3}}{2}i\right)\left(\frac{\sqrt{2}}{2} + \frac{\sqrt{2}}{2}i\right)(1+i) = \sqrt{2}\left(-\frac{\sqrt{3}}{2} + \frac{1}{2}i\right) = \sqrt{2}\left(\cos\frac{5\pi}{6} + i\sin\frac{5\pi}{6}\right).$

2. Find trigonometric form of complex number:

   (1) $1 + \cos\theta + i\sin\theta \ (0 \leq \theta \leq \pi)$;
   (2) $\frac{1-i\tan\alpha}{1+i\tan\alpha}$;
   (3) $1 + i\tan\alpha$;
   (4) $i\sin\alpha$;
   (5) $\frac{1-\sin\theta+i\cos\theta}{1-\sin\theta-i\cos\theta}$.

   Solution:

   (1) When $0 \leq \theta \leq \pi$,

   $$1 + \cos\theta + i\sin\theta = 2\cos^2\frac{\theta}{2} + 2\sin\frac{\theta}{2}\cos\frac{\theta}{2}$$

   $$= 2\cos\frac{\theta}{2}\left(\cos\frac{\theta}{2} + i\sin\frac{\theta}{2}\right).$$

   (2) $\dfrac{1-i\tan\alpha}{1+i\tan\alpha} = \dfrac{\cos\alpha - i\sin\alpha}{\cos\alpha + i\sin\alpha}$

   $$= \frac{\cos(-\alpha) + i\sin(-\alpha)}{\cos\alpha + i\sin\alpha} = \cos(-2\alpha) + i\sin(-2\alpha).$$

(3) $1 + i\tan\alpha = \dfrac{\cos\alpha + i\sin\alpha}{\cos\alpha}$

$$= \begin{cases} \dfrac{1}{\cos\alpha}(\cos\alpha + i\sin\alpha), & \alpha \in \left(2k\pi - \dfrac{\pi}{2}, 2k\pi + \dfrac{\pi}{2}\right), \\ & k \in Z, \\ -\dfrac{1}{\cos\alpha} & \alpha \in \left(2k\pi + \dfrac{\pi}{2}, 2k\pi + \dfrac{3\pi}{2}\right), \\ \quad [\cos(\pi + \alpha) + i\sin(\pi + \alpha)], & k \in Z. \end{cases}$$

(4) $i\sin\alpha$

$$= \begin{cases} \sin\alpha\left(\cos\dfrac{\pi}{2} + i\sin\dfrac{\pi}{2}\right), & \alpha \in [2k\pi, 2k\pi + \pi),\ k \in Z, \\ -\sin\alpha\left(\cos\dfrac{3\pi}{2} + i\sin\dfrac{3\pi}{2}\right), & \alpha \in (2k\pi + \pi, 2k\pi + 2\pi), \\ & k \in Z. \end{cases}$$

(5) $\dfrac{1 - \sin\theta + i\cos\theta}{1 - \sin\theta - i\cos\theta} = \dfrac{1 - \cos\left(\frac{\pi}{2} - \theta\right) + i\sin\left(\frac{\pi}{2} - \theta\right)}{1 - \cos\left(\frac{\pi}{2} - \theta\right) - i\sin\left(\frac{\pi}{2} - \theta\right)}$

$$= \dfrac{2\sin^2\left(\frac{\pi}{4} - \frac{\theta}{2}\right) + 2i\sin\left(\frac{\pi}{4} - \frac{\theta}{2}\right)\cos\left(\frac{\pi}{4} - \frac{\theta}{2}\right)}{2\sin^2\left(\frac{\pi}{4} - \frac{\theta}{2}\right) - 2i\sin\left(\frac{\pi}{4} - \frac{\theta}{2}\right)\cos\left(\frac{\pi}{4} - \frac{\theta}{2}\right)}$$

$$= \dfrac{\sin\left(\frac{\pi}{4} - \frac{\theta}{2}\right) + i\cos\left(\frac{\pi}{4} - \frac{\theta}{2}\right)}{\sin\left(\frac{\pi}{4} - \frac{\theta}{2}\right) - i\cos\left(\frac{\pi}{4} - \frac{\theta}{2}\right)}$$

$$= \dfrac{\cos\left(\frac{\pi}{4} + \frac{\theta}{2}\right) + i\sin\left(\frac{\pi}{4} + \frac{\theta}{2}\right)}{\cos\left(\frac{\pi}{4} + \frac{\theta}{2}\right) - i\sin\left(\frac{\pi}{4} + \frac{\theta}{2}\right)}$$

$$= \dfrac{\cos\left(\frac{\pi}{4} + \frac{\theta}{2}\right) + i\sin\left(\frac{\pi}{4} + \frac{\theta}{2}\right)}{\cos\left[-\left(\frac{\pi}{4} + \frac{\theta}{2}\right)\right] + i\sin\left[-\left(\frac{\pi}{4} + \frac{\theta}{2}\right)\right]}$$

$$= \cos\left(\dfrac{\pi}{2} + \theta\right) + i\sin\left(\dfrac{\pi}{2} + \theta\right).$$

3. If $z = 3\cos\theta + i\sin\theta$, find the maximum of $y = \tan(\theta - \arg z)$ $(0 < \theta < \frac{\pi}{2})$ and its θ.

Solution:

$$\because \ z = 3\cos\theta + i\sin\theta, \ \therefore \ \tan(\arg z) = \frac{\sin\theta}{3\cos\theta} = \frac{1}{3}\tan\theta.$$

$$\therefore \ y = \tan(\theta - \arg z) = \frac{\tan\theta - \frac{1}{3}\tan\theta}{1 + \frac{1}{3}\tan^2\theta} = \frac{2}{\frac{3}{\tan\theta} + \tan\theta}.$$

$$\because \ 0 < \theta < \frac{\pi}{2}, \ \tan\theta > 0, \ \therefore \ \frac{3}{\tan\theta} + \tan\theta \geq 2\sqrt{3},$$

$$\therefore \ y = \frac{2}{\frac{3}{\tan\theta} + \tan\theta} \leq \frac{\sqrt{3}}{3}.$$

If and only if $\frac{3}{\tan\theta} = \tan\theta(0 < \theta < \frac{\pi}{2})$, then $\theta = \sqrt{3}, \theta = \frac{\pi}{3}$, $y_{\max} = \frac{\sqrt{3}}{3}$.

4. If $z = \left(\frac{2}{1+\alpha}\right)^2$, and $\frac{1-\alpha}{1+\alpha}$ is purely complex number, find the locus of complex number z.

Solution:

$\because \ \dfrac{1-\alpha}{1+\alpha}$ is purely imaginary, $\therefore \ |a| = 1$ and $a \neq \pm 1$.

$\therefore$ Suppose $a = \cos\theta + i\sin\theta, \ \theta \in (0, \pi) \cup (\pi, 2\pi)$,

$z = x + yi \, (x, y \in R)$;

$$x + yi = \frac{1}{\cos^2\frac{\theta}{2}}(\cos\theta - i\sin\theta) = \left(1 - \tan^2\frac{\theta}{2}\right) + \left(-2\tan\frac{\theta}{2}\right)i,$$

$$\therefore \ \begin{cases} x = 1 - \tan^2\dfrac{\theta}{2}, \\ y = -2\tan\dfrac{\theta}{2}, \end{cases} \quad \theta \in (0, \pi) \cup (\pi, 2\pi).$$

We cancel the θ, which implies $y^2 = -4(x - 1)(x \neq 1)$.
So the track is the focus of the origin, the x-axis is the axis of symmetry, open to the left and not containing $(1, 0)$ of a parabola.

5. If $0 < \theta < 2\pi$, $z = 1 - \cos\theta + i\sin\theta$, $\mu = a^2 + ai$, and $z\mu$ is purely imaginary number, $a \in R$,

(1) find the $\arg\mu$ of complex number μ (express with θ);

(2) if $\omega = z^2 + \mu^2 + 2z\mu$, is ω a positive real number? Why?

Solution:

(1) $\because$ $z\mu = (1 - \cos\theta + i\sin\theta)(a^2 + ai)$

$$= [a^2(1 - \cos\theta) - a\sin\theta] + [a^2\sin\theta + a(1 - \cos\theta)]i,$$

and $z\mu$ is purely imaginary,

$$\therefore \begin{cases} a^2\sin\theta + a(1 - \cos\theta) \neq 0, & \text{①} \\ a^2(1 - \cos\theta) = a\sin\theta. & \text{②} \end{cases}$$

According to ①, $a \neq 0$, and $0 < \theta < 2\pi$, $\therefore$ $1 - \cos\theta \neq 0$.

$\therefore$ According to ②,

$$a = \frac{\sin\theta}{1 - \cos\theta} = \cot\frac{\theta}{2};$$

$$\therefore \tan(\arg\mu) = \frac{a}{a^2} = \frac{1}{a} = \tan\frac{\theta}{2}.$$

(i) When $0 < \theta < \pi$, $\mu = \cot^2\frac{\theta}{2} + i\cot\frac{\theta}{2}$, corresponding points in the first quadrant, $\therefore$ $\arg\mu = \frac{\theta}{2}$.

(ii) When $\pi < \theta < 2\pi$, $\frac{\pi}{2} < \frac{\theta}{2} < \pi$, $\mu = \cot^2\frac{\theta}{2} + i\cot\frac{\theta}{2}$, in the fourth quadrant, $\therefore$ $\arg\mu = \pi + \frac{\theta}{2}$.

(2) ω cannot be a real number, actually, according to the condition,

$$\omega = (z + \mu)^2, \quad z + \mu = (1 - \cos\theta + a^2) + (a + \sin\theta)i;$$

if $\omega \in R^+$ then $a + \sin\theta = 0$,

$$\therefore a = -\sin\theta, \quad \because a = \frac{\sin\theta}{1 - \cos\theta}, \quad \sin\theta \neq 0,$$

$$\therefore -\sin\theta = \frac{\sin\theta}{1 - \cos\theta}, \therefore \cos\theta = 2 \text{ is impossible.}$$

Multiplication and Exponentiation of Complex Number

1. Operation law: If $z_1 = r_1(\cos\theta_1 + i\sin\theta_1)$, $z_2 = r_2(\cos\theta_2 + i\sin\theta_2)$, then $z_1z_2 = r_1r_2[\cos(\theta_1 + \theta_2) + i\sin(\theta_1 + \theta_2)]$.

2. **Geometrical meaning:** $\overrightarrow{OP_1}, \overrightarrow{OP_2}$ are two vectors corresponding to $z_1 = r_1(\cos\theta_1 + i\sin\theta_1)$ and $z_2 = r_2(\cos\theta_2 + i\sin\theta_2)$, $\overrightarrow{OP_1}$ rotates θ_2 in counterclockwise direction, modulus is r_2 times to the original, get $\overrightarrow{OP}$, which expresses $z_1 z_2$.

3. **Exponentiation of complex number — Dymov theorem**
   If $z = r(\cos\theta + i\sin\theta)$, then $z^n = r^n(\cos n\theta + i\sin n\theta)(n \in N^*)$.

Exercise

6. In the complex plane, four vertices of a square are Z_1, Z_2, Z_3, O (O is origin) in a counterclockwise direction. If $z_1 = 1 + \sqrt{3}i$ corresponds to Z_2, find the complex numbers corresponding to Z_1 and Z_3.

   Solution: If the complex numbers z_1 and z_2 correspond to Z_1 and Z_3,

   $$z_1 = z_2 \frac{1}{\sqrt{2}}\left[\cos\left(\frac{\pi}{4}\right) + i\sin\left(\frac{\pi}{4}\right)\right] = \frac{\sqrt{3}+1}{2} + i\frac{\sqrt{3}-1}{2},$$

   $$z_3 = z_2 \frac{1}{\sqrt{2}}\left[\cos\left(\frac{\pi}{4}\right) + i\sin\left(\frac{\pi}{4}\right)\right] = \frac{1-\sqrt{3}}{2} + i\frac{\sqrt{3}+1}{2}.$$

7. (1) If $z = \frac{\sqrt{3}}{2} - \frac{1}{2}i$, $w = \frac{\sqrt{2}}{2} + \frac{\sqrt{2}}{2}i$, P and Q correspond to complex numbers $\overline{zw}$, $z^2 w^3$, prove $\triangle OPQ$ is isosceles and right-angled triangle (O is origin).

   (2) If $|z_1| = 2$, $|z_2| = 3$, $3z_1 - 2z_2 = \frac{3}{2} - i$, find $z_1 z_2$.

   Proof:

   $$(1)\ z = \frac{\sqrt{3}}{2} - \frac{1}{2}i = \cos\left(-\frac{\pi}{6}\right) + i\sin\left(-\frac{\pi}{6}\right),$$

   $$w = \frac{\sqrt{2}}{2} + \frac{\sqrt{2}}{2}i = \cos\frac{\pi}{4} + i\sin\frac{\pi}{4},$$

$$\therefore zw = \cos\left(\frac{\pi}{12}\right) + i\sin\left(\frac{\pi}{12}\right).$$

$$\therefore \overline{zw} = \cos\left(-\frac{\pi}{12}\right) + i\sin\left(-\frac{\pi}{12}\right),$$

$$\therefore z^2 w^3 = \cos\left(-\frac{\pi}{3} + \frac{3\pi}{4}\right) + i\sin\left(-\frac{\pi}{3} + \frac{3\pi}{4}\right)$$

$$= \cos\left(\frac{5\pi}{12}\right) + i\sin\left(\frac{5\pi}{12}\right).$$

Included angle of OP and OQ is $\frac{5\pi}{12} - (-\frac{\pi}{12}) = \frac{\pi}{2}$.

$$\therefore OP \perp OQ.$$

$$\because |OP| = |\overline{zw}| = 1, \quad |OQ| = |z^2 w^3| = 1.$$

$$\therefore \triangle OPQ \text{ is isosceles and right-angled triangle.} \qquad \square$$

(2) If $z_1 = 2(\cos\alpha + i\sin\alpha)$ and $z_2 = 3(\cos\beta + i\sin\beta)$, $3z_1 - 2z_2 = \frac{3}{2} - i$ and the necessary and sufficient condition of the complex number is given by

$$6(\cos\alpha - \cos\beta) = \frac{3}{2} \quad \text{and} \quad 6(\sin\alpha - \sin\beta) = -1,$$

$$\therefore -12\sin\frac{\alpha+\beta}{2}\sin\frac{\alpha-\beta}{2} = \frac{3}{2} \tag{1}$$

$$\text{and } 12\cos\frac{\alpha+\beta}{2}\sin\frac{\alpha-\beta}{2} = -1. \tag{2}$$

By (1) $\div$ (2), we have

$$\tan\frac{\alpha+\beta}{2} = \frac{3}{2}.$$

According to universal formula: $\sin(\alpha + \beta) = \frac{12}{13}$, $\cos(\alpha + \beta) = -\frac{5}{13}$.

$$\therefore z_1 \cdot z_2 = 6[\cos(\alpha + \beta) + i\sin(\alpha + \beta)] = -\frac{30}{13} + \frac{72}{13}i.$$

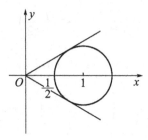

Comment: Question (2) appeared the National High School Math League in 2001, and reference answer is given skill strong solution. According to the characteristics of the problem, triangle form of complex number seems to be more conform with the student's thinking characteristics, but also not complicated. It can also use the geometric meaning of complex number.

8. If $|z_1| = |z_1 + z_2| = 3$, $|z_1 - z_2| = 3\sqrt{3}$, find $\log_3 |(z_1 \bar{z_2})^{2000} + (\bar{z_1} z_2)^{2000}|$.

Solution: According to the condition

$$|z_1 + z_2|^2 = (z_1 + z_2)(\bar{z_1} + \bar{z_2}) = 9,$$

$$\therefore |z_1|^2 + |z_2|^2 + z_1\bar{z_2} + \bar{z_1} z_2 = 9,$$

$$|z_1 - z_2|^2 = (z_1 - z_2)(\bar{z_1} - \bar{z_2}) = 27,$$

$$\therefore |z_1|^2 + |z_2|^2 - (z_1\bar{z_2} + \bar{z_1} z_2) = 27,$$

$$\therefore |z_1|^2 + |z_2|^2 = 18,$$

$$\because |z_1|^2 = 9, \therefore |z_2| = 3,$$

and $z_1\bar{z_2} + \bar{z_1} z_2 = -9$, $|z_1\bar{z_2}| = |\bar{z_1} z_2| = 9$.

Suppose $z_1\bar{z_2} = 9(\cos\theta + i\sin\theta)$,

$$\therefore \bar{z_1} z_2 = \overline{z_1\bar{z_2}} = 9[\cos(-\theta) + i\sin(-\theta)].$$

$$\therefore \cos\theta = \frac{-1}{2}, \therefore z_1\bar{z_2} = 9w \text{ or } 9w^2, \text{ where } w = -\frac{1}{2} - \frac{\sqrt{3}}{2}i.$$

When $z_1 \bar{z}_2 = 9w, \bar{z}_1 z_2 = 9w^2$,

$\therefore (z_1 \bar{z}_2)^{2000} + (\bar{z}_1 z_2)^{2000} = -9^{2000}$;

When $z_1 z_2 = 9w^2, z_1 z_2 = 9w$,

$\therefore (z_1 \bar{z}_2)^{2000} + (\bar{z}_1 z_2)^{2000} = -9^{2000}$.

$\therefore$ The original type is equal to $\log_3 9^{2000} = 4000$.

9. If $n = 2000$, find $\frac{1}{2^n}(1 - 3C_n^2 + 3^2 C_n^4 - 3^3 C_n^6 + \cdots - 3^{999} C_n^{1998} + 3^{1000} C_n^{2000})$.

Solution:

$\because$ The original type $= \left(\frac{1}{2}\right)^n - C_n^2 \left(\frac{1}{2}\right)^{n-2} \left(\frac{\sqrt{3}}{2}\right)^2 + \cdots$

$\qquad + C_n^{n-2} \left(\frac{1}{2}\right)^2 \left(\frac{\sqrt{3}}{2}\right)^{n-2} - C_n^n \left(\frac{\sqrt{3}}{2}\right)^n.$

$\therefore$ The original type is $\left(\frac{1}{2} + \frac{\sqrt{3}}{2} i\right)^n$,

real part of expanded type.

$\because \left(\frac{1}{2} + \frac{\sqrt{3}}{2} i\right)^n = \left(\cos \frac{\pi}{3} + i \sin \frac{\pi}{3}\right)^n = \left(\cos \frac{n}{3}\pi + i \sin \frac{n}{3}\pi\right),$

$\because \cos \frac{n}{3}\pi = \cos \frac{2000}{3}\pi = \cos \left(666\pi + \frac{2}{3}\pi\right) = \cos \frac{2}{3}\pi = -\frac{1}{2}.$

$\therefore$ The original type is equal to $-\frac{1}{2}$.

10. If the 20 vertices of the unit circle inscribed 20-sided regular are $z_1, z_2, \ldots, z_{20}$, how many different points are corresponding to $z_1^{1995}, z_2^{1995}, \ldots, z_{20}^{1995}$.

Solution: The 20 vertices of the unit circle inscribed 20-sided regular are $\frac{2\pi}{20}$; if $Z_1 = \cos\theta + i\sin\theta$, by complex multiplication,

$$Z_k = (\cos\theta + i\sin\theta)\left[\cos\frac{2(k-1)\pi}{20} + i\sin\frac{2(k-1)\pi}{20}\right],$$

$$k = 1, 2, \ldots, 20,$$

$$\therefore Z_k^{1995} = (\cos 1995\theta + i\sin 1995\theta)$$

$$\times\left[\cos\frac{2(k-1)\pi}{20}\cdot 1995 + i\sin\frac{2(k-1)\pi}{20}\cdot 1995\right]$$

$$= (\cos 1995\theta + i\sin 1995\theta)$$

$$\times\left(\cos\frac{2\pi}{20}\cdot 1995 + i\sin\frac{2\pi}{20}\cdot 1995\right)^{k-1}$$

$$= (\cos 1995\theta + i\sin 1995\theta)\left[\cos\left(99\times 2\pi + \frac{15\times 2\pi}{20}\right)\right.$$

$$\left.+ i\sin\left(99\times 2\pi + \frac{15\times 2\pi}{20}\right)\right]^{k-1}$$

$$= (\cos 1995\theta + i\sin 1995\theta)\left(\cos\frac{3\pi}{2} + i\sin\frac{3\pi}{2}\right)^{k-1}$$

$$= (\cos 1995\theta + i\sin 1995\theta)\left(\cos\frac{3\pi}{2} + i\sin\frac{3\pi}{2}\right)^{k-1}$$

$$= (\cos 1995\theta + i\sin 1995\theta)(-i)^{k-1}, \quad k = 1, 2, \ldots, 20.$$

By periodic imaginary unit i, four different values are known.

11. If $z = x + yi$ (x, y are rational numbers) and $|z| = 1$, prove: to any natural numbers n, $|z^{2n} - 1|$ is rational number.

Solution: Let $z = \cos\theta + i\sin\theta$,

$$\therefore \begin{cases} x = \cos\theta, \\ y = \sin\theta, \end{cases}$$

$\therefore$ $\cos\theta, \sin\theta$ are rational numbers,

$\because z^{2n} = \cos 2n\theta + i\sin 2n\theta$,

$$\therefore |z^{2n} - 1| = \sqrt{(\cos 2n\theta - 1)^2 + \sin^2 2n\theta}$$

$$= \sqrt{2 - 2\cos 2n\theta} = 2|\sin n\theta|.$$

Here we prove $\sin n\theta$, $\cos n\theta$ are rational numbers.
(1) When $n = 1$, $\sin\theta$ and $\cos\theta$ are rational numbers.
And they are clearly established by the subject.
(2) Suppose when $n = k$, $\sin k\theta$, $\cos k\theta$ are rational numbers.

$\because$ $\sin(k+1)\theta = \sin\theta \cdot \cos k\theta + \cos\theta \cdot \sin k\theta$,

$\cos(k+1)\theta = \cos\theta \cdot \cos k\theta - \sin\theta \cdot \sin k\theta$.

$\therefore$ when $n = k+1$, $\sin(k+1)\theta, \cos(k+1)\theta$ are rational numbers.

$\therefore$ For all $n \in N, \sin n\theta$ is rational number.

$\therefore$ We obtain the original proposition.

12. If $\sin A + \sin 3A + \sin 5A = a$, $\cos A + \cos 3A + \cos 5A = b$, prove:
(1) when $b \neq 0$, $\tan 3A = \frac{a}{b}$; (2) $(1 + 2\cos 2A)^2 = a^2 + b^2$.

Solution:
(1) According to the condition

$$b + ai = (\cos A + \cos 3A + \cos 5A) + i(\sin A + \sin 3A + \sin 5A)$$

$$= (\cos A + i\sin A) + (\cos 3A + i\sin 3A)$$

$$+ (\cos 5A + i\sin 5A)$$

$$= (\cos A + i\sin A) + (\cos A + i\sin A)^3 + (\cos A + i\sin A)^5$$

$$= (\cos A + i \sin A)^3 + [(\cos A + i \sin A)^{-2}$$
$$+ 1 + (\cos A + i \sin A)]$$
$$= (\cos 3A + i \sin 3A) + (1 + 2\cos 2A). \qquad (*)$$

$\because b \neq 0.$

$\therefore$ Condition by the plural,

$$\frac{a}{b} = \frac{[(1 + 2\cos 2A)\sin 3A]}{[(1 + 2\cos 2A)\cos 2A]}, \quad \therefore \ \tan 3A = \frac{a}{b}.$$

(2) Take modulus on the both sides of $(*)$, we get $\sqrt{a^2 + b^2} = |(1 + 2\cos 2A)|$.

On both sides of the upper type, the proof is complete.

13. For given $n \in N^*$, prove $\sin 1 + \sin 2 + \sin 3 + \cdots + \sin n \leq \frac{1}{\sin \frac{1}{2}}$.

Proof: Suppose $z = \cos 1 + i \sin 1$,

$$\therefore |\sin 1 + \sin 2 + \sin 3 + \cdots + \sin n|$$
$$\leq |(\cos 1 + i \sin 1) + (\cos 2 + i \sin 2) + (\cos 3 + i \sin 3)$$
$$+ \cdots + (\cos n + i \sin n)|$$
$$= |(\cos 1 + i \sin 1) + (\cos 1 + i \sin 1)^2 + (\cos 1 + i \sin 1)^3$$
$$+ \cdots + (\cos 1 + i \sin 1)^n|$$
$$= |z + z^2 + z^3 + \cdots + z^n|$$
$$= \left| z \cdot \frac{1 - z^n}{1 + z} \right|$$
$$= \frac{|1 - \cos n - i \sin n|}{2 \sin \frac{1}{2}} \leq \frac{1 + |\cos n - i \sin n|}{2 \sin \frac{1}{2}} = \frac{1}{\sin \frac{1}{2}}.$$

14. In the complex plane, complex number z_1 moves on segment connecting $1 + i$ and $1 - i$. Complex number z_2 moves on the circle, where the center is origin and 1 is the radius.

Find: (1) the locus of z_1^2; (2) are of the moving rang of $z_1 \cdot z_2$; (3) area of the moving range of $z_1 + z_2$.

Solution: Suppose $z_1 = 1 + ti(-1 \le t \le 1)$, $z_2 = \cos\theta + i\sin\theta\,(0 \le \theta < 2\pi)$; then $z_1^2 = (1 + ti)^2 = (1 - t^2) + 2ti$.

(1) If let $z_1^2 = x + yi(x, y \in R)$,

$$\therefore \begin{cases} x = 1 - t^2, \\ y = 2t, \end{cases} \qquad \therefore y^2 = -4(x - 1)\,(0 \le x \le 1).$$

(2) Because $z_1 \cdot z_2 = (1 + ti)(\cos\theta + i\sin\theta)$

$$= (\cos\theta - t\sin\theta) + i(\sin\theta + t\cos\theta).$$

If let $z_1 \cdot z_2 = x + yi\,(x, y \in R)$,

$$\therefore \begin{cases} x = \cos\theta - t\sin\theta, \\ y = \sin\theta + t\cos\theta, \end{cases} \qquad \therefore x^2 + y^2 = 1 + t^2.$$

$\therefore$ The image of $z_1 \cdot z_2$ is the area of concentric circle

where the center is origin and radius is $\sqrt{1 + t^2}$.

$\because |t| \le 1$, $\therefore r_{\min} = 1$, $r_{\max} = \sqrt{2}$.

$\therefore S = 2\pi - \pi = \pi$.

$\therefore$ The area of $z_1 \cdot z_2$ moving range is π.

(3) $\because z_1 + z_2 = 1 + \cos\theta + i(t + \sin\theta)$,

If let $z_1 + z_2 = x + yi\,(x, y \in R)$,

then

$$\begin{cases} x = 1 + \cos\theta, \\ y = t + \sin\theta, \end{cases} \text{ therefore } (x - 1)^2 + (y - t)^2 = 1.$$

$\therefore$ The image of $z_1 + z_2$ is the area of the concentric circle,

where center is $(1, t)$ and radius is 1.

$\because |t| \le 1$.

$\therefore$ The area is $4 + \pi$.

15. If complex numbers $z_1, z_2, \ldots, z_n$ are geometric progression, where $|z_1| \ne 1$ and $q\,(|q| = 1$ and $q \ne \pm 1)$ is the common ratio.

$\omega_1, \omega_2, \ldots, \omega_n$ meet: $\omega_k = z_k + \frac{1}{z_k} + h$, where $k = 0, 1, \ldots, n, h$ is known complex number, prove: points $P_1, P_2, \ldots, P_n$ corresponding to $\omega_1, \omega_2, \ldots, \omega_n$ are all on the ellipse where focal length is 4.

Solution: Suppose

$$z_1 = r(\cos\alpha + i\sin\alpha), \quad q = \cos\theta + i\sin\theta, \quad \text{and } r \neq 1, \theta \neq k\pi$$

$$\therefore z_k = z_1 \cdot q^{k-1} = r[\cos[\alpha + (k-1)\theta] + i\sin[\alpha + (k-1)\theta],$$

$$\therefore \omega_k - h = z_k + \frac{1}{z_k}$$

$$= \left(r + \frac{1}{r}\right)\cos[\alpha + (k-1)\theta] + i\left(r - \frac{1}{r}\right)\sin[\alpha + (k-1)\theta].$$

Let $\omega_k = x + y_i$ $(x, y \in R)$; according to the conditions of the plural:

$$\begin{cases} x - h = \left(r + \frac{1}{r}\right)\cos[\alpha + (k-1)\theta], \\ y = \left(r - \frac{1}{r}\right)\sin[\alpha + (k-1)\theta]. \end{cases}$$

Eliminate α, θ we can get $\dfrac{(x-h)^2}{\left(r + \frac{1}{r}\right)^2} + \dfrac{y^2}{\left(r - \frac{1}{r}\right)^2} = 1.$

Explain the point P_i $(i = 1, 2, \ldots, n)$ both on the same ellipse. The focal length is

$$2\sqrt{\left(r + \frac{1}{r}\right)^2 - \left(r - \frac{1}{r}\right)^2} = 4.$$

Division and Square of Complex Number

1. Division law

If $z_1 = r_1(\cos\theta_1 + i\sin\theta_1)$, $z_2 = r_2(\cos\theta_2 + i\sin\theta_2)$, then $\frac{z_1}{z_2} = \frac{r_1}{r_2}[\cos(\theta_1 - \theta_2) + i\sin(\theta_1 - \theta_2)]$.

2. Geometrical meaning

$\overrightarrow{OP_1}, \overrightarrow{OP_2}$ are two vectors corresponding to $z_1 = r_1(\cos\theta_1 + i\sin\theta_1)$ and $z_2 = r_2(\cos\theta_2 + i\sin\theta_2)$, $\overrightarrow{OP_1}$ rotates θ_2, in clockwise direction, modulus is $\frac{1}{r_2}$ times to the original, get $\overrightarrow{OP_1}$, which expresses $\frac{z_1}{z_2}$.

3. Square of complex number

If $z = r(\cos\theta + i\sin\theta)$, then $\sqrt[n]{z} = \sqrt[n]{r}(\cos\frac{2k\pi+\theta}{n} + i\sin\frac{2k\pi+\theta}{n})$ ($k = 0, 1, 2, \ldots, n-1$).

Exercise

16. If $z_1 = \cos\frac{2\pi}{3} + i\sin\frac{2\pi}{3}$, $z_2 = \cos\frac{11\pi}{6} + i\sin\frac{11\pi}{6}$, find $\arg\left(\frac{2z_1^2}{z_2}\right)$.

Solution:

$$\frac{2z_1^2}{z_2} = \frac{2\left(\cos\frac{4}{3}\pi + i\sin\frac{4\pi}{3}\right)}{\cos\frac{11\pi}{6} + i\sin\frac{11\pi}{6}}$$

$$= 2\left[\cos\left(\frac{4\pi}{3} - \frac{11\pi}{6}\right) + i\sin\left(\frac{4\pi}{3} - \frac{11\pi}{6}\right)\right]$$

$$= 2\left[\cos\left(\frac{\pi}{2}\right) + i\sin\left(-\frac{\pi}{2}\right)\right] = 2\left(\cos\frac{3\pi}{2} + i\sin\frac{3\pi}{2}\right)$$

$$\therefore \arg\left(\frac{2z_1^2}{z_2}\right) = \frac{3\pi}{2}.$$

17. If $z_1, z_2 \in C$, $|z_1| = 1$, $|z_2| = 4$, $z_1 - z_2 = 1 - 2\sqrt{3}i$, find $\frac{z_1}{z_2}$.

Proof: Suppose $z_1 = \cos\alpha + i\sin\alpha$, $z_2 = 4(\cos\beta + i\sin\beta)$

$\because z_1 - z_2 = 1 - 2\sqrt{3}i$,

$$\therefore \begin{cases} \cos\alpha - 4\cos\beta = 1, & \text{①} \\ \sin\alpha - 4\sin\beta = -2\sqrt{3}, & \text{②} \end{cases}$$

Now $①^2 + ②^2 \Rightarrow 1 + 16 - 8\cos(\alpha - \beta) = 13$, $\cos(\alpha - \beta) = \frac{1}{2}$;

$$\therefore \sin(\alpha - \beta) = \pm\frac{\sqrt{3}}{2}.$$

$$\therefore \frac{z_1}{z_2} = \left|\frac{z_1}{z_2}\right| \cdot [\cos(\alpha - \beta) + i\sin(\alpha - \beta)]$$

$$= \frac{1}{4}\left(\frac{1}{2} \pm \frac{\sqrt{3}}{2}i\right) = \frac{1}{8} \pm \frac{1}{8}\sqrt{3}i.$$

□

18. If $z = \cos\theta + i\sin\theta(0 < \theta < \pi)$, $\omega = \frac{1-(\bar{z})^4}{1+z^4}$, and $|\omega| = \frac{\sqrt{3}}{3}$, $\arg\omega < \frac{\pi}{2}$, find θ.

Solution:

$$\because \ w = \frac{1 - [\cos(-\theta) + i\sin(-\theta)]^4}{1 + (\cos\theta + i\sin\theta)^4} = \frac{1 - \cos(-4\theta) - i\sin(-4\theta)}{1 + \cos 4\theta + i\sin 4\theta}$$

$$= \frac{2\sin^2 2\theta + 2\sin 2\theta \cos 2\theta \cdot i}{2\cos^2 2\theta - 2\sin 2\theta \cos 2\theta \cdot i} = \tan 2\theta \cdot (\sin 4\theta + i\cos 4\theta).$$

$$\because \ |w| = |\tan 2\theta| = \frac{\sqrt{3}}{3}, \ \therefore \ \tan 2\theta = \pm\frac{\sqrt{3}}{3}.$$

$$\because \ 0 < \theta < \pi, \therefore \ \text{when } \tan 2\theta = \frac{\sqrt{3}}{3}, \ \therefore \ \theta = \frac{\pi}{12} \text{ or } \theta = \frac{7}{12}\pi,$$

$$\therefore \ w = \frac{\sqrt{3}}{3}\left(\sin\frac{\pi}{3} + i\cos\frac{\pi}{3}\right) = \frac{\sqrt{3}}{3}\left(\cos\frac{\pi}{6} + i\sin\frac{\pi}{6}\right),$$

$$\therefore \arg w = \frac{\pi}{6} < \frac{\pi}{2}, \ \therefore \ \theta = \frac{\pi}{12} \text{ or } \theta = \frac{7\pi}{12}.$$

When $\tan 2\theta = -\frac{\sqrt{3}}{3}$, $\therefore \ \theta = \frac{5}{12}\pi$ or $\theta = \frac{11}{12}\pi$, where

$$w = -\frac{\sqrt{3}}{3}\left(\sin\frac{5}{3}\pi - i\cos\frac{5}{3}\pi\right) = \frac{\sqrt{3}}{3}\left(\cos\frac{11\pi}{6} + i\sin\frac{11\pi}{6}\right)$$

$$\therefore \ \arg w = \frac{11}{6}\pi > \frac{1}{2}\pi.$$

$$\therefore \theta = \frac{5}{12}\pi \text{ or } \theta = \frac{11}{12}\pi \quad \text{should be rejection,}$$

$$\therefore \theta = \frac{\pi}{12} \text{ or } \frac{7}{12}\pi.$$

19. If $z = 1 + i$, find $\sqrt[n]{z}$.

Solution:

$$\because z = 1 + i = \sqrt{2}\left(\cos\frac{\pi}{4} + i\sin\frac{\pi}{4}\right),$$

$$\therefore \sqrt[n]{z} = \sqrt[2n]{2}\left[\cos\frac{2k\pi + \frac{\pi}{4}}{n} + i\sin\frac{2k\pi + \frac{\pi}{4}}{n}\right]$$

$$(k = 0, 1, 2, \ldots, n-1).$$

Exponent Form of Complex Number

1. Triangle form of complex number can also be expressed by exponent form of complex number: $z = re^{i\theta}$. In the multiplication and division of complex number, we have certain advantages.

2. Euler formula: $e^{i\theta} = \cos\theta + i\sin\theta$.

3. Some solutions

If $z = \cos\theta + i\sin\theta = e^{i\theta}$, $n \in N^*$, then

$$\cos n\theta = \text{Re}(z^n) = \frac{z^{2n} + 1}{2z^n};$$

$$\sin n\theta = \text{Im}(z^n) = \frac{z^{2n} - 1}{2z^n i};$$

$$\tan n\theta = \frac{z^{2n} - 1}{(z^{2n} + 1)i};$$

$$1 - z = -2i\sin\frac{\theta}{2}\left(\cos\frac{\theta}{2} + i\sin\frac{\theta}{2}\right) = -2i\sin\frac{\theta}{2}\cdot e^{i\cdot\frac{\theta}{2}};$$

$$1 + z = 2\cos\frac{\theta}{2}\left(\cos\frac{\theta}{2} + i\sin\frac{\theta}{2}\right) = 2\cos\frac{\theta}{2}\cdot e^{i\cdot\frac{\theta}{2}}.$$

Exercise

20. If $1, z_1, z_2, \ldots, z_{n-1}$ are n roots of equation $z^n = 1 \, (n \geq 2, n \in N)$, find $\frac{1}{1-z_1} + \frac{1}{1-z_2} + \cdots + \frac{1}{1-z_{n-1}}$.

Analysis:

(1) According to the condition: $z_k = \cos \frac{2k\pi}{n} + i \sin \frac{2k\pi}{n}$ $(k = 1, 2, \ldots, n-1)$.

(2) Using complex division to simplify: $\frac{1}{1-z_k}$ $(k = 1, 2, \ldots, n-1)$.

(3) The solution of this case can be got by using the triangle knowledge:

Solution: $\because z_1, z_2, \ldots z_{n-1}$ are roots of equation $z^n = 1$ $(n \geq 2, n \in N^*)$.

$$\therefore z_k = \cos \frac{2k\pi}{n} + i \sin \frac{2k\pi}{n} \quad (k = 1, 2, \ldots, n-1).$$

$$\frac{1}{1-z_k} = \frac{1}{(1 - \cos \frac{2k\pi}{n}) - i \sin \frac{2k\pi}{n}}$$

$$= \frac{1}{2 \sin \frac{k\pi}{n} (\sin \frac{k\pi}{n} - i \cos \frac{k\pi}{n})}$$

$$= \frac{\sin \frac{k\pi}{n} + i \cos \frac{k\pi}{n}}{2 \sin \frac{k\pi}{n}} \quad (k = 1, 2, \ldots, n-1).$$

$$\therefore \frac{1}{1-z_1} + \frac{1}{1-z_2} + \cdots + \frac{1}{1-z_{n-1}}$$

$$= \sum_{k=1}^{n-1} \left(\frac{1}{2} + \frac{1}{2} \text{ctg} \frac{k\pi}{n} \cdot i \right).$$

Whether n is odd or even, it has

$$\sum_{i=1}^{n=1} \text{ctg} \frac{k\pi}{n} = 0,$$

$$\therefore \frac{1}{1-z_1} + \frac{1}{1-z_2} + \cdots + \frac{1}{1-z_{n-1}} = \frac{n-1}{2}.$$

21. Prove: $\cos^n \theta = \frac{1}{2^n} \sum_{k=0}^{n} C_n^k \cos(n-2k)\theta, n \in N^*$.

Analysis: According to exponent form of complex number: $\cos^n \theta = \left(\frac{e^{i\theta} + e^{-i\theta}}{2} \right)^n$. Expand type by using the binomial theorem, and then get right type.

Proof: According to exponent form of complex number and binomial theorem:

$$\cos^n \theta = \left(\frac{e^{i\theta} + e^{-i\theta}}{2} \right)^n$$

$$= \frac{1}{2^n} \sum_{k=0}^{n} C_n^k (e^{i\theta})^{n-k} (e^{-i\theta})^k$$

$$= \frac{1}{2^n} \sum_{k=0}^{n} C_n^k e^{i(n-2k)\theta}$$

$$= \frac{1}{2^n} \sum_{k=0}^{n} C_n^k \cos(n-2k)\theta.$$

$\square$

Comment: In the final step, the establishment is due to Euler's formula. Sum is real numbers, so each item remains real parts. Because $C_n^k = C_n^{n-k}$ and $\sin(-x) = -\sin x$, sum of imaginary parts are zero.

When $n = 2, 3, 4$, this identity can be simplified to $\cos^2 \theta = \frac{1+\cos 2\theta}{2}$, $\cos^3 \theta = \frac{3\cos\theta + \cos 3\theta}{4}$, $\cos^4 \theta = \frac{3+4\cos 2\theta + \cos 4\theta}{2}$ by using $\cos(-x) = \cos x$. These formulas are of great use.

22. If n is a prime number and more than 3, find $(1 + 2\cos \frac{2\pi}{n})(1 + 2\cos \frac{4\pi}{n})(1 + 2\cos \frac{6\pi}{n}) \ldots (1 + 2\cos \frac{2k\pi}{n})$.

Solution: Suppose

$$w = e^{\frac{2\pi i}{n}}, \; \therefore \; w^\pi = 1, \; w^{-\frac{n}{2}} = e^{-\pi}$$

$$= -1, \; 2\cos \frac{2k\pi}{n} = w^k + w^{-k}.$$

$$\therefore \prod_{k=1}^{\pi}\left(1+2\cos\frac{2k\pi}{n}\right)=\prod_{k=1}^{T}(1-w^k+w^k)$$

$$=\prod_{k=1}^{n}w^{-k}(w^{2k}+w^k+1)$$

$$=w^{\frac{\pi(n-1)}{2}}\cdot 3\prod_{k=1}^{n-1}\frac{1-w^{3k}}{1-w^k}=(-1)^{n+}\cdot 3\prod_{k=1}^{n+1}\frac{1-w^{3k}}{1-w^k}.$$

Because n is more than 3 of prime numbers,

$\therefore (-1)^{n-1}=1$, and $3, 3\times 2, \ldots, 3(n-1)$ was n except. The remainder solutions are not the same

$$\therefore \prod_{k=1}^{n-1}(1-w^{3k})=\prod_{k=1}^{n-1}(1-w^k), \therefore \prod_{k=1}^{n}\left(1+2\cos\frac{2k\pi}{n}\right)=3.$$

23. n is a positive integer, $a_j(j=1,2,\ldots,n)$ is complex number, and there are $|\prod_{j\in I}(1+a_j)-1|\leq\frac{1}{2}$ to any non-empty set I of assemblage $\{1,2,\ldots,n\}$. Prove: $\sum_{j=1}^{n}|a_j|\leq 3$.

Proof: Suppose $1+a_j=r_je^{i\theta_j}$, $r_j\geq 0, |\theta_j|\leq\pi, j=1,2,\ldots,n$;

$$\therefore \left|\prod_{j\in I}r_j\cdot e^{i\sum_{j\in I}\theta_j}-1\right|\leq\frac{1}{2} \tag{i}$$

suppose r, θ are real numbers, $r\geq 0, |\theta\leq\pi|, |re^{i\theta}-1|\leq\frac{1}{2}$;

$$\therefore \frac{1}{2}\leq r\leq\frac{3}{2}, |\theta|\leq\frac{\pi}{6}, |re^{i\theta}-1|\leq|r-1|+|\theta|.$$

From the geometrical meaning of complex number:

$$\frac{1}{2}\leq r\leq\frac{3}{2}, |\theta|\leq\frac{\pi}{6}$$

$$\because |re^{i\theta}-1|=|r(\cos\theta+\sin\theta)-1|$$

$$=|(r-1)(\cos\theta+i\sin\theta)+((\cos\theta-1)+i\sin\theta)|$$

$$\leq |r - 1| + \sqrt{(\cos\theta - 1)^2 + \sin\theta}$$

$$= |r - 1| + \sqrt{2(1 - \cos\theta)}$$

$$= |r - 1| + 2\left|\sin\frac{\theta}{2}\right| \leq |r - 1| + |\theta|.$$

Using mathematical induction to the assemblage I , from (i) and lemma, we have

$$\therefore \quad \frac{1}{2} \leq \prod_{j\in I} r_j \leq \frac{3}{2}, \quad \left|\sum_{j=I}\theta_j\right| \leq \frac{\pi}{6}, \tag{ii}$$

$$|a_j| = |r_j e^{i\theta_j} - 1| \leq |r - 1| + |\theta_j|,$$

$$\therefore \quad \sum_{j=1}^{n}|a_j| \leq \sum_{j=1}^{n}|r_j - 1| + \sum_{j=1}^{n}|\theta_j|$$

$$= \sum_{r_j\geq 1}|r_j - 1| + \sum_{r_j<1}|r_j - 1| + \sum_{\theta_j\geq 0}|\theta_j| + \sum_{\theta_j<0}|\theta_j|.$$

From (ii),

$$\sum_{r_j\geq 1}|r_j - 1| = \sum_{r_j\geq 1}(r_j - 1) \leq \prod_{r_j\geq 1}(1 + (r_j - 1)) - 1$$

$$= \frac{3}{2} - 1 = \frac{1}{2},$$

$$\sum_{r_j<1}|r_j - 1| = \sum_{r_j<1}(1 - r_j) \leq \prod_{r_j<1}(1 - (1 - r_j))^{-1} - 1$$

$$\leq 2 - 1 = 1,$$

$$\sum_{\theta_j\geq 0}|\theta_j| + \sum_{\theta_j<0}|\theta_j| = \sum_{\theta_j\geq 0}\theta_j - \sum_{\theta_j<0}\theta_j \leq \frac{\pi}{6} - \left(-\frac{\pi}{6}\right) = \frac{\pi}{3},$$

$$\therefore \quad \sum_{j=1}^{n}|a_j| \leq \frac{1}{2} + 1 + \frac{\pi}{3} < 3.$$

$\square$

Exercise 9

1. If $|z^2 + \frac{1}{z^2}| = 1$, find the range of $\arg z$.

2. In complex plane, A and B are vertexes of $\triangle OAB$, and its corresponding complex numbers are z_1, z_2 (O is origin). If $z_1 - (1+i)z_2 = 0$ and $|z_2 - 2 - 2i| = 2$, find the maximum and minimum of area of $\triangle OAB$.

3. If $|z| = 1$, find the maximum of $u = \frac{(z+4)^2 - (\bar{z}+4)^2}{4i}$.

4. One of the vertex of the quadrilateral $OABC$ is at the origin. Point A moves on $y = x + 2(|x| \leq 1)$, and C moves on the unit circle. Find the area of moving area of B.

5. The first item of geometric progression $\{z_n\}$ is 48, common ratio is $\frac{\sqrt{6}}{4} + \frac{\sqrt{2}}{4}i$. $\{a_n\}$ is the sequence of real parts of its original sequence, and it do not change the original order. Find $\sum_{k=1}^{+\infty} a_k$.

6. α, β, γ are arguments of z_1, z_2, z_3. If $|z_1| = 1$, $|z_2| = k$, $|z_3| = 2 - k$, and $z_1 + z_2 + z_3 = 0$, find the range of $\cos(\beta - \gamma)$.

7. ϑ_1, ϑ_2 are arguments of non-zero complex number z_1, z_2. If $z_1 + z_2 = 5i$, $|z_1 z_2| = 14$, find the minimum of $\cos(\vartheta_1 - \vartheta_2)$ and its z_1, z_2.

8. If $\alpha^{2012} + \beta^{2012}$ expresses the binary polynomial about $\alpha + \beta, \alpha\beta$, find the sum of coefficient of this polynomial.

9. If complex sequence $a_n = (1+i)(1 + \frac{i}{\sqrt{2}})(1 + \frac{i}{\sqrt{3}}) \cdots (1 + \frac{i}{\sqrt{n}})$, find $|a_n - a_{n+1}|$.

10. If $|z| = 1$, $w = z^4 - z^3 - 3z^2 i - z + 1$, find z where $|w|$ is the maximum.

11. If $|z - z_0| = \sqrt{2}$, $\arg z_1 = \frac{5}{12}\pi$, $z_0 - (1+i)z_1 = 0$, (1) find z_1 and z_0; (2) prove: if $|z - z_0| = \sqrt{2}$, the argument of z_1 is minimum.

12. Difference of two roots of equation $z^2 - (2\cos\theta + i\sin^2\theta)z + i\cos\theta(1 - \cos 2\theta) = 0$ ($\theta \in R$) is $b + ci$ ($b, c \in R$), find the locus of focus of parabola $y = -x^2 + bx + c$.

13. For $x \in R$, find the range of $y = \sqrt{x^2 + x + 1} - \sqrt{x^2 - x + 1}$.

14. If $x, y, A, B \in r$, $A = x\cos^2\theta + y\sin^2\theta$, $B = x\sin^2\theta + y\cos^2\theta$, prove: $x^2 + y^2 \geq A^2 + B^2$.

15. For $n \in N$, S_n is the minimum of $\sum_{k=1}^{n} \sqrt{(2k-1)^2 + a_k^2}$ ($a_1, a_2, \ldots, a_n \in R^+$), and its sum is 17. If only one n makes S_n an integer number, find n.

16. In an acute triangle ABC, if $\tan A = m$, $\tan B = m - 2$, prove: $m > \sqrt{2} + 1$.

17. Complex numbers z_1, z_2, z_3, z_4, z_5 meet:

$$\begin{cases} |z_1| \leq 1, \\ |z_2| \leq 1, \\ |2z_3 - (z_1 + z_2)| \leq |z_1 - z_2|, \\ |2z_4 - (z_1 + z_2)| \leq |z_1 - z_2|, \\ |2z_5 - (z_3 + z_4)| \leq |z_3 - z_4|. \end{cases}$$

Find the maximum of $|z_5|$.

Chapter 10

Complex Number and Equation

In field F, $f(x) = 0$ ① is called equation of nth degree with one unknown, where $f(x)$ is a polynomial of nth degree with one unknown in field F. If the number α in field F meets ①, α is root in field F.

Roots of Complex Coefficient Polynomial of Nth Degree with One Unknown

In 1799, German mathematician Gauss proved a very important theorem in polynomial theory.

Theorem 1 (Fundamental Theorem of Algebra): Every non-constant single-variable polynomial with complex coefficients has at least one complex root.

Deduction 1: Every polynomial $f(x)$ of degree $n \geq 1$ has exactly n complex roots where k multiple roots have k roots.

Deduction 2 (Unique Decomposition Theorem): If do not consider the order of the factors, $f(x)$ with complex coefficients is uniquely decomposed into the form of

$$f(x) = A(x - \alpha_1)^{m_1}(x - \alpha_2)^{m_2} \cdots (x - \alpha_s)^{m_s},$$

where α_i $(i = 1, 2, \ldots, s)$ are the different complex roots of $f(x)$ and m_j $(j = 1, 2, \ldots, s)$ are its multiple roots, $\sum_{j=1}^{s} m_j = n$.

The root of a polynomial with real coefficients:

To every polynomial of degree $n \geq 1$ with real coefficients, according to fundamental theorem of algebra, it has at most n complex roots.

Theorem 2: If imaginary number α is a root of a real polynomial $f(x)$, then its complex conjugate is also a root.

Binomial equation: Like the equation $a_n x^n + a_0 = 0$ $(a_0, a_n \in C, a_n \neq 0)$ is called binomial equation. Every binomial equation can be translated into the form $x^n = b$ $(b \in C)$, which can be solved by extracting the complex number.

The geometric meaning of nth complex root:
If $x^n = b$ $(b \in C)$, it is called the nth root of complex number b. This nth root corresponds to n points in complex plane and these points are evenly distributed in the origin as the center, with $\sqrt[n]{|b|}$ as the radius of the circle.

Viete Theorem:

Theorem 3: Suppose that $x_1, x_2, \ldots, x_n$ are n roots of equation $a_n x^n + a_{n-1} x^{n-1} + \cdots + a_1 x + a_0 = 0$ $(a_n \neq 0)$, so

$$
\begin{cases}
\displaystyle\sum_{i-1}^{n} x_i = -\frac{a_{n-1}}{a_n} \quad \text{(sum of every root)} \\[4mm]
\displaystyle\sum_{1 \leq i \leq j \leq n} x_i x_j = \frac{a_{n-1}}{a_n} \quad \text{(sum of the product of every two roots)} \\[4mm]
\displaystyle\sum_{1 \leq i \leq j \leq k \leq n} x_i x_j x_k = -\frac{a_{n-3}}{a_n} \quad \text{(sum of the product of every} \\[4mm]
\hspace{8cm} \text{three roots)} \\[2mm]
\vdots \\[2mm]
\displaystyle\prod_{i-1}^{n} x_i = (-1)^n \frac{a_0}{a_n} \quad \text{(product of every roots).}
\end{cases}
$$

Specially, a quadratic equation with real coefficients is $ax^2 + bx + c = 0$ $(a, b, c \in R, a \neq 0)$, so

$$
\begin{cases}
x_1 + x_2 = -\dfrac{b}{a}, \\[4mm]
x_1 x_2 = \dfrac{c}{a}.
\end{cases}
$$

Roots of Polynomial with Rational Coefficient or Integral Coefficient:

Theorem 4: If reduced fraction $\frac{r}{s}$ is the root of polynomial with integral coefficient $f(x) = a_n x^n + a_{n-1} x^{n-1} + \cdots + a_1 x + a_0$, then

(1) $r | a_0, s | a_n$;

(2) $f(x) = \left(x - \frac{r}{s}\right) g(x)$, where $g(x)$ is polynomial with integral coefficient.

Discrimination of Equations with Real Coefficient:

1. A quadratic equation with real coefficients is $ax^2 + bx + c = 0$ $(a, b, c \in R, a \neq 0)$, and discriminant is $\Delta = b^2 - 4ac$.

- When $\Delta > 0$, the equation has two different real numbers $x_{1,2} = \frac{-b \pm \sqrt{b^2 - 4ac}}{2a}$.
- When $\Delta = 0$, the equation has two same real numbers $x_{1,2} = -\frac{b}{2a}$.
- When $\Delta < 0$, the equation has a pair of conjugate imaginary roots $x_{1,2} = \frac{b}{2a} \pm \frac{\sqrt{4ac - b^2}}{2a} i$.

2. Discriminant of higher order equation with real coefficients:

Theorem 5: Suppose $x_1, x_2, \ldots, x_n$ are roots of higher order equation with real coefficients: $a_n x^n + a_{n-1} x^{n-1} + \cdots + a_1 x + a_0 = 0$ $(n > 1, a_n a_0 \neq 0)$.

$$\Delta_1 = (n-1) a_{n-1}^2 - 2n a_{n-2} a_n;$$

$$\Delta_2 = (n-1) a_1^2 - 2n a_2 a_0.$$

$\therefore$ (1) When $x_1, x_2, \ldots, x_n$ are real numbers, $\Delta_1 \geq 0$, and $\Delta_2 \geq 0$.

(2) When $\Delta_1 < 0$ or $\Delta_2 < 0$, $x_1, x_2, \ldots, x_n$ are not all real numbers.

Proof: According to Viete theorem:

$$\sum_{i=1}^{n} x_i = -\frac{a_{n-1}}{a_n}, \qquad \sum_{1\leq i<j\leq n} x_i x_j = \frac{a_{n-2}}{a_n}$$

$$I_1 = \sum_{1\leq i\leq j\leq n} (x_i - x_j)^2$$

$$= \sum_{1\leq i\leq j\leq n} (x_i^2 + x_i^2 - 2x_i x_j)$$

$$= \sum_{1\leq i\leq j\leq n} (x_i^2 + x_j^2) - 2\sum_{1\leq i<j\leq n} x_i x_j$$

$$= (n-1)\sum_{i=1}^{n} x_i^2 - 2\cdot \sum_{1\leq i<j\leq n} x_i x_j$$

$$= (n-1)\left[\left(\sum_{i=1}^{n} x_i\right)^2 - 2\sum_{1\leq i<j\leq n} x_i x_j\right] - 2\sum_{1\leq i<j\leq n} x_i x_j$$

$$= (n-1)\left[\left(\sum_{i=1}^{n} x_i\right)^2 - 2n\sum_{1\leq i<j\leq n} x_i x_j\right]$$

$$= (n-1)\left(-\frac{a_{n-1}}{a_n}\right)^2 - 2n\cdot\frac{a_{n-2}}{a_n}$$

$$= \frac{\Delta_1}{a_n^2}$$

$$\therefore \Delta_1 = I_1 \cdot a_n^2.$$

Let $y = \frac{1}{x}$, so original equations are translated into $a_0 y^n + a_1 y^{n-1} + a_2 y^{n-2} + \cdots + a_{n-1}y + a_n = 0$, and its roots are $\frac{1}{x_1}, \frac{1}{x_2}, \ldots, \frac{1}{x_n}$.

$$I_2 = \sum_{1\leq i<j\leq n} \left(\frac{1}{x_i} - \frac{1}{x_j}\right)^2 = \frac{\Delta_2}{a_0^2}.$$

Similarly,

$$\therefore \Delta_2 = a_0^2 \cdot I_2.$$

∴ When $x_1, x_2, \ldots, x_n$ are real numbers, $\Delta_1 = a_0^2 \cdot I_1 \geq 0$ and $\Delta_2 = a_0^2 \cdot I_2 \geq 0$.

Conclusion (1) holds.

∴ Conclusions (1) and (2) are each converse-negative proposition.

∴ Conclusion (2) holds.

Criterion of Divisibility of Polynomial:

Theorem 6: If n roots of polynomial $f(x)$ are different, and these roots are roots of another polynomial $g(x)$, then $f(x)|g(x)$.

Lagrange's Interpolation Formula:

In complex number sets, every polynomial of degree m $(m \leq n)$ can be only represented as follows:

$$f(x) = \frac{(x - x_1)(x - x_2)\cdots(x - x_n)}{(x_0 - x_1)(x_0 - x_2)\cdots(x_0 - x_n)} f(x_0)$$

$$+ \frac{(x - x_0)(x - x_2)\cdots(x - x_n)}{(x_1 - x_1)(x_1 - x_2)\cdots(x_1 - x_n)} f(x_1)$$

$$+ \cdots + \frac{(x - x_0)(x - x_1)\cdots(x - x_{n-1})}{(x_n - x_1)(x_n - x_2)\cdots(x_n - x_{n-1})} f(x_n).$$

Namely,

$$f(x) = \sum_{i=1}^{n} \left(f(x_i) \cdot \prod_{\substack{0 \leq j \leq n \\ j \neq i}} \frac{x - x_j}{x_i - x_j} \right).$$

Property of n-order Unit Root of 1:

Property 1: The n-order unit root of 1 has n roots, which are $\varepsilon_k = \cos\frac{2k\pi}{n} + i\sin\frac{2k\pi}{n} = e^{\frac{2k\pi}{n}i}$ $(k = 0, 1, 2, \ldots, n - 1)$.

Property 2: $(\varepsilon_k)^n = 1, \varepsilon_k = (\varepsilon_1)^n, |\varepsilon_k| = 1$ $(k = 0, 1, 2, \ldots, n - 1)$.

Property 3: When n is even number, $\varepsilon_0 = 1$ is only the real root; when n is odd number, $\varepsilon_0 = 1, \varepsilon_{\frac{n}{2}} = 1$ are two real numbers.

Other imaginary roots are pair of pair of conjugates, where ε_k and ε_{n-k} are conjugates and $\varepsilon_k \cdot \varepsilon_{n-k} = 1$ $(k = 0, 1, 2, \ldots, n - 1)$.

Property 4: $\{\varepsilon_k\}$ is closed for multiplication and division, where product of several roots of equation $x^n - 1 = 0$ is also the root of this equation and quotient of two roots (or reciprocal of root) is also the root of this equation.

Property 5:

$$1+\varepsilon_k^p+\varepsilon_k^{2p}+\cdots+\varepsilon_k^{(n-1)p} = \begin{cases} n & \text{when } p \text{ is an integer multiple of} \\ & \quad n \text{ or } k = 0, \\ 0 & \text{when } p \text{ is not an integer multiple of} \\ & \quad n \text{ or } k = 0, \end{cases}$$

Especially, when $1 + \varepsilon_1 + \varepsilon_2 + \cdots + \varepsilon_{n-1} = 0$ or $k \neq 0, 1 + \varepsilon_k + \varepsilon_k^2 + \cdots + \varepsilon_k^{n-1} = 0$.

Property 6: If $p_1, p_2, \ldots, p_m$ are coprime numbers to each other and $n = p_1, p_2, \ldots, p_m$, so n-order unit roots of 1 are product of $P_1 p_1$-order unit root and P_2 p_2-order unit root,$\ldots$, P_m p_m-order unit root.

Property 7:

$$\sum_{k=0}^{n-1} x^k = \prod_{k=1}^{n-1} (x - \varepsilon_k),$$

especially, when $x = 1$, $n = \prod_{k=1}^{n-1}(1 - \varepsilon_k)$.

Property 8: ε_k represents the N equal points of the unit circle in the complex plane (or vertices of inscribed regular n polygon of unit circle), where $\varepsilon_0 = 1$ is the intersect of unit circle and positive real axis.

Some n-order unit root of 1 is called primitive n-order unit root of 1 (primitive root for short). One is not m-order unit root of 1 if and only if $m < n$.

For example, 3-order unit root of 1 is $\omega = -\frac{1}{2} + \frac{\sqrt{3}}{2}i$ and $\omega^2 = -\frac{1}{2} - \frac{\sqrt{3}}{2}i$, 4-order unit root of 1 is i and $-i$.

Property 9: All primitive root of 1 can be found in all unit roots ε_k $(k = 0, 1, 2, \ldots, n - 1)$ where k are all integers which are less than

n and coprime with n. The number of primitive n-order root of 1 is equal or less than the number of these numbers which are coprime with n and note it $\phi(n)$. When p and q are coprime, $\phi(p \cdot q) = \phi(p) \cdot \phi(q)$.

Property 10: All n-order unit root of 1 equal to n successive integer power of its any primitive root; if $n = p_1 p_2 \dots p_m$ and $P_1, P_2, \dots, P_m$ are coprime to each other, all prime n-order unit root equal to product of prime P_1-order unit root, prime P_2-order unit root, $\dots$, prime P_n-order unit root.

For example, prime 12-order unit root of 1 is equal to product of prime 3-order unit roots of 1 which are ω , ω^2 and prime 4-order unit roots of 1 which are i, -1. Its prime roots are $\pm\frac{\sqrt{3}}{2} \pm \frac{1}{2}i$.

Property 11: In n-order unit root of 1, each pair of conjugate prime roots corresponds to inscribed regular n polygon, and vice versa. Namely, the number of inscribed regular n polygon is equal the number of pair of conjugate prime n-order roots of 1, which is $1\frac{1}{2}\phi(n)$. (Inscribed regular n polygon means dividing the circle n parts equally, and, every p (where p and n are coprime and $p < \frac{n}{2}$) points of division connect to regular n polygon.)

The proof of the above is not difficult, as a practice for the reader.

Because of these properties, unit root of 1 in the problem solving (especially the math contest) has a wide range of applications.

Exercise

1. If $z \in C$ and equation $x^2 - zx + 4 + 3i = 0$ has real roots, find the minimum of $|z|$.

   Analysis: Suppose $x_0 = a + bi$ $(a, b \in R)$ is the real root of equation. Consider from the equality of complex number.

Solution: Suppose $x_0 = a + bi$ $(a, b \in R)$ is the real root of equation;

$$\therefore \ x_0^2 - (a + bi)x_0 + 4 + 3i = 0,$$

$$\therefore \ (x_0^2 - ax_0 + 4) + (3 - bx_0)i = 0, \ \therefore \ \begin{cases} x_0^2 - ax_0 + 4 = 0, \\ 3 - bx_0 = 0. \end{cases}$$

$$\because \ x_0 \neq 0, \ \therefore \ a = x_0 + \frac{4}{x_0}, b = \frac{3}{x_0},$$

$$\therefore \ |z| = \sqrt{a^2 + b^2} = \sqrt{\left(x_0 + \frac{4}{x_0}\right)^2 + \left(\frac{3}{x_0}\right)^2}$$

$$= \sqrt{x_0^2 + \frac{25}{x_0^2} + 8} \geq \sqrt{2\sqrt{x_0^2 \cdot \frac{25}{x_0^2}} + 8} = 3\sqrt{2}.$$

If and only if $x_0^2 = \frac{25}{x_0^2}$, namely, $x_0 = \pm\sqrt{5}$, the equality holds.

$\therefore$ the minimum of $|z|$ is $3\sqrt{2}$.

Comment: The imaginary coefficient quadratic equation has real roots. "$\Delta > 0$" cannot be used. We should suppose the real root and consider from the perspective of the equality of complex number.

2. If z_1, z_2 are two imaginary roots of real coefficient quadratic equation, $\omega = \frac{a(\sqrt{3}+i)z_1}{z_2}$ $(a \in R)$, and $|\omega| \leq 2$, find the range of $|(a - 4) + ai|$.

Analysis: Take care that two imaginary roots are conjugate complex numbers in real coefficient quadratic equation.

Solution: $|\omega| = |\frac{a(\sqrt{3}+i)z_1}{z_2}|$, $\because \ z_1 = \overline{z_2}$, $\therefore \ |z_1| = |z_2|$,

$|a\sqrt{3} + i)| \leq 2, |a| \leq 1, \ \therefore \ -1 \leq a \leq 1,$

$(a - 4) + ai| = \sqrt{a^2 - 8a + 16 + a^2} = \sqrt{2(a - 2)^2 + 8},$

$-3 \leq a - 2 \leq -1,$

$(a - 2)^2 \in [1, 9], 2(a - 2)^2 \in [2, 18], 2(a - 2)^2 + 8 \in [10, 26],$

$\therefore \ |(a - 4) + ai| \in [\sqrt{10}, \sqrt{26}].$

3. (1) $1 + 2i$ is imaginary root of equation $x^2 - 2x + k = 0$, find k.

(2) $1 + 2i$ is imaginary root of equation $x^2 - 4x + k = 0$, find k.

Solution: (1) Another root is $1 - 2i$, $\therefore$ $k = (1+2i)(1-2i) = 5$.

(2) $(1 + 2i)^2 - 4(1 + 2i) + k = 0 \Rightarrow k = 7 - 4i$.

Comment: The difference between the former and the latter: the former: $k \in R$, $x^2 - 2x + k = 0$ is an equation with real coefficient; the latter $x^2 - 4x + k = 0$ is not an equation with real coefficient (an equation with complex coefficient). So the latter cannot use imaginary pairs theorem of equation with real coefficients. When solving the problem, see the conditions clearly.

4. (1) Solve $x^2 - 3x + 5m = 0$ $(m \in R)$.

(2) Suppose p is a given even prime number, $k \in R$ and $\sqrt{k^2 - pk} \in R$, find k.

Solution: (1) $\Delta = 9 - 20m$.

- When $\Delta = 9 - 20m > 0$, namely, $m < \dfrac{9}{20}$, $x = \dfrac{3 \pm \sqrt{9 - 20m}}{2}$.

- When $\Delta = 9 - 20m = 0$, namely, $m = \dfrac{9}{20}$, $x = \dfrac{3}{2}$.

- When $\Delta = 9 - 20m < 0$, namely, $m > \dfrac{9}{20}$, $x = \dfrac{3 \pm \sqrt{20m - 9}\,i}{2}$.

(2) Suppose $\sqrt{k^2 - pk} = n, n \in N^*$, $\therefore$ $k^2 - pk - n^2 = 0$, $k = \dfrac{p \pm \sqrt{p^2 + 4n^2}}{2}$, so $p^2 + 4n^2$ is square number. Suppose it $m^2, m \in N^*$, $\therefore$ $(m - 2n)(m + 2n) = p^2$,

$\because$ p is a prime number, and $p \geq 3$,

$$\therefore \begin{cases} m - 2n = 1, \\ m + 2n = p^2, \end{cases} \therefore \begin{cases} m = \dfrac{p^2 + 1}{2}, \\ n = \dfrac{p^2 - 1}{4}, \end{cases}$$

$$\therefore k = \frac{p \pm m}{2} = \frac{2p \pm (p^2 + 1)}{4},$$

$$\therefore k = \frac{(p+1)^2}{4} \quad \text{(Give off negative)}.$$

5. If one of the root of equation with real coefficient $x^2 + kx + k^2 - 3k = 0$ is an imaginary number which modulus is 2, find k.

Analysis: According to imaginary pairs theorem of equation with real coefficients, suppose this pair conjugate imaginary roots is $z, \overline{z}$, so $z \cdot \overline{z} = 2^2 = 4$. The equation can be constructed by Viete theorem.

Solution: $\because$ Original equation has imaginary roots, $\therefore$ $\Delta = k^2 - 4(k^2 - 3k) < 0 \Rightarrow k > 4$ or $k < 0$. Suppose one root of equation is z, so another root is $\overline{z}$, and $z\overline{z} = |z|^2 = 4 \Rightarrow k^2 - 3k = z\overline{z} = 4 \Rightarrow k = 4$ (rejection) or $k = -1$. $\therefore$ When $k = -1$, the equation has an imaginary number which modulus is 2.

6. If x_1, x_2 are two roots of equation with real coefficients $x^2 + x + p = 0$ and $|x_1 - x_2| = 3$, find p.

Solution: $\Delta = 1 - 4p$,

(1) When $\Delta \geq 0$, namely, when $p \leq \frac{1}{4}$, x_1, x_2 are real roots, so $|x_1 - x_2| = \sqrt{(x_1 + x_2)^2 - 4x_1x_2} = 3$, namely, $\sqrt{1 - 4p} = 3 \Rightarrow p = -2$.

(2) When $\Delta < 0$, namely, when $p > \frac{1}{4}$, x_1, x_2 are imaginary roots, suppose $x_1 = a + bi \, (a, b \in R)$, so $x_2 = a - bi$.

$$\therefore |x_1 - x_2| = |2bi| = 2|b| = 3 \Rightarrow b = \pm\frac{3}{2}$$

$$\because x_1 + x_2 = 2a = -1$$

$$\therefore a = -\frac{1}{2}$$

$$\therefore p = x_1 x_2 = |x_1|^2 = \frac{5}{2}$$

$$\therefore p = -2 \text{ or } \frac{5}{2}.$$

Comment: This example does not clearly indicate that the roots are the real or imaginary roots, so must be discussed. The wrong solution: $|x_1 - x_2| = \sqrt{(x_1 + x_2)^2 - 4x_1x_2} = 3$, so $\sqrt{1 - 4p} = 3 \Rightarrow p = -2$.

7. If α, β are two roots of equation with real coefficients, $x^2 - mx + 3 = 0$, find $|\alpha| + |\beta|$.

Analysis: α, β are two roots of equation with real coefficients, $x^2 - mx + 3 = 0$, so we only need to discuss whether two roots are real roots or imaginary conjugate roots. Then according to Viete theorem and modulus of complex numbers, find $|\alpha| + |\beta|$.

Solution: $\Delta = m^2 - 4 \times 3 = m^2 - 12$,

(1) When $\Delta \geq 0$, namely, when $m \geq 2\sqrt{3}$ or $m \leq -2\sqrt{3}$, $\alpha\beta = 3 > 0$, so $|\alpha| + |\beta| = |\alpha + \beta| = |m|$.

(2) When $\Delta < 0$, namely, when $-2\sqrt{3} < m < 2\sqrt{3}$,

$$|\alpha| + |\beta| = 2|\alpha| = 2\sqrt{\alpha\beta} = 2\sqrt{3};$$

$$\therefore \ |\alpha| + |\beta| = \begin{cases} m, & m \geq 2\sqrt{3}, \\ -m, & m \leq -2\sqrt{3}, \\ 2\sqrt{3}, & -2\sqrt{3} < m < 2\sqrt{3}. \end{cases}$$

8. If α, β are two imaginary roots of equation with real coefficients $ax^2 + bx + c = 0$ and $\frac{\alpha^2}{\beta} \in R$, find $\frac{\alpha}{\beta}$.

Solution: $\because \ \alpha, \beta$ are two imaginary roots of equation with real coefficients $ax^2 + bx + c = 0$

$$\therefore \ \overline{\alpha} = \beta, \overline{\beta} = \alpha$$

$$\because \ \frac{\alpha^2}{\beta} \in R$$

$$\therefore \ \overline{\left(\frac{\alpha^2}{\beta}\right)} = \frac{\alpha^2}{\beta}, \text{ namely, } \alpha^3 = \beta^3,$$

$$\text{so } (\alpha - \beta)(\alpha^2 + \alpha\beta + \beta^2) = 0$$

$$\because \ \alpha \neq \beta$$

$$\therefore \ \alpha^2 + \alpha\beta + \beta^2 = 0 \Rightarrow \frac{\alpha}{\beta} = -\frac{1}{2} \pm \frac{\sqrt{3}i}{2}.$$

9. If $z + \frac{5}{z} \in R$, and the real part and imaginary part of $z + 3$ are opposite numbers. Is the imaginary is exist? If exist, find z; if not, give the reason.

Analysis: This is an exploration problem. Firstly, assume the existence, then find it according to the known conditions. If it does not exist, then give the contradiction.

Solution: Assume exist imaginary number $z = a + bi$ $(a, b \in R, b \neq 0)$,

so $\begin{cases} a + bi + \dfrac{5}{a + bi} \in R, \\ a + 3 + b = 0, \end{cases}$ $\therefore \ \begin{cases} b - \dfrac{5b}{a^2 + b^2} = 0, \\ a + b = -3, \\ b \neq 0, \end{cases}$ so $\begin{cases} a = -1 \\ b = -2 \end{cases}$

or $\begin{cases} a = -2, \\ b = -1. \end{cases}$

$\therefore \ z = -1 - 2i$ or $z = -2 - i$.

10. If i is a root of equation $x^3 + 2x + k = 0$, find two roots and k.

Solution: $\because \ i$ is a root of equation

$$\therefore \ i^3 + 2i + k = 0, \text{ so } k = -i.$$

$$\therefore \ \text{The original equation is } x^3 + 2x - i = 0$$

$$\therefore \ (x - i)(x^2 + ix + 1) = 0$$

$$\therefore \ x - i = 0, \text{ namely, } x = i.$$

$$\because \ x^2 + ix + 1 = 0$$

$$\therefore \ x = \frac{-1 \pm \sqrt{5}i}{2}.$$

$$\therefore \ \text{Other two roots are } x = \frac{-1 \pm \sqrt{5}i}{2} \text{ and } k = -i.$$

11. (1) If equation $x^3 \sin\theta - (\sin\theta + 2)x^2 + 6x - 4 = 0$ has three positive roots, find the minimum of

$$u = \frac{9\sin^2\theta - 4\sin\theta + 3}{(1 - \cos\theta)(2\cos\theta - 6\sin\theta - 3\sin 2\theta + 2)}.$$

(2) Suppose a, b, c are roots of equation $x^3 - k_1 x - k_2 = 0$ ($k_1 + k_2 \neq 1$), find $\frac{1+a}{1-a} + \frac{1+b}{1-b} + \frac{1+c}{1-c}$.

(3) Find the number of rational root (x, y, z) of

$$\begin{cases} x + y + z = 0, \\ xyz + z = 0, \\ xy + yz + xz + y = 0. \end{cases}$$

(4) $P(x) = x^5 + a_1 x^4 + a_2 x^3 + a_3 x^2 + a_4 x + a_5$. When $k = 1, 2, 3, 4, P(k) = k \cdot 2007$, find $P(10) - P(-5)$.

(5) Find the number of different non-zero integer solution of equation

$$\frac{(x-1)(x-4)(x-9)}{(x+1)(x+4)(x+9)} + \frac{2}{3}\left(\frac{x^3+1}{(x+1)^3} + \frac{x^3+4^3}{(x+4)^3} + \frac{x^3+9^3}{(x+9)^3}\right) = 1.$$

(6) Solve $x^5 - 11x^4 + 36x^3 - 36x^2 + 11x - 1 = 0$.

(7) Solve $x^6 + x^5 + 2x^4 + x^3 + 2x^2 + x + 1 = 0$.

(8) Solve $x^6 + x^5 + x^3 + x - 1 = 0$.

Solution: (1) $\because$ Equation $(x - 1)(x^2 \sin\theta - 2x^2 + 4) = 0$ has three positive roots.

$\therefore$ Equation $x^2 \sin\theta - 2x^2 + 4 = 0$ has two positive roots.

$$\therefore \begin{cases} \Delta = 4 - 16\sin\theta \geq 0 \\ \sin\theta > 0 \end{cases} \Leftrightarrow 0 < \sin\theta \leq \frac{1}{4},$$

$\because \ 9\sin^2\theta - 4\sin\theta + 3 = 9\left(\sin\theta - \dfrac{2}{9}\right)^2 + \dfrac{23}{9} \geq \dfrac{23}{9},$

$0 < (1 - \cos\theta)(2\cos\theta - 6\sin\theta - 3\sin 2\theta + 2)$

$\qquad = 2(1 - \cos\theta)(1 + \cos\theta)(1 - 3\sin\theta)$

$\qquad = 2\sin^2\theta(1 - 3\sin\theta) = \dfrac{8}{9} \times \dfrac{3}{2}\sin\theta \times \dfrac{3}{2}\sin\theta(1 - 3\sin\theta)$

$\qquad \leq \dfrac{8}{9}\left(\dfrac{1}{3}\right)^3$

$\therefore \ u \geq \dfrac{\frac{23}{9}}{\frac{8}{9\times27}} = \dfrac{621}{8}.$

If and only if $\sin\theta = \frac{2}{9}$, the equality holds. $\therefore \ u \geq \frac{621}{8}$.

(2) $x^3 - k_1 x - k_2 = (x - a)(x - b)(x - c)$

$\therefore \ a + b + c = 0$, $ab + bc + ca = -k_1$, $abc = k_2$ and $1 - k_1 - k_2 = (1 - a)(1 - b)(1 - c),$

$\dfrac{1+a}{1-a} + \dfrac{1+b}{1-b} + \dfrac{1+c}{1-c} = \dfrac{3-(a+b+c) - (ab+bc+ca) + 3abc}{(1-a)(1-b)(1-c)}$

$\qquad\qquad\qquad\qquad = \dfrac{3 + k_1 + 3k_2}{1 - k_1 - k_2}.$

(3) If $z = 0$, so $\begin{cases} x + y = 0, \\ xy + y = 0. \end{cases}$

$\therefore \begin{cases} x = 0, \\ y = 0 \end{cases}$ or $\begin{cases} x = -1, \\ y = 1. \end{cases}$

If $z \neq 0$, $xyz + z = 0$, so $xy = -1$. $\qquad\qquad$ (1)

$\because \ x + y + z = 0$, $\therefore \ z = -x - y$. $\qquad\qquad$ (2)

Substitute (2) into $xy + yz + xz + y = 0$, so

$$x^2 + y^2 + xy - y = 0. \qquad\qquad (3)$$

According to (1): $x = -\frac{1}{y}$. Substitute it into (3):

$(y - 1)(y^3 - y - 1) = 0.$

$\therefore y^3 - y - 1 = 0$ has no rational root, $\therefore y = 1$.

According to (1): $x = -1$. According to (2) $z = 0$ (contradict $z \neq 0$).

$\therefore$ This equation has two rational roots:

$$
\begin{cases} x = 0, \\ y = 0, \\ z = 0 \end{cases} \text{or} \quad \begin{cases} x = -1, \\ y = 1, \\ z = 0. \end{cases}
$$

(4) Let $Q(x) = P(x) - 2007x$, so when $k = 1, 2, 3, 4$, $Q(k) = P(k) - 2007k = 0$.

$\therefore$ 1, 2, 3, 4 are roots of $Q(x) = 0$.

$\because Q(x)$ is 5-order equation, so suppose

$Q(x) = (x - 1)(x - 2)(x - 3)(x - 4)(x - r)$

$P(10) = Q(10) + 2007 \times 10$

$= 9 \times 8 \times 7 \times 6(10 - r) + 2007 \times 10$

$P(-5) = Q(-5) + 2007 \times)(-5)$

$= (-6)(-7)(-8)(-9)(-5 - r) + 2007 \times (-5)$

$\therefore P(10) - P(-5) = 9 \times 8 \times 7 \times 6 \times 15 + 2007 \times 15 = 75465$.

(5) Using $a^3 + b^3 = (a + b)(a^2 - ab + b^2)$, the original equation

$$
\frac{(x - 1)(x - 4)(x - 9)}{(x + 1)(x + 4)(x + 9)} + 1
$$

$$
+ \frac{2}{3} \left(\frac{x^3 + 1}{(x + 1)^3} - 1 + \frac{x^3 + 4^3}{(x + 4)^3} - 1 + \frac{x^3 + 9^3}{(x + 9)^3} - 1 \right) = 0
$$

is equal to

$$
\frac{x^3 + 49x}{(x + 1)(x + 4)(x + 9)} - \left(\frac{x}{(x + 1)^2} + \frac{4x}{(x + 4)^2} + \frac{9x}{(x + 9)^2} \right) = 0.
$$

Divide x by two sides of equation, and after arrangement:

$$x(x^4 - 98x^2 - 288x + 385) = 0$$

Then divide x: $(x^2 - 31)^2 - (6x + 24)^2 = 0$.

Namely, $(x^2 + 6x - 7)(x^2 - 6x - 55) = 0$, so

$(x + 7)(x - 1)(x + 5)(x - 11) = 0$.

After examine: $x_1 = -7, x_2 = 1, x_3 = -5, x_4 = 11$ are roots of original equation.

$\therefore$ Original equation has four integer roots.

(6) $x = 1$ is the root of original equation, so

$(x - 1)(x^4 - 10x^3 + 26x^2 - 10x + 1) = 0$.

Find roots of $x^4 - 10x^3 + 26x^2 - 10x + 1 = 0$ as follows:

Divide x^2 by two sides of original equation:

$$x^2 - 10x + 26 - 10 \cdot \frac{1}{x} + \frac{1}{x^2} = 0,$$

$$\left[\left(x + \frac{1}{x}\right)^2 - 2\right] - 10\left(x + \frac{1}{x}\right) + 26 = 0,$$

$$\left(x + \frac{1}{x}\right)^2 - 10\left(x + \frac{1}{x}\right) + 24 = 0,$$

$$x + \frac{1}{x} = 4 \quad \text{or} \quad x + \frac{1}{x} = 6.$$

$\therefore$ Five roots of original equation are $1, 2 + \sqrt{3}, 2 - \sqrt{3}, 3 + 2\sqrt{2}, 3 - 2\sqrt{2}$.

(7) Left $= x^6 + x^5 + x^4 + x^4 + x^3 + x^2 + x^2 + x + 1$

$\therefore (x^4 + x^2 + 1)(x^2 + x + 1) = 0$

$\therefore x_{1,2} = \dfrac{-1 \pm \sqrt{5}i}{2}, \quad x_{3,4} = \sqrt{\dfrac{-1 \pm \sqrt{5}i}{2}},$

$$x_{5,6} = -\sqrt{\dfrac{-1 \pm \sqrt{5}i}{2}}.$$

(8) Left $= x^6 + x^5 - x^4 + x^4 + x^3 - x^2 + x^2 + x - 1$

$\therefore (x^4 + x^2 + 1)(x^2 + x - 1) = 0$

$$\therefore x_{1,2} = \dfrac{-1 \pm \sqrt{5}}{2}, \quad x_{3,4} = \sqrt{\dfrac{-1 \pm \sqrt{5}i}{2}},$$

$$x_{5,6} = -\sqrt{\dfrac{-1 \pm \sqrt{5}i}{2}}.$$

12. In complex set C, solve $z^n = (\bar{z})^m$ $(m, n \in N)$.
    Solution: (1) When $m \neq n$, suppose $n > m$.

① When $z = 0$, it meets original equation.

$\therefore z = 0$ is solution of original equation.

② When $z \neq 0$, take modulus by both sides.

$\therefore |z|^n = |\bar{z}|^n$, namely, $|z|^{n-m} = 1$.

Substitute $\bar{z} = \dfrac{1}{z}$ into original equation, so $z^{m+n} = 1$.

$\therefore z = \cos\dfrac{2l\pi}{m+n} + i\sin\dfrac{2l\pi}{m+n}$ $(l = 0, 1, 2, \ldots, m+n-1)$.

(2) When $m = n$, we have the following.

① For arbitrary $z \in R$, they meet the original equation.

③ If z is imaginary number, let $z = re^{i\theta}$ $(r > 0)$

and substitute it into original equation.

$\therefore (e^{i\theta})^{2n} = 1$, namely, $e^{i\theta} = \pm e^{i \cdot \frac{k\pi}{n}}$ $(k = 1, 2, \ldots, n-1)$.

$\therefore$ The solution of original equation is

$$z = \pm r \left(\cos \frac{k\pi}{n} + i \sin \frac{k\pi}{n} \right) \quad (k = 1, 2, \ldots, n-1, r > 0).$$

$\therefore$ When $m = n$, $z = \pm r \left(\cos \dfrac{k\pi}{n} + i \sin \dfrac{k\pi}{n} \right)$

$(k = 1, 2, \ldots, n-1, r > 0)$ or $z \in R.$

When $m \neq n$, $z = \cos \dfrac{2l\pi}{m+n} + i \sin \dfrac{2l\pi}{m+n}$

$(l = 0, 1, 2, \ldots, m+n-1)$ or $z = 0.$

13. Suppose $n \in N, 0 < r \in R$. Prove: equation $x^{n+1} + rx^n - r^{n+1} = 0$ has no complex root which modulus is r.

Proof: (reductio ad absurdum): Suppose z is the root where r is the modulus of original equation.

$$z^n(z+r) = r^{n+1} \tag{4}$$

Take modulus by two sides: $|z|^n|z+r| = r^{n+1}$.

$$\because |z| = r$$
$$\therefore |z+r| = r$$
$$\therefore z \text{ meets}$$
$$\begin{cases} |z| = r, \\ |z+r| = r, \end{cases}$$

which is the intersect of circle $|z| = r$ and circle $|z + r| = r$.

$$z_1 = r \left(\cos \frac{2\pi}{3} + i \sin \frac{2\pi}{3} \right), \quad z_2 = r \left(\cos \frac{4\pi}{3} + i \sin \frac{4\pi}{3} \right).$$

Substitute z_1 into (4):

$$\cos \frac{(2n+1)\pi}{3} + i \sin \frac{(2n+1)\pi}{3} = 1.$$

According to equality of complex number:

$$\cos\frac{(2n+1)\pi}{3} = 1, \quad \sin\frac{(2n+1)\pi}{3} = 0$$

$$\therefore \frac{(2n+1)\pi}{3} = 2k\pi \ (k \in Z), \text{ namely,}$$

$2n + 1 = 6k \ (k, n \in Z)$, which is a contradiction.

Substitute z_2 into (4), and similarly, which is a contradiction.

$\therefore z_1$ and z_2 are not roots of original equation.

$\therefore$ The original equation has no complex root which modulus is r.

$\square$

14. In complex set C: solve

$$\begin{cases} \arg(x^2 - 2) = \dfrac{3\pi}{4}, \\ \arg(x^2 + 2\sqrt{3}) = \dfrac{\pi}{6}. \end{cases}$$

Solution: Suppose

$$x^2 - 2 = r_1\left(\cos\frac{3}{4}\pi + i\sin\frac{3}{4}\pi\right), \tag{5}$$

$$x^2 + 2\sqrt{3} = r_1\left(\cos\frac{3}{4}\pi + i\sin\frac{3}{4}\pi\right), \tag{6}$$

Subtracting two equations:

$$2 + 2\sqrt{3} = \left(\frac{\sqrt{3}}{2}r_2 + \frac{\sqrt{2}}{2}r_1\right) + i\left(\frac{1}{2}r_2 - \frac{\sqrt{2}}{2}r_1\right).$$

According to equality of complex numbers:

$$\frac{1}{2}r_2 - \frac{\sqrt{2}}{2}r_1 = 0, \quad \frac{\sqrt{3}}{2}r_2 + \frac{\sqrt{2}}{2}r_1 = 2 + 2\sqrt{3},$$

$$\therefore r_1 = 2\sqrt{2}, \quad r_2 = 4.$$

Substitute $r_1 = 2\sqrt{2}, r_2 = 4$ into (5) and (6), so $x = \pm(1 + i)$.

15. In complex plane, find the area of polygon whose vertices are points corresponding to roots of equation $x^6 - \sqrt{2} + \sqrt{7}i = 0$.

Solution: According to geometric meaning of quadratic equation: six points corresponding to roots of $x^6 = \sqrt{2} - \sqrt{7}i$ construct inscribed hexagon of circle whose radius is $|x| = \sqrt[12]{(\sqrt{2})^2 + (-\sqrt{7})^2} = \sqrt[6]{3}$.

∴ Side of regular hexagon is $\sqrt[6]{3}$, so the area of this regular hexagon is $S = 6 \times \frac{\sqrt{3}}{4} \times (\sqrt[6]{3})^2 = \frac{3}{2}\sqrt[6]{243}$.

16. Suppose $A = \{z | z^{18} = 1, z \in C.\}$, $B = \{w | w^{48} = 1, w \in C\}$, $D = \{zw | z \in A, w \in B\}$, find the number of elements of D.

Solution:

Element of A is $z = \cos\dfrac{2k}{18}\pi + i\sin\dfrac{2k}{18}\pi$.

Element of B is $w = \cos\dfrac{2t}{48}\pi + i\sin\dfrac{2t}{48}\pi$.

Element of D is $zw = \cos\dfrac{2(8k+3t)}{144}\pi + i\sin\dfrac{2(8k+3t)}{144}\pi$.

∵ 8 and 3 are coprime and $8k + 3t$ can equal to any integer.

∵ $8k + 3t$ can be equal to $0, 1, 2, \ldots, 143$, namely, zw

has 144 different numbers.

∴ Set D has 144 elements.

17. If q is the 7-order extraction root of 1, find $\frac{q}{1+q^2} + \frac{q^2}{1+q^4} + \frac{q^3}{1+q^6}$.

Solution: $q^7 = 1$.

If $q = 1$, original type is equal to $\dfrac{3}{2}$.

If $q \neq 1$, original type is equal to

$$\frac{q}{1+q^2} + \frac{q^5}{1+q^3} + \frac{q^4}{1+q} = \frac{q}{1+q^2} + \frac{q^4(1+q^2)}{1+q^3}$$

$$= \frac{2 \cdot (q^4 + q + q^6)}{(1 + q^2)(1 + q^3)}.$$

$\because q \neq 1$ and $q^7 = 1$

$\therefore q^6 + q^5 + q^4 + q^3 + q^2 + q + 1 = 0,$

namely, $q^6 + q^4 + q = -(1 + q^2 + q^3 + q^5)$

$\qquad = -(1 + q^2) \cdot (1 + q^3).$

$\therefore$ The original type is equal to -2.

$\therefore$ The original type is equal to $\dfrac{3}{2}$ or -2.

18. If the modulus of complex number A is 1, prove: the roots of equation $\left(\frac{1+ix}{1-ix}\right)^n = A$ are all different.

Proof:

$\qquad \because |A| = 1,$ let $A = \cos\alpha + i\sin\alpha.$

$$\because \left(\frac{1 + ix}{1 - ix}\right)^n = \cos\alpha + i\sin\alpha$$

$$\therefore \frac{1 + ix}{1 - ix} = \cos\frac{\alpha + 2k\pi}{n} + i\sin\frac{\alpha + 2k\pi}{n}$$

$$(k = 0, 1, 2, \ldots, n - 1).$$

According to componendo and dividendo method:

$$-ix = \frac{1 - \cos\frac{\alpha + 2k\pi}{n} - i\sin\frac{\alpha + 2k\pi}{n}}{1 + \cos\frac{\alpha + 2k\pi}{n} + i\sin\frac{\alpha + 2k\pi}{n}}$$

$$x = \frac{2\sin^2\frac{\alpha + 2k\pi}{2n} - 2i\sin\frac{\alpha + 2k\pi}{2n}\cos\frac{\alpha + 2k\pi}{2n}}{2\cos^2\frac{\alpha + 2k\pi}{2n} + 2i\sin\frac{\alpha + 2k\pi}{2n}\cos\frac{\alpha + 2k\pi}{2n}}i$$

$$= \tan\frac{\alpha + 2k\pi}{2n} \cdot \frac{i\sin\frac{\alpha + 2k\pi}{2n} + \cos\frac{\alpha + 2k\pi}{2n}}{\cos\frac{\alpha + 2k\pi}{2n} + i\sin\frac{\alpha + 2k\pi}{2n}} = \tan\frac{\alpha + 2k\pi}{2n}.$$

Let $k = 0, 1, 2, \ldots, n - 1$, take n different real roots of original equation which are

$$\tan \frac{\alpha}{2n}, \ \tan \frac{\alpha + 2\pi}{2n}, \ldots, \tan \frac{\alpha + 2(n-1)\pi}{2n}.$$

□

19. Prove: in complex plane, all points outside a certain point belonging to $S = \{z | z^3 + z + 1 = 0, z \in C\}$ are in annulus $\frac{\sqrt{13}}{3} < |z| < \frac{5}{4}$.

Proof: Suppose

$$f(x) = x^3 + x + 1, \ \text{so} \ f\left(-\frac{9}{13}\right) < 0, \ f\left(-\frac{16}{25}\right) > 0.$$

$\therefore$ one real root x_0 of $f(x) = 0$ belongs to $\left(-\frac{9}{13}, -\frac{16}{25}\right)$.

$\because$ $f(x)$ is increasing function.

$\therefore$ $f(x) = 0$ has only one real root x_0.

Suppose other two complex roots of $f(x) = 0$ are z_1 and $\overline{z_1}$.

According to Viete theorem: $z_1 \cdot \overline{z_1} \cdot x_0 = -1$.

$$\therefore |z_1| = |\overline{z_1}| = \sqrt{-\frac{1}{x_0}} \in \left(\frac{\sqrt{13}}{3}, \frac{5}{4}\right).$$

$\therefore$ Beside x_0, other two roots belonging to

S are in the annulus $\dfrac{\sqrt{13}}{3} < |z| < \dfrac{5}{4}$.

20. If a, b, c are real numbers and complex numbers, z_1, z_2, z_3 meet

$$\begin{cases} |z_1| = |z_2| = |z_3| = 1, \\[2mm] \dfrac{z_1}{z_2} + \dfrac{z_2}{z_3} + \dfrac{z_3}{z_1} = 1, \end{cases}$$

find $|az_1 + bz_2 + cz_3|$.

□

Solution: Use triangular form and exponential form of complex number to solve the following.

Suppose $\frac{z_1}{z_2} = e^{i\theta}$, $\frac{z_2}{z_3} = e^{i\phi}$, so $\frac{z_3}{z_1} = e^{-i(\theta+\phi)}$.

$$\because e^{i\theta} + e^{i\phi} + e^{-i(\theta+\phi)} = 1.$$

Take the imaginary part of two sides:

$$\theta = \sin\theta + \sin\varphi - \sin(\theta + \varphi)$$

$$= 2\sin\frac{\theta+\varphi}{2}\cos\frac{\theta-\varphi}{2} - 2\sin\frac{\theta+\varphi}{2}\cos\frac{\theta+\varphi}{2}$$

$$= 2\sin\frac{\theta+\varphi}{2}\left(\cos\frac{\theta-\varphi}{2} - \cos\frac{\theta+\varphi}{2}\right)$$

$$= 4\sin\frac{\theta+\varphi}{2}\sin\frac{\theta}{2}\sin\frac{\varphi}{2}.$$

$\therefore$ $\theta = 2k\pi$ or $\varphi = 2k\pi$ or $\theta + \varphi = 2k\pi$, $k \in Z$.

$\therefore$ $z_1 = z_2$ or $z_2 = z_3$ or $z_3 = z_1$.

If $z_1 = z_2$, then $1 + \frac{z_1}{z_3} + \frac{z_3}{z_1} = 1$.

$\therefore$ $\left(\frac{z_3}{z_1}\right)^2 + 1 = 0$, $\frac{z_3}{z_1} = \pm i$

$\therefore$ $|az_1 + bz_2 + cz_3| = |z_1| \cdot |a + b \pm ci| = \sqrt{(a+b)^2 + c^2}$.

Similarly, if $z_2 = z_3$, $|az_1 + bz_2 + cz_3| = \sqrt{(b+c)^2 + a^2}$.

If $z_3 = z_1$, then $|az_1 + bz_2 + cz_3| = \sqrt{(a+c)^2 + b^2}$.

Comment: According to $|z_1| = |z_2| = |z_3| = 1$ and properties of conjugate complex numbers

$$\overline{z_k} = \frac{1}{z_k}, \quad k = 1, 2, 3;$$

$$\because \frac{z_1}{z_2} + \frac{z_2}{z_3} + \frac{z_3}{z_1} = 1 \therefore z_1^2 z_3 + z_2^2 z_1 + z_3^2 z_2 = z_1 z_2 z_3, \quad (7)$$

and $\dfrac{\overline{z_1}}{\overline{z_2}} + \dfrac{\overline{z_2}}{\overline{z_3}} + \dfrac{\overline{z_3}}{\overline{z_1}} = 1$, namely, $\dfrac{z_2}{z_1} + \dfrac{z_3}{z_2} + \dfrac{z_1}{z_3} = 1$

$$\therefore \ z_2^2 z_3 + z_3^2 z_1 + z_1^2 z_2 = z_1 z_2 z_3. \tag{8}$$

According to (7) and (8): $(z_1 - z_2)(z_2 - z_3)(z_3 - z_1) = 0$.
Namely, $z_1 = z_2$ or $z_2 = z_3$ or $z_1 = z_3$ and then solve the problem.

21. Find the maximum integer n which makes all non-zero solutions of equation $(z + 1)^n = z^n + 1$ into the unit circle. (Chinese National Team Selection Test in 1989).

Solution: According to binomial theorem:

$$z(C_n^1 z^{n-2} + C_n^2 z^{n-3} + C_n^3 z^{n-4} + \cdots$$
$$+ C_n^{n-2} z + C_n^{n-1}) = 0 \ (n > 3).$$

Suppose that non-zero solutions are z_i $(i = 1, 2, \ldots, n-2)$.
According to Viete theorem:

$$S_1 = \sum_{i=1}^{n-2} z_i = -\frac{C_n^2}{C_n^1} = -\frac{n-1}{2},$$

$$S_2 = \sum_{1 < i < j < n-2} z_i z_j = \frac{C_n^3}{C_n^1} = \frac{1}{6}(n-1)(n-2).$$

Suppose $n > 4$ which meets the condition, so $z_i \cdot \overline{z_i} = |z_i|^2 = 1$ and $x_i = z_i + \overline{z_i}$ are real numbers $(i = 1, 2, \ldots, n-2)$.

$\because$ The coefficients of the equation are real numbers and the root is in the form of a conjugate number.

$\therefore$ Non-zero solutions can be expressed as $\overline{z_i}$ $(i = 1, 2, \ldots, n-2)$, so

$$t_1 = \sum_{i=1}^{n-2} x_i = \sum_{i=1}^{n-2} (z_i + \overline{z}i) = 2s_1 = 1 - n,$$

$$t_2 = \sum_{1 < i < j < n-2} x_i x_j = \frac{1}{2}\left(t_1^2 - \sum_{i=1}^{n-2} x_i^2\right),$$

$$\sum_{i=1}^{n-2} x_i^2 = \sum_{i=1}^{n-2}(z_i^2 + \bar{z}_i^2 + 2z_i \cdot \bar{z}_i)$$

$$= \sum_{i=1}^{n-2}(z_i^2 + \bar{z}_i^2) + 2\sum_{i=1}^{n-2} z_i \cdot \bar{z}_i$$

$$= 2\sum_{i=1}^{n-2} z_i^2 + 2\sum_{i=1}^{n-2} 1$$

$$= 2(S_1^2 - 2S_2) + 2(n - 2),$$

$$\therefore \ t_2 = \frac{1}{2}[(2S_1)^2 - 2(S_1^2 - 2S_2) - 2(n - 2)]$$

$$= \frac{1}{12}(7n^2 - 30n + 35).$$

According to Viete theorem: real numbers x_i ($i = 1, 2, \ldots,$ $n - 2$) are $n - 2$ roots of equation with real coefficient $x^{n-2} - t_1 x^{n-3} + t_2 x^{n-4} + \cdots + t_{n-3}x + t_{n-2} = 0$. According to Theorem 5,

$$\Delta_1 = (n - 3)(-t_1)^2 - 2(n - 2)t_2 \geq 0$$

$$\therefore \ (n - 3)(n - 1)^2 - 2(n - 2)\frac{7n^2 - 30n + 35}{12} \geq 0.$$

$$\therefore \ (n - 4)[(n - 5)^2 - 12] \leq 0$$

$$\therefore \ n \leq 5 + \sqrt{12} < 9, \ \therefore \ n \leq 8.$$

When $n = 8$ the equation is given by

$$\therefore \ (8z^6 + 28z^5 + 56z^4 + 70z^3 + 56z^2 + 28z + 8)z = 0.$$

Non-zero solutions are six roots of the following equation

$$4z^6 + 14z^5 + 28z^4 + 35z^3 + 28z^2 + 14z + 4 = 0. \tag{9}$$

Equation (9) can be translated:

$$4(z^3 + z^{-3}) + 14(z^2 + z^{-2}) + 28(z + z^{-1}) + 35 = 0$$

$$\therefore \ 4(z + z^{-1})^3 + 14(z + z^{-1})^2 + 16(z + z^{-1}) + 7 = 0$$

$$\because \ \Delta_2 = (3 - 1) \cdot 16^2 - 2 \cdot 3 \cdot 14 \cdot 7 = -76 < 0.$$

According to Theorem 5, roots of $4x^3 + 14x^2 + 16x + 7 = 0$ are not all real numbers.

$\therefore$ There is root z_i of Eq. (9), which make $z_i + z_i^{-1}$ is not a real number.

$\therefore \ z_i + z_i^{-1} \neq z_i + \overline{z_i}$, $\therefore \ |z_i| \neq 1$, $\therefore \ n = 8$ does not meet the problem set conditions.

When $n = 7$, the equation is given by

$$0 = (z + 1)^7 - (z^7 + 1)$$

$$= (z + 1) \sum_{i=0}^{6} C_6^i z^i - (z + 1) \sum_{i=0}^{6} (-1)^i z^i$$

$$= (z + 1) \sum_{i=0}^{6} [C_6^i - (-1)^i] z^i$$

$$= 7(z + 1)z(z^2 + z + 1)^2.$$

Non-zero solutions are -1 and $\cos 120° \pm i \sin 120°$, and they are on the unit circle.

$\therefore$ The maximum number n of integer is 7.

22. Suppose a, b, c are three sides of the non-isosceles triangle ABC, where area is $S_\triangle$,
Prove:

$$\frac{a^3}{(a - b)(a - c)} + \frac{b^3}{(b - c)(b - a)} + \frac{c^3}{(c - a)(c - b)} > 2 \cdot 3^{\frac{3}{4}} \cdot S_\triangle^{\frac{1}{2}}.$$

Proof: The quadratic polynomial is given by
$$f(x) = x^3 - (x - x_1)(x - x_2)(x - x_3).$$

Take different figures: $x_1 = a$, $x_2 = b$, $x_3 = c$,

$$\therefore f(a) = a^3, \ f(b) = b^3, \ f(c) = c^3.$$

According to Lagrange formula,

$$\frac{(x-b)(x-c)}{(a-b)(a-c)} \cdot a^3 + \frac{(x-c)(x-a)}{(b-c)(b-a)} \cdot b^3 + \frac{(x-a)(x-b)}{(c-a)(c-b)} \cdot c^3$$

$$= x^3 - (x-a)(x-b)(x-c).$$

Compare the coefficients of the x^2 on both sides of the formula.

$$\frac{a^3}{(a-b)(a-c)} + \frac{b^3}{(b-c)(b-a)} + \frac{c^3}{(c-a)(c-b)} = a+b+c.$$

By the Helen formula, $S_\triangle = \sqrt{p(p-a)(p-b)(p-c)}$, and $p = \frac{1}{2}(a+b+c)$,

$$\because S_\triangle \leq \sqrt{p \cdot \left(\frac{p}{3}\right)^3} = \frac{p^2}{3\sqrt{3}}, \ \therefore p \geq (3\sqrt{3}S_\triangle)^{\frac{1}{2}}.$$

Obviously it cannot be established,

$$\therefore a+b+c = 2p > 2(3\sqrt{3}S_\triangle)^{\frac{1}{2}} = 2 \cdot 3^{\frac{3}{4}} \cdot S_\triangle^{\frac{1}{2}}$$

$$\therefore \frac{a^3}{(a-b)(a-c)} + \frac{b^3}{(b-c)(b-a)} + \frac{c^3}{(c-a)(c-b)} > 2 \cdot 3^{\frac{3}{4}} \cdot S_\triangle^{\frac{1}{2}}. \quad \square$$

Comment: According to Lagrange formula, find one of the general formulas of any finite sequence $a_1, a_2, a_3, \ldots, a_m$:

$$a_n = f(n) = b_1 \cdot n^{m-1} + b_2 \cdot n^{m-2} + \cdots + b_m.$$

This shows: general formulas of $a_1, a_2, a_3, \ldots, a_m$ equal to the value of $g(x)$ when $x = n$, and b_1, b_2, b_m are constants.

For example, suppose $a_1 = 1$, $a_2 = 5$, $a_3 = 11$, find one general formulas of this sequence.

Suppose $a_n = f(n) = b_1 \cdot n^2 + b_2 \cdot n + \cdots + b_3$.

Then, $x = n$, $x_1 = 1$, $x_2 = 2$, $x_3 = 3$ and $f(1) = 1$, $f(2) = 5$, $f(3) = 11$.

According to Lagrange formula:

$$\frac{(n-2)(n-3)}{(1-2)(1-3)} \cdot f(1) + \frac{(n-1)(n-3)}{(2-1)(2-3)} \cdot f(2)$$

$$+ \frac{(n-1)(n-2)}{(3-1)(3-2)} \cdot f(3) = f(n).$$

Simplify, $n^2 + n - 1 = f(n)$.

So the general formulas is $a_n = n^2 + n - 1$ $(n = 1, 2, 3)$.

23. Find the remainder term of $g(x) = x^3 - x$ which is divided by $f(x) = x^{81} + x^{49} + x^{25} + x^9 + x$.

Solution:

$$\because\ g(x) = x^3 - x = x(x-1)(x+1)$$

$$\therefore\ x_1 = 0,\ x_2 = 1,\ x_3 = -1$$

$$\because\ f(0) = 0,\ f(1) = 5,\ f(-1) = -5$$

$$\therefore\ r(x) = \frac{(x-1)(x+1)}{(0-1)(0+1)} \cdot f(0) + \frac{(x-0)(x+1)}{(1-0)(1+1)} \cdot f(1)$$

$$+ \frac{(x-0)(x+1)}{(-1-0)(-1-1)} \cdot f(-1)$$

$$= \frac{5x(x+1)}{1\cdot 2} + \frac{-5x(x-1)}{(-1)(-2)} = 5x.$$

$\therefore$ The complement form of $f(x) = x^{81} + x^{49} + x^{25} + x^9 + x$ divides $g(x) = x^3 - x$ is $r(x) = 5x$.

Comment: According to Lagrange formula, find the residue of polynomial division.

According to division formula with remainder, polynomial $f(x)$ and $g(x)(g(x) \neq 0)$ must exist polynomial $q(x)$ and $r(x)$ $(\deg r(x) < \deg q(x)$ or $r(x) = 0)$ where $f(x) = g(x) \cdot q(x) + r(x)$.

Suppose $\deg f(x) \geq n$, $\deg g(x) = n$, and $g(x) = (x - x_1)(x - x_2) \cdots (x - x_n)$.

According to Lagrange formula, $r(x) = \sum_{i=1}^{n} f(x_i) \cdot R_i(x)$, $R_i(x) = \prod_{\substack{j=1 \\ j \neq i}}^{n} \frac{x - x_i}{x_i - x_j}$ and $r(x_i) = f(x_i)$.

According to factorization, $(x - x_i) | (f(x) - r(x))$, so $g(x) | (f(x) - r(x))$.

$$\because \deg g(x) = n$$

$$\therefore \deg r(x) \leq n - 1 \leq n$$

$$\therefore r(x) \text{ is the residue of } g(x) \text{ divided by } f(x).$$

24. The given integer: $x_0 < x_1 < \cdots < x_n$. Prove: among x_0, $x_1, \ldots, x_n$ in $x^n + a_1 x^{n-1} + \cdots + a_n$, one of the absolute is not less than $\frac{n!}{2^n}$ (Nineteenth IMO pre-selection)

Prove: According to Lagrange's formula:

$$P(x) = \sum_{j=1}^{n} \left(\prod_{\substack{i \neq k \\ -n \leq i \leq n}} \frac{x - x_i}{x_j - x_i} \right) \cdot P(x_j) \equiv x^n + a_1 x^{n-1} + \cdots + a_n.$$

Assume that the conclusion is wrong, namely, when $j = 1, 2, \ldots, n$, $P(x_j) < \frac{n!}{2^n}$.

$\therefore$ 1 (first coefficient of $P(x)$) equal to sum of first coefficient of $\prod_{i \neq j} \frac{x - x_i}{x_j - x_i}$ and its modulus is no more than

$$\sum_{j=0}^{n} P(x_j) \prod_{\substack{i \neq j \\ 1 \leq i \leq n}} \frac{1}{|x_j - x_i|}$$

$$< \frac{n!}{2^n} \prod_{\substack{i \neq j \\ 1 \leq i \leq n}} \frac{1}{|x_j - x_i|} \leq \sum_{j=0}^{n} \frac{n!}{2^n} \cdot \frac{1}{\prod_{i<j}(j-i)} \cdot \frac{1}{\prod_{i<j}(j-i)}$$

$$= \frac{1}{2^n} \sum_{j=0}^{n} \frac{n!}{j!(n-j)!} = \frac{1}{2^n} \sum_{j=0}^{n} C_n^i = 1.$$

Contradiction, so the original conclusion is established.

Comment: Obviously, a special case in this example is as follows: to polynomial with real coefficient $f(x) = x^n +$

$a_1 x^{n-1} + \cdots + a_{n-1}x + a_n$, the conclusion is follows: Among $|f(1)|, |f(2)|, \ldots, |f(n+1)|$, at least one of them is not less than $\frac{n!}{2^n}$.

Exercise 10

1. Find the number of roots of equation $x^2 - 3|x| + 2 = 0$ in complex set.

2. If quadratic equation $x^2 + zx + 4 + 3i = 0$ (i is imaginary unit) has real root, find the minimum of $|z|$.

3. If equation $x^2 - (2i - 1)x + 3m - i = 0$ has real root, find the range of m.

4. If $z_1, z_2, z_3, \ldots, z_6$ is 6-order extraction root of $2 + i$ and their arg's are $\alpha_1, \alpha_2, \alpha_3, \ldots, \alpha_6$, find $\tan \alpha_1 \tan \alpha_2 + \tan \alpha_2 \tan \alpha_3 + \cdots + \tan \alpha_6 \tan \alpha_1$.

5. If α is the root of $x + \frac{1}{x} = 2 \cos \theta$, find $\alpha^n + \frac{1}{\alpha^n}$.

6. Suppose $z = x + yi$ ($x, y \in R$) and $|z + 2| - |z - 2| = 4$, find the locus of point (x, y) corresponding to z.

7. Factorization in complex range: $x^3 - 2x^2 + 4x - 8$.

8. Solve complex equation: $z^3 = \bar{z}$.

9. If $\frac{1}{2} + \frac{\sqrt{3}}{2}i$ is one root of equation $x^2 + px + q = 0$ ($p, q \in R$), find the arg of $p + qi$.

10. If z_1 and z_2 are imaginary roots of $x^2 + 5x + m = 0$ and $|z_1 - z_2| = 3$, find m.

11. If quadratic equation $x^2 - (i + \tan \theta)x - (2 + i) = 0$ has a real root, find θ.

12. In complex plane, M is the set of points corresponding to roots of $x^{10} = 1$. How many right triangles among the triangles are vertices from M.

13. If quadratic equation $x^2 + 2(p - q)x + 2(p^2 + q^2) = 0$ ($p, q \in R$) has imaginary root and x^3 is real number, find $\frac{p}{q}$.

14. If equation $x^2 + (4 + i)x + 4 + ai = 0$ ($a \in R$) has real root b and $z = a + bi$, find the range of arg of $\bar{z}(1 - ci)(c > 0)$.

15. Prove: points corresponding to complex roots of $z^n \cos\theta_n + z^{n-1}\cos\theta_{n-1} + z^{n-2}\cos\theta_{n-2} + \cdots + z\cos\theta_1 + \cos\theta_0 = 2$ are all outside of curve $|z| = \frac{1}{2}$ ($\theta_n, \theta_{n-1}, \ldots, \theta_1, \theta_0$ are real numbers).

16. Suppose $z_1, z_2, \ldots, z_n \in C$, $n \geq 2$. Prove: $\sum_{k=1}^n |z_k| \leq |\sum_{k=1}^n z_k| + \sum_{1 \leq i < j \leq n} |z_i - z_j|$.

17. If $x, y \in R$, prove: $\sqrt{x^2 + y^2} + \sqrt{(x-1)^2 + y^2} + \sqrt{x^2 + (y-1)^2} \geq \frac{\sqrt{2}}{2}(\sqrt{3} - 1)$.

18. Suppose a, b are real numbers, and equation $x^4 + ax^3 + bx^2 + ax + 1 = 0$ at least has one real root, find the minimum of $a^2 + b^2$.

19. Suppose $A = \{z | \bar{z} = \frac{1}{1+ti}, t \in R\}$, $A = \{z' | \arg(z' - i) = \frac{5\pi}{4}\}$, find $\min_{z \in A, z' \in B} |\overrightarrow{ZZ'}|$.

20. If complex numbers z_1, z_2 meet $|z_1 + z_2| = 2, |z_1 z_2| = 3$, and $z = \frac{z_1}{z_2}$, find the maximum and minimum of $\arg(z)$.

21. Suppose a and b are real numbers. If A is the solution set of equation $x^2 + a|x| + b = 0$ in C, and $B = \{n | n = |A|, \forall a, b \in R\}$, find set B.

22. $r_1, \overline{r_1}, r_2, \overline{r_2}, r_3, \overline{r_3}, r_4, \overline{r_4}, r_5, \overline{r_5}$ are 10 complex roots of equation $x^{10} + (13x - 1)^{10} = 0$, find $\frac{1}{r_1 \overline{r_1}} + \frac{1}{r_2 \overline{r_2}} + \cdots + \frac{1}{r_5 \overline{r_5}}$.

23. Prove: $C_n^1 + C_n^5 + \cdots + C_n^{4m-3} = \frac{1}{2}(2^{n-1} + 2^{\frac{n}{2}} \cdot \sin\frac{n\pi}{4})$ where the maximum integer is given by $4m - 3 \leq n$.

24. Prove: $\sin n\theta = 2^{n-1} \prod_{k=0}^{n-1} \sin(\theta + \frac{k\pi}{n})$.

25. (1) If $n > 3$ and n is a prime number,
$$\text{find } \left(1 + 2\cos\frac{2\pi}{n}\right)\left(1 + 2\cos\frac{4\pi}{n}\right) \cdots \left(1 + 2\cos\frac{2n\pi}{n}\right).$$

(2) If $n > 3$ and n is a natural number,
$$\text{find } \left(1 + 2\cos\frac{\pi}{n}\right)\left(1 + 2\cos\frac{2\pi}{n}\right)\left(1 + 2\cos\frac{3\pi}{n}\right)$$
$$\cdots \left(1 + 2\cos\frac{(n-1)\pi}{n}\right).$$

26. If x, y, z and w meet

$$
\begin{cases}
\dfrac{x^2}{2^2 - 1^2} + \dfrac{y^2}{2^2 - 3^2} + \dfrac{z^2}{2^2 - 5^2} + \dfrac{w^2}{2^2 - 7^2} = 1, \\[2mm]
\dfrac{x^2}{4^2 - 1^2} + \dfrac{y^2}{4^2 - 3^2} + \dfrac{z^2}{4^2 - 5^2} + \dfrac{w^2}{4^2 - 7^2} = 1, \\[2mm]
\dfrac{x^2}{6^2 - 1^2} + \dfrac{y^2}{6^2 - 3^2} + \dfrac{z^2}{6^2 - 5^2} + \dfrac{w^2}{6^2 - 7^2} = 1, \\[2mm]
\dfrac{x^2}{8^2 - 1^2} + \dfrac{y^2}{8^2 - 3^2} + \dfrac{z^2}{8^2 - 5^2} + \dfrac{w^2}{8^2 - 7^2} = 1,
\end{cases}
$$

find $x^2 + y^2 + z^2 + w^2$.

27. If $f(z) = C_0 z^n + C_1 z^{n-1} + C_2 z^{n-2} + \cdots + C_{n-1} z + C_n$ is n-order polynomial with complex coefficient, prove: there must be a complex number z_0 where $|z_0| \leq 1$ and $|f(z_0)| \geq |C_0| + |C_n|$.

Answers

Exercise 1

1. We have

$$\sin^2\theta + \sin\theta\cos\theta - 2\cos^2\theta = \frac{\sin^2\theta + \sin\theta\cos\theta - 2\cos^2\theta}{\sin^2\theta + \cos^2\theta}$$

$$= \frac{\tan^2\theta + \tan\theta - 2}{\tan^2\theta + 1}$$

$$= \frac{4+2-2}{4+1} = \frac{4}{5}.$$

2. We have

$$\because \begin{cases} \sin\theta + \cos\theta = \dfrac{\sqrt{3}+1}{2}, \\[2mm] \sin\theta\cos\theta = \dfrac{m}{2}, \end{cases}$$

$$\therefore \frac{\sin\theta}{1-\cot\theta} + \frac{\cos\theta}{1-\tan\theta} = \cdots = \sin\theta + \cos\theta = \frac{\sqrt{3}+1}{2}.$$

3. We have

$$3\sin^2\alpha + 2\sin^2\beta = 2\sin\alpha \Rightarrow \sin^2\beta = \sin\alpha - \frac{3}{2}\sin^2\alpha$$

$$\Rightarrow \sin\alpha - \frac{3}{2}\sin^2\alpha \geq 0$$

$$\Rightarrow 0 \le \sin \alpha \le \frac{2}{3}, \quad \therefore \ \sin^2 \alpha + \sin^2 \beta = -\frac{1}{2}(\sin \alpha - 1)^2 + \frac{1}{2}$$

$$\because \ 0 \le \sin \alpha \le \frac{2}{3}, \quad \therefore \ 0 \le \sin^2 \alpha + \sin^2 \beta \le \frac{4}{9}.$$

4. The question is equivalent to find the range of $m = -\cos^2 x - \sin x$,

$$m = \left(\sin x - \frac{1}{2}\right)^2 - \frac{5}{4}, \quad \because \ \sin x \in [-1, 1], \quad \therefore \ m \in \left[-\frac{5}{4}, 1\right].$$

5. Suppose $\sin x - \sin y = t$, $\therefore \ \cos x \cos y - \sin x \sin y = \frac{t^2 - 1}{2}$

$$\therefore \ \cos(x + y) = \frac{t^2 - 1}{2}$$

$$\because \ -1 \le \cos(x + y) \le 1$$

$$\therefore \ -1 \le \frac{t^2 - 1}{2} \le 1$$

$$\therefore \ -\sqrt{3} \le t \le \sqrt{3}.$$

6. The origin type is equal to $\frac{1 + \sin \alpha}{|\cos \alpha|} - \frac{1 - \sin \alpha}{|\cos \alpha|} = \frac{2 \sin \alpha}{|\cos \alpha|} = 2 \tan \alpha$.

(1) When $\cos \alpha > 0$, namely, α is the first or fourth quadrant angel or on the positive x-axis,

$$\frac{2 \sin \alpha}{|\cos \alpha|} = \frac{2 \sin \alpha}{\cos \alpha} = 2 \tan \alpha \text{ always holds,}$$

$$\therefore \ 2k\pi - \frac{\pi}{2} < \alpha < \frac{\pi}{2} + 2k\pi.$$

(2) When $\cos \alpha < 0$, namely, α is the second or third quadrant angel or on the negative x-axis,

$$\frac{2 \sin \alpha}{|\cos \alpha|} = \frac{2 \sin \alpha}{\cos \alpha} = -2 \tan \alpha = 2 \tan \alpha \text{ if and only if } \tan \alpha = 0,$$

$$\therefore \ \alpha = (2k + 1)\pi.$$

$$\therefore \ \text{The set is } \{(2k + 1)\pi, k \in Z\} \cup \left(2k\pi - \frac{\pi}{2}, 2k\pi + \frac{\pi}{2}\right) \ (k \in Z).$$

7. $\because \ \sin A(\sin B + \cos B) - \sin C = 0$

$\therefore \ \sin A \sin B + \sin A \cos B - \sin(A + B) = 0$

$\therefore \ \sin A \sin B + \sin A \cos B - \sin A \cos B - \cos A \sin B = 0$

$\therefore \ \sin B(\sin A - \cos A) = 0$

$\because \ B \in (0, \pi)$

$\therefore \ \sin B \neq 0, \ \therefore \ \cos A = \sin A$

$\because \ A \in (0, \pi), \ \therefore \ A = \dfrac{\pi}{4}, \ \therefore \ B + C = \dfrac{3}{4}\pi$

$\because \ \sin B + \cos 2C = 0 \ \therefore \ \sin B + \cos 2\left(\dfrac{3}{4}\pi - B\right) = 0$

$\because \ \sin B - \sin 2B = 0. \ \therefore \ \sin B - 2\sin B \cos B = 0$

$\therefore \ \cos B = \dfrac{1}{2}, \ B = \dfrac{\pi}{3}, \ C = \dfrac{5\pi}{12}$

$\therefore \ A = \dfrac{\pi}{4}, \ B = \dfrac{\pi}{3}, \ C = \dfrac{5\pi}{12}.$

8. $\because \ \sin x + \cos x = \dfrac{1}{5} \Rightarrow \sin^2 x + 2\sin x \cos x + \cos^2 x = \dfrac{1}{25},$

$\therefore \ 2\sin x \cos x = -\dfrac{24}{25}.$

$\because \ (\sin x - \cos x)^2 = 1 - 2\sin x \cos x = \dfrac{49}{25}.$

$\because \ -\dfrac{\pi}{2} < x < 0, \ \therefore \ \sin x < 0, \ \cos x > 0, \ \sin x - \cos x < 0,$

$\therefore \ \sin x - \cos x = -\dfrac{7}{5}.$

$$\therefore \quad \frac{3\sin^2\frac{x}{2} - \sin\frac{x}{2} + \cos^2\frac{x}{2}}{\tan x + \cot x} = \frac{2\sin^2\frac{x}{2} - \sin x + 1}{\frac{\sin x}{\cos x} + \frac{\cos x}{\sin x}}$$

$$= \sin x \cos x (2 - \cos x - \sin x) = \left(-\frac{12}{25}\right) \times \left(2 - \frac{1}{5}\right) = -\frac{108}{125}.$$

9. (1) $g(x) = \cos x \cdot \sqrt{\dfrac{1 - \sin x}{1 + \sin x}} + \sin x \cdot \sqrt{\dfrac{1 - \cos x}{1 + \cos x}}$

$$= \cos x \cdot \frac{1 - \sin x}{|\cos x|} + \sin x \cdot \frac{1 - \cos x}{|\sin x|}.$$

$$= \cos x \cdot \frac{\sqrt{(1 - \sin x)^2}}{\cos^2 x} + \sin x \cdot \sqrt{\frac{(1 - \cos x)^2}{\sin^2 x}}.$$

$$\because x \in \left(\pi, \frac{17\pi}{12}\right], \ \therefore \ |\cos x| = -\cos x, |\sin x| = -\sin x,$$

$$\therefore g(x) = \cos x \cdot \frac{1 - \sin x}{-\cos x} + \sin x \cdot \frac{1 - \cos x}{-\sin x}$$

$$= \sin x + \cos x - 2 = \sqrt{2}\sin\left(x + \frac{\pi}{4}\right) - 2.$$

(2) $\because \pi < x \le \dfrac{17\pi}{12}, \ \therefore \ \dfrac{5\pi}{4} < x + \dfrac{\pi}{4} \le \dfrac{5\pi}{3}.$

$\because \ \sin t$ is decreasing function in $\left(\dfrac{5\pi}{4}, \dfrac{3\pi}{2}\right]$

and increasing function in $\left(\dfrac{3\pi}{2}, \dfrac{5\pi}{3}\right]$

$$\because \ \sin\frac{5\pi}{3} < \sin\frac{5\pi}{4},$$

$$\therefore \ \sin\frac{3\pi}{2} \le \sin\left(x + \frac{\pi}{4}\right) < \sin\frac{5\pi}{4} \ \left(\text{when } x \in \left(\pi, \frac{17\pi}{2}\right]\right)$$

$$\therefore \ -1 \le \sin\left(x + \frac{\pi}{4}\right) < -\frac{\sqrt{2}}{2},$$

$$\therefore \ -\sqrt{2} - 2 \leq \sqrt{2}\sin\left(x + \frac{\pi}{4}\right) - 2 < -3,$$

$\therefore$ The range of $g(x)$ is $[-\sqrt{2} - 2, -3)$.

Exercise 2

1. $$\sin\alpha + \cos\alpha = -\frac{1}{5} \Rightarrow (\sin\alpha + \cos\alpha)^2 = \frac{1}{25}$$

$$\Rightarrow \sin\alpha\cos\alpha = -\frac{12}{25}.$$

$\therefore$ $\sin\alpha, \cos\alpha$ are two roots of $x^2 + \dfrac{1}{5}x - \dfrac{12}{25} = 0,$

and its solution is $x = -\dfrac{4}{5}, \dfrac{3}{5},$

$\because \alpha \in \left(\dfrac{3\pi}{2}, 2\pi\right)$, $\therefore$ $\sin\alpha < 0, \cos\alpha > 0,$

$\therefore$ $\sin\alpha = -\dfrac{4}{5}, \cos\alpha = \dfrac{3}{5}.$

$\therefore$ The original type

$$= \frac{2\sin\alpha}{\sin 3\alpha - \sin\alpha + \cos\alpha - \cos 3\alpha}$$

$$= \frac{2\sin\alpha}{2\cos 2\alpha \sin\alpha + 2\sin 2\alpha \sin\alpha}$$

$$= \frac{1}{\cos 2\alpha + \sin 2\alpha} = -\frac{25}{31}.$$

2. $\because A + B + C = 180°$, $\sin B = \dfrac{3}{5}$

$\therefore 0° < B < 45°$ or $135° < B < 180°$

$\because \cos A = \dfrac{12}{13} > \dfrac{\sqrt{2}}{2} = \cos 45°$

$\therefore 0° < A < 45°,$

$$\therefore\ 0° < B < 45° \text{ or } 135° < B < 180°,$$

$$\therefore\ \cos B = \cdots = \pm\frac{4}{5},$$

$$\therefore\ \sin C = \sin(A + B) = \cdots = \frac{56}{65}, \frac{16}{65}.$$

3. $\quad \because\ \alpha, \beta \in \left(0, \frac{\pi}{2}\right),\ \therefore\ \cos\alpha = \frac{5}{13},\ 0 < \alpha + \beta < \pi$

When $0 < \alpha + \beta < \dfrac{\pi}{2},\ \because\ \sin(\alpha + \beta) < \sin\alpha,$

$\therefore\ \alpha + \beta < \alpha,$ it is impossible

$\therefore\ \dfrac{\pi}{2} < \alpha + \beta < \pi,$

$$\cos(\alpha + \beta) = -\frac{3}{5},\ \therefore\ \cos\beta = \cos[(\alpha - \beta) - \alpha] = \cdots = \frac{33}{65},$$

$$\because\ 0 < \beta < \frac{\pi}{2} \Rightarrow 0 < \frac{\beta}{2} < \frac{\pi}{4}$$

$$\therefore\ \cos\frac{\beta}{2} = \sqrt{\frac{1 + \cos\beta}{2}} = \frac{7\sqrt{65}}{65}.$$

4. Let

$$A = \cos^2 x + \cos^2\left(x + \frac{2\pi}{3}\right) + \cos^2\left(x + \frac{4\pi}{3}\right),$$

$$B = \sin^2 x + \sin^2\left(x + \frac{2\pi}{3}\right) + \sin^2\left(x + \frac{4\pi}{3}\right),$$

$$A + B = 3,\ A - B = \cos 2x + \cos\left(2x + \frac{4\pi}{3}\right) + \cos\left(2x + \frac{8\pi}{3}\right)$$

$$= \cos 2x + 2\cos(2x + 2\pi)\cos\frac{2\pi}{3}$$

$$= \cos 2x - \cos 2x = 0,$$

$$\therefore\ \text{the original type is equal to } \frac{3}{2}.$$

5. Method 1:

$$\cos\left(2a + \frac{\pi}{4}\right) = \cos 2a \cos\frac{\pi}{4} - \sin 2a \sin\frac{\pi}{4}$$

$$= \frac{\sqrt{2}}{2}(\cos 2a - \sin 2a).$$

$$\because \frac{3}{4}\pi \le a + \frac{\pi}{4} \le \frac{7}{4}\pi, \text{ and } \cos\left(a + \frac{\pi}{4}\right) = \frac{3}{5},$$

$$\therefore \sin\left(a + \frac{\pi}{4}\right) = -\frac{4}{5},$$

$$\therefore \cos 2a = \sin\left(2a + \frac{\pi}{2}\right) = 2\sin\left(a + \frac{\pi}{4}\right)\cos\left(a + \frac{\pi}{4}\right)$$

$$= 2 \times \left(-\frac{4}{5}\right) \times \frac{3}{5} = -\frac{24}{25},$$

$$\sin 2a = -\cos\left(2a + \frac{\pi}{2}\right) = 1 - 2\cos^2\left(a + \frac{\pi}{4}\right)$$

$$= 1 - 2 \times \left(\frac{3}{5}\right)^2 = \frac{7}{25},$$

$$\therefore \cos\left(2a + \frac{\pi}{4}\right) = \frac{\sqrt{2}}{2}\left(-\frac{24}{25} - \frac{7}{25}\right) = -\frac{31\sqrt{2}}{50}.$$

Method 2:

$$\because \frac{3}{4}\pi \le a + \frac{\pi}{4} \le \frac{7}{4}\pi, \text{ and } \cos\left(a + \frac{\pi}{4}\right) = \frac{3}{5},$$

$$\therefore \sin\left(a + \frac{\pi}{4}\right) = -\frac{4}{5}.$$

$$\therefore \sin 2\left(a + \frac{\pi}{4}\right) = 2\sin\left(a + \frac{\pi}{4}\right)\cos\left(a + \frac{\pi}{4}\right) = -\frac{24}{25},$$

$$\cos 2\left(a + \frac{\pi}{4}\right) = 2\cos^2\left(a + \frac{\pi}{4}\right) - 1 = -\frac{7}{25},$$

$$\therefore \cos\left(2a + \frac{\pi}{4}\right) = \cos\left[2\left(a + \frac{\pi}{4}\right) - \frac{\pi}{4}\right]$$

$$= \frac{\sqrt{2}}{2}\left[\cos 2\left(a + \frac{\pi}{4}\right) + \sin 2\left(a + \frac{\pi}{4}\right)\right] = -\frac{31\sqrt{2}}{50}.$$

6. $\quad \because \alpha \in (0, \pi)$ and $\tan\alpha = \dfrac{1}{3} > 0, \therefore 0 < \alpha < \dfrac{\pi}{2}.$ (1)

$\quad \because \tan\beta = -\dfrac{1}{7}$ and $\beta \in (0, \pi),$

$\quad \therefore \dfrac{\pi}{2} < \beta < \pi, \ -\pi < -\beta < -\dfrac{\pi}{2}, -\pi < \alpha - \beta < 0.$

$\quad \because \tan(\alpha - \beta) = \dfrac{1}{2} > 0, \ -\pi < \alpha - \beta < -\dfrac{\pi}{2};$ (2)

according to (1) and (2), $-\pi < 2\alpha - \beta < 0,$

$\quad \because \tan(2\alpha - \beta) = 1$

$\quad \therefore 2\alpha - \beta = -\frac{3\pi}{4}.$

7. Suppose E is the midpoint of $BC, \therefore DE//AB,$ and $DE = \frac{2\sqrt{6}}{3}.$
Suppose $BE = x, \therefore$ In $\triangle BDE, \cos\angle BED = -\cos\angle ABC = -\frac{\sqrt{6}}{6}.$

According to cosine law, $5 = x^2 + \frac{8}{3} + 2 \times \frac{2\sqrt{6}}{3} \times \frac{\sqrt{6}}{6}x,$ the solution
is $x = 1, \ x = -\frac{7}{3}$ (rejection), $\therefore \quad BC = 2. \therefore \quad AC^2 = AB^2 + BC^2 - 2AB \cdot BC \cos B = \frac{28}{3}, \therefore \ AC = \frac{2\sqrt{21}}{3}.$

According to sine law, $\sin A = \frac{\sqrt{70}}{14}.$

8. We have

$$y = \sin\left(x + \frac{\pi}{3}\right)\sin\left(x + \frac{\pi}{2}\right)$$

$$= \left(\sin x \cos\frac{\pi}{3} + \cos x \sin\frac{\pi}{3}\right)\cos x$$

$$= \frac{1}{2} \sin x \cos x + \frac{\sqrt{3}}{2} \cos^2 x = \frac{1}{4} \sin 2x + \frac{\sqrt{3}}{2} \cdot \frac{1 + \cos 2x}{2}$$

$$= \frac{\sqrt{3}}{4} + \frac{1}{2} \sin\left(2x + \frac{\pi}{3}\right)$$

$\therefore T = \pi.$

9. $\because \tan\alpha + \cot\alpha = -\dfrac{10}{3} \quad \therefore 3\tan^2\alpha + 10\tan\alpha + 3 = 0,$

$\therefore \tan\alpha = -3 \text{ or } \tan\alpha = -\dfrac{1}{3},$

$\because \dfrac{3\pi}{4} < \alpha < \pi, \therefore \tan\alpha = -\dfrac{1}{3}$

$\therefore \dfrac{5\sin^2\frac{\alpha}{2} + 8\sin\frac{\alpha}{2}\cos\frac{\alpha}{2} + 11\cos^2\frac{\alpha}{2} - 8}{\sqrt{2}\sin\left(\alpha - \frac{\pi}{2}\right)}$

$$= \frac{5\frac{1-\cos\alpha}{2} + 4\sin\alpha + 11\frac{1+\cos\alpha}{2} - 8}{-\sqrt{2}\cos\alpha}$$

$$= \frac{5 - 5\cos\alpha + 8\sin\alpha + 11 + 11\cos\alpha - 16}{-2\sqrt{2}\cos\alpha}$$

$$= \frac{8\sin\alpha + 6\cos\alpha}{-2\sqrt{2}\cos\alpha}$$

$$= \frac{8\tan\alpha + 6}{-2\sqrt{2}} = -\frac{5\sqrt{2}}{6}.$$

10. **Method 1:**

$\because A + B + C = \pi, \cot A = -\cot(B+C) = -\dfrac{\cot B \cot C - 1}{\cot B + \cot C},$

$\therefore -\dfrac{\cot B \cot C - 1}{\cot B + \cot C} + \cot B + \cot C = \sqrt{3},$

$\therefore \cot^2 B + (\cot C - \sqrt{3})\cot B + (\cot^2 C - \sqrt{3}\cot C + 1) = 0$

$\because \cot B \in R, \therefore \Delta \geq 0,$

$\because \Delta = (\cot C - \sqrt{3})^2 - 4(\cot^2 C - \sqrt{3}\cot C + 1)$

$\quad = -(\sqrt{3}\cot C - 1)^2 \leq 0$

$\therefore \sqrt{3}\cot C - 1 = 0 \Rightarrow C = 60°.$

Similarly, $A = B = 60°$, $\therefore \triangle ABC$ is equilateral triangle.

Method 2:

$(\cot A + \cot B + \cot C)^2 = (\sqrt{3})^2 \Rightarrow \cot^2 A + \cot^2 B + \cot^2 C$

$\quad + 2(\cot A \cot B + \cot A \cot C + \cot B \cot C) = 3,$ $\qquad$ (*)

$\because A + B + C = \pi,$

$\therefore \tan A + \tan B + \tan C = \tan A \tan B \tan C.$

Multipling $\cot A \cot B \cot C$ by both sides

$\therefore \cot A \cot B + \cot B \cot C + \cot C \cot A = 1.$

Substitute it for type (*), we get $\cot^2 A + \cot^2 B + \cot^2 C - 1 = 0$

$\therefore \cot^2 A + \cot^2 B + \cot^2 C - (\cot A \cot B$

$\qquad + \cot B \cot C + \cot C \cot A) = 0,$

$\Rightarrow (\cot A - \cot B)^2 + (\cot A - \cot C)^2 + (\cot B - \cot C)^2 = 0$

$\therefore \triangle ABC$ is equilateral triangle.

11. Suppose E is tangent point of circle and AD and G is tangent point of circle and AB, so $OG \perp AB$, $OE \perp AD$

$\because AD // BC, \therefore EO \perp BC$, it and BC intersect at F.

Connect OB, and suppose $\angle OBG = \alpha$, $\angle OBF = \beta$, $\therefore \phi = \alpha + \beta.$

Finding $\sin \phi$ is equal to finding $\sin(\alpha + \beta)$,

$\because OF \perp BF, OE = 1, EF = 5, OF = 4,$

$\because BC = 5, \therefore BF = \dfrac{5}{2}.$

In the right-angled triangle OFB, $OB = \sqrt{16 + \dfrac{25}{4}} = \dfrac{\sqrt{89}}{2}$,

$$\sin\beta = \frac{OF}{OB} = \frac{8}{\sqrt{89}}, \cos\beta = \frac{BF}{OB} = \frac{5}{\sqrt{89}}.$$

In the right-angled triangle OBG, $OB = \dfrac{\sqrt{89}}{2}, OG = 1$,

$$BG = \sqrt{OB^2 - OG^2} = \frac{\sqrt{85}}{2}, \cos\alpha = \frac{BG}{OB} = \sqrt{\frac{85}{89}},$$

$$\sin\alpha = \frac{OG}{OB} = \frac{2}{\sqrt{89}},$$

$$\therefore \ \sin\phi = \sin(\alpha + \beta) = \cdots = \frac{10 + 8\sqrt{85}}{89}.$$

12. $y = 7 - 4\sin x \cos x + 4\cos^2 x - 4\cos^4 x$

$$= 7 - 2\sin 2x + 4\cos^2 x(1 - \cos^2 x)$$

$$= 7 - 2\sin 2x + 4\cos^2 x \sin^2 x$$

$$= 7 - 2\sin 2x + \sin^2 2x$$

$$= (1 - \sin 2x)^2 + 6.$$

$\because$ The maximum of $z = (u-1)^2 + 6$ in $[-1, 1]$ is given by

$z_{\max} = (-1 - 1)^2 + 6 = 10.$

The minimum of $z = (u-1)^2 + 6$ in $[-1, 1]$ is given by

$z_{\min} = (1 - 1)^2 + 6 = 6.$

$\therefore$ When $\sin 2x = -1$, $y_{\max} = 10$; When $\sin 2x = 1$, $y_{\min} = 6$.

13. $f(x) = \sqrt{3}\sin x + \cos x = 2\sin\left(x + \dfrac{\pi}{6}\right)$, $\because -\dfrac{\pi}{2} \le x \le \dfrac{\pi}{2}$,

$$\therefore \ -\frac{\pi}{3} \le x + \frac{\pi}{6} \le \frac{2\pi}{3}.$$

According to trigonometric function line:

$$-\frac{\sqrt{3}}{2} \le \sin\left(x + \frac{\pi}{6}\right) \le 1 \Rightarrow -\sqrt{3} \le f(x) \le 2.$$

$f(x)_{min} = -\sqrt{3}$ if and only if $x + \frac{\pi}{6} = -\frac{\pi}{3}$, namely, $x = -\frac{\pi}{2}$.

$f(x)_{max} = 2$ if and only if $x + \frac{\pi}{6} = \frac{\pi}{2}$, namely, $x = \frac{\pi}{3}$.

14. (1)
$$A = 2\sqrt{3}, \ \frac{T}{4} = 3, \ T = \frac{2\pi}{\omega},$$

$$\therefore \ \omega = \frac{\pi}{6} . \ \therefore \ y = 2\sqrt{3}\sin\frac{\pi}{6}x$$

$$\therefore \ y = 2\sqrt{3}\sin\frac{2\pi}{3} = 3 \text{ when } x = 4.$$

$$\therefore \ M(4,3).$$

$$\because \ p(8,0).$$

$$\therefore \ MP = \sqrt{4^2 + 3^2} = 5.$$

(2) Method 1:

In $\triangle MNP$, $\angle MNP = 120°, MP = 5$.

Suppose $\angle PMN = \theta$, $\therefore \ 0° < \theta < 60°$.

According to sine law: $\dfrac{MP}{\sin 120°} = \dfrac{NP}{\sin\theta} = \dfrac{MN}{\sin(60° - \theta)}.$

$$\therefore \ NP = \frac{10\sqrt{3}}{3}\sin\theta, \ \therefore \ MN = \frac{10\sqrt{3}}{3}\sin(60° - \theta)$$

$$\therefore NP + MN = \frac{10\sqrt{3}}{3}\sin\theta + \frac{10\sqrt{3}}{3}\sin(60° - \theta)$$

$$= \frac{10\sqrt{3}}{3}\left(\frac{1}{2}\sin\theta + \frac{\sqrt{3}}{3}\cos\theta\right) = \frac{10\sqrt{3}}{3}\sin(\theta + 60°)$$

$$\because \ 0° < \theta < 60°,$$

$\therefore$ broken line track MNP is the longest when $\theta = 30°$,

namely, broken line track MNP is the longest when $\angle PMN = 30°$.

Method 2:

In $\triangle MNP$, $\angle MNP = 120°$, $MP = 5$,
According to cosine law:

$$MN^2 + NP^2 - 2MN \cdot NP \cdot \cos \angle MNP = MP^2$$

$$\therefore MN^2 + NP^2 + MN \cdot NP = 25$$

$$\therefore (MN + NP)^2 - 25 = MN \cdot NP \leq \left(\frac{MN + NP}{2} \right)^2$$

$$\therefore \frac{3}{4}(MN + NP)^2 \leq 25, \text{ namely, } MN + NP \leq \frac{10\sqrt{3}}{3}$$

broken line track MNP is the longest if and

only if $MN = NP$.

Comment: Answer and presentation mode of question (2) is not unique.
Besides methods 1 and 2, it can also be designed as:
① $N \left(\frac{12+\sqrt{3}}{2}, \frac{9+4\sqrt{3}}{6} \right)$; ② $N \left(\frac{12-\sqrt{3}}{2}, \frac{9-4\sqrt{3}}{6} \right)$; ③ point N is on the perpendicular bisector of segment MP.

15.

$$\because \frac{\sin 1°}{\cos k° \cos(k+1)°} = \frac{\sin((k+1)° - k°)}{\cos k° \cos(k+1)°}$$

$$= \frac{\sin(k+1)° \cos k° - \cos(k+1)° \sin k°}{\cos k° \cos(k+1)°}$$

$$= \tan(k+1)° - \tan k°$$

$$\therefore \sin 1° \cdot \sum_{k=0}^{88} \frac{1}{\cos k° \cos(k+1)°} = \sum_{k=0}^{88} \frac{\sin 1°}{\cos k° \cos(k+1)°}$$

$$= \sum_{k=0}^{88} (\tan(k+1)^\circ - \tan k^\circ) = \tan 89^\circ - \tan 0^\circ$$

$$= \tan 89^\circ = \cot 1^\circ = \frac{\cos 1^\circ}{\sin 1^\circ}$$

$$\therefore \sum_{k=0}^{88} \frac{1}{\cos k^\circ \cos(k+1)^\circ} = \frac{\cos 1^\circ}{\sin 1^\circ}.$$

16. Suppose

$$M = \cos \frac{\pi}{2n+1} \cos \frac{2\pi}{2n+1} \cdots \cos \frac{n\pi}{2n+1},$$

$$N = \sin \frac{\pi}{2n+1} \sin \frac{2\pi}{2n+1} \cdots \sin \frac{n\pi}{2n+1}$$

$$\therefore MN = \left(\sin \frac{\pi}{2n+1} \cos \frac{\pi}{2n+1} \right) \left(\sin \frac{2\pi}{2n+1} \cos \frac{2\pi}{2n+1} \right)$$

$$\cdots \left(\sin \frac{n\pi}{2n+1} \cos \frac{n\pi}{2n+1} \right)$$

$$= \frac{1}{2^n} \sin \frac{2\pi}{2n+1} \sin \frac{4\pi}{2n+1} \cdots \sin \frac{2n\pi}{2n+1}$$

$$\because \sin \frac{2n\pi}{2n+1} = \sin \frac{\pi}{2n+1}, \quad \sin \frac{(2n-2)\pi}{2n+1} = \sin \frac{3\pi}{2n+1}, \cdots$$

When n is an even number, $\sin \dfrac{(n+2)\pi}{2n+1} = \sin \dfrac{(n-1)\pi}{2n+1}$.

When n is an odd number, $\sin \dfrac{(n+1)\pi}{2n+1} = \sin \dfrac{n\pi}{2n+1}$.

$$\therefore MN = \frac{1}{2^n} \sin \frac{\pi}{2n+1} \sin \frac{2\pi}{2n+1} \cdots \sin \frac{n\pi}{2n+1} = \frac{1}{2^n} N$$

$$\therefore M = \frac{1}{2^n}.$$

17. Suppose

$$a_n = \frac{1}{2^n} \cot \frac{x}{2^n} - \cot x,$$

$$\therefore \ a_{k+1} - a_k = \left(\frac{1}{2^{k+1}} \cot \frac{x}{2^{k+1}} - \cot x \right) - \left(\frac{1}{2^k} \cot \frac{x}{2^k} - \cot x \right)$$

$$= \frac{1}{2^{k+1}} \left(\cot \frac{x}{2^{k+1}} - 2 \cot \frac{x}{2^k} \right)$$

$$= \frac{1}{2^{k+1}} \left(\frac{1}{\tan \frac{x}{2^{k+1}}} - \frac{1 - \tan^2 \frac{x}{2^{k+1}}}{\tan \frac{x}{2^k}} \right) = \frac{1}{2^{k+1}} \tan \frac{x}{2^{k+1}}$$

$$\therefore \ \text{left} = \sum_{k=1}^{n} \left(\frac{1}{2^k} \tan \frac{x}{2^k} \right)$$

$$= a_1 + (a_2 - a_1) + (a_3 - a_2) + \cdots (a_n - a_{n-1})$$

$$= a_n = \text{Right}.$$

18.

$$\because \ \sin 3\alpha = 3 \sin \alpha - 4\sin^3 \alpha$$

$$\therefore \ 6 \sin 50° - 8\sin^3 50° = 2 \sin 150° = 1$$

$$\therefore \ (3 \sin 50° - 1)^2 = 9 \sin^2 50° - 6 \sin 50° + 1$$

$$= 9 \sin^2 50° - 8\sin^3 50° = \sin^2 50° (9 - 8 \sin 50°)$$

$$\therefore \ \sin^2 50° (9 - 8 \sin 50°) = (3 \sin 50° - 1)^2$$

$$\therefore \ \sqrt{9 - 8 \sin 50°} = \frac{3 \sin 50° - 1}{\sin 50°} = 3 - \csc 50°$$

$$\therefore \ a = 1, \ b = -1.$$

19. In $\triangle ABC$,

$$S = 2R^2 \sin A \sin B \ \sin C = \frac{1}{2}(a + b + c)r$$

$$\therefore \ \frac{r}{R} = \frac{4R \sin A \sin B \sin C}{a + b + c} = \frac{2 \sin A \sin B \ \sin C}{\sin A + \sin B + \sin C}.$$

$\therefore$ Only to prove : $\dfrac{\sin A \sin B \sin C}{\sin A + \sin B + \sin C} = 2 \sin \dfrac{A}{2} \sin \dfrac{B}{2} \sin \dfrac{C}{2}$,

namely, $4 \cos \dfrac{A}{2} \cos \dfrac{B}{2} \cos \dfrac{C}{2} = \sin A + \sin B + \sin C$

$\because \sin A + \sin B + \sin C$

$= 2 \sin \dfrac{A+B}{2} \cos \dfrac{A-B}{2} + 2 \cos \dfrac{C}{2} \sin \dfrac{C}{2}$

$= 2 \cos \dfrac{C}{2} \left(\cos \dfrac{A-B}{2} + \cos \dfrac{A+B}{2} \right) = 4 \cos \dfrac{A}{2} \cos \dfrac{B}{2} \cos \dfrac{C}{2}$

$\therefore$ Equation holds.

20. Suppose

$$B_n = (a^2 + b^2)^n \cos n\theta.$$

$\because \sin \theta = \dfrac{2ab}{a^2 + b^2}$

$\therefore \cos \theta = \dfrac{a^2 - b^2}{a^2 + b^2} \quad (a > b > 0)$

$\therefore (a^2 + b^2) \sin \theta = 2ab \quad$ and

$\quad = (a^2 + b^2) \cos \theta$ are integer numbers

$\therefore A_1, B_1 \in Z.$

Suppose $A_k, B_k \in Z$

$\therefore A_{k+1} = (a^2 + b^2)^{k+1} \sin(k+1)\theta = A_k(a^2 - b^2) + B_k \cdot 2ab$

$B_{k+1} = (a^2 + b^2)^{k+1} \cos(k+1)\theta = B_k(a^2 - b^2) - A_k \cdot 2ab$

$\because A_k, B_k \in Z$

$\therefore A_{k+1}, B_{k+1} \in Z.$

According to mathematical induction, $\forall \, n \in N^*, A_n \in Z.$

Exercise 3

1. $\because$ The image of function $y = 3\cos(2x + \phi)$ is central symmetry about point $(\frac{4\pi}{3}, 0)$.

$$\therefore 2 \cdot \frac{4\pi}{3} + \phi = k\pi \therefore \phi = k\pi - 2 \cdot \frac{4\pi}{3}(k \in Z) \therefore |\phi|_{\min} = \frac{\pi}{3}.$$

2. $f(x) = 2\sin\left(\omega x + \frac{\pi}{6}\right)$

$\because T = \pi, \therefore \omega = 2$

$$\because 2k\pi - \frac{\pi}{2} \leq 2x + \frac{\pi}{6} \leq 2k\pi + \frac{\pi}{2}$$

$$\therefore k\pi - \frac{\pi}{3} \leq x \leq k\pi + \frac{\pi}{6}, k \in z.$$

$\therefore$ Increasing interval of the function is $\left[k\pi - \frac{\pi}{3}, k\pi + \frac{\pi}{6}\right]$,

$k \in Z.$

3. $y = \tan(\omega x + \frac{\pi}{4}) \xrightarrow{\text{right translation } \frac{\pi}{6} \text{ unit}} y = \tan[\omega(x - \frac{\pi}{6}) + \frac{\pi}{4}]$
$= \tan(\omega x + \frac{\pi}{6})$

$$\therefore \frac{\pi}{4} - \frac{\pi}{6}\omega + k\pi = \frac{\pi}{6} \therefore \omega = 6k + \frac{1}{2} \quad (k \in Z)$$

$$\therefore \omega > 0 \therefore \omega_{\min} = \frac{1}{2}.$$

4. Make the image of $y_1 = \sin\frac{\pi x}{2}$ and $y_2 = kx$. If inequality $\sin\frac{\pi x}{2} \geq kx$ holds, $k \leq 1$.

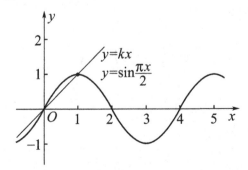

5. $\therefore f(x)$ is decreasing function

$$\therefore \begin{cases} \sin x > 0 \\ \log_2 \dfrac{1}{\sin x} \geq 1 \\ 2k\pi - \dfrac{\pi}{2} \leq x \leq 2k\pi + \dfrac{\pi}{2} \end{cases} \Rightarrow \begin{cases} 0 < \sin x \leq \dfrac{1}{2} \\ 2k\pi - \dfrac{\pi}{2} \leq x \leq 2k\pi + \dfrac{\pi}{2} \end{cases}$$

$$\Rightarrow \begin{cases} 2k\pi < x \leq 2k\pi + \dfrac{\pi}{6} \text{ or } 2k\pi + \dfrac{5\pi}{6} \leq x < 2k\pi + \pi \\ 2k\pi - \dfrac{\pi}{2} \leq x \leq 2k\pi + \dfrac{\pi}{2}. \end{cases}$$

$$\therefore 2k\pi < x \leq 2k\pi + \frac{\pi}{6}, \quad k \in Z.$$

6. $y = -1 + \frac{4}{2+\sin 2x}$, $\because x \in [0, \pi]$, $\therefore 2x \in [0, 2\pi]$,

$\therefore \sin 2x \in [-1, 1]$

implies $\frac{1}{3} \leq -1 + \frac{4}{2+\sin 2x} \leq 3$.

7. $\because f(2+x) = f(2-x)$, $f(x)$ is axial symmetry about $x = 2$.

$\because$ The first intersection of image and x-axis at the right

of the origin is $N(6, 0)$

$\therefore \dfrac{T}{4} = 6 - 2 = 4$, namely, $T = 16$

$\therefore \omega = \dfrac{2\pi}{T} = \dfrac{\pi}{8}$.

Substitute $N(6, 0)$ into

$$f(x) = \sin\left(\frac{\pi}{8}x + \varphi\right) : \sin\left(\frac{3\pi}{4} + \varphi\right) = 0,$$

$\therefore \varphi = 2k\pi + \dfrac{\pi}{4}$ or $\varphi = 2k\pi + \dfrac{5\pi}{4}$ $(k \in Z)$

$\because f(0) < 0$

$$\therefore \; \varphi = 2k\pi + \frac{5\pi}{4} \quad (k \in Z).$$

$$\therefore \quad \text{the minimum positive number of } \varphi = \frac{5\pi}{4}$$

$$\therefore \; f(x) = \sin\left(\frac{\pi}{8}x + \frac{5\pi}{4}\right).$$

8. $f(x) = 2a\sin^2 x - 2\sqrt{3}a\sin x \cos x + a + b - 1$

$= a(1 - \cos 2x) - \sqrt{3}a\sin 2x + a + b - 1$

$= -2a\;\sin(2x + \frac{\pi}{6}) + 2a + b - 1$

$$\because \; 0 \le x \le \frac{\pi}{2} \quad \therefore \; \frac{\pi}{6} \le 2x + \frac{\pi}{6} \le \frac{7}{6}\pi$$

$$\therefore \; -\frac{1}{2} \le \sin\left(2x + \frac{\pi}{6}\right) \le 1$$

$$\because \; a < 0$$

$$\therefore \; a \le -2a\sin\left(2x + \frac{\pi}{6}\right) \le -2a,$$

$$\therefore \; 3a + b - 1 \le -2a\sin\left(2x + \frac{\pi}{6}\right) + 2a + b - 1 \le b - 1$$

$$\because \quad \text{range is } [-3, 1]$$

$$\therefore \; \begin{cases} b - 1 = 1 \\ 3a + b - 1 = -3 \end{cases} \quad \therefore \; \begin{cases} a = -\dfrac{4}{3} \\ b = 2 \end{cases}$$

$$\therefore \; a + b = \frac{2}{3}.$$

9. (1) The lowest point $M(\frac{2\pi}{3}, -2) \therefore A = 2.$

$\therefore$ Distance between two adjacent intersects

on the x-axis is $\dfrac{\pi}{2}$,

$$\therefore \frac{T}{2} = \frac{\pi}{2}, \text{ namely, } T = \pi, \omega = \frac{2\pi}{T} = \frac{2\pi}{\pi} = 2$$

Point $M\left(\frac{2\pi}{3}, -2\right)$ on the image

$$\therefore 2\sin\left(2 \times \frac{2\pi}{3} + \varphi\right) = -2, \text{ namely, } \sin\left(\frac{4\pi}{3} + \varphi\right) = -1$$

$$\therefore \frac{4\pi}{3} + \varphi = 2k\pi - \frac{\pi}{2}, k \in Z \quad \therefore \varphi = 2k\pi - \frac{11\pi}{6}$$

$$\therefore \varphi \in \left(0, \frac{\pi}{2}\right), \quad \therefore \varphi = \frac{\pi}{6}, \text{ so } f(x) = 2\sin\left(2x + \frac{\pi}{6}\right).$$

(2) $\quad \because x \in \left[\frac{\pi}{12}, \frac{\pi}{2}\right], \quad 2x + \frac{\pi}{6} \in \left[\frac{\pi}{3}, \frac{7\pi}{6}\right]$

$$f(x)_{\max} = 2 \text{ when } 2x + \frac{\pi}{6} = \frac{\pi}{2}, \text{ namely, } x = \frac{\pi}{6};$$

$$f(x)_{\min} = -1 \text{ when } 2x + \frac{\pi}{6} = \frac{7\pi}{6}, \text{ namely, } x = \frac{\pi}{2};$$

$\therefore$ the range of $f(x)$ is $[-1, 2]$.

10. (1)

$$f(x) = \sin\frac{\pi}{4}x\cos\frac{\pi}{6} - \cos\frac{\pi}{4}x\sin\frac{\pi}{6} - \cos\frac{\pi}{4}x$$

$$= \frac{\sqrt{3}}{2}\sin\frac{\pi}{4}x - \frac{3}{2}\cos\frac{\pi}{4}x = \sqrt{3}\sin\left(\frac{\pi}{4}x - \frac{\pi}{3}\right)$$

$$\therefore T = \frac{2\pi}{\frac{\pi}{4}} = 8.$$

(2) $(x, g(x))$ is an arbitrarily point on the image of $y = g(x)$ and its symmetric point is $(2 - x, g(x))$ about $x = 1$.

$\because (2 - x, g(x))$ is on the image of $y = f(x)$

$$\therefore g(x) = f(2 - x) = \sqrt{3}\sin\left[\frac{\pi}{4}(2 - x) - \frac{\pi}{3}\right]$$

$$= \sqrt{3} \sin \left(\frac{\pi}{2} - \frac{\pi}{4} x - \frac{\pi}{3} \right) = \sqrt{3} \cos \left(\frac{\pi}{4} x + \frac{\pi}{3} \right)$$

When $0 \le x \le \frac{3}{4}$, $\frac{\pi}{3} \le \frac{\pi}{4} x + \frac{\pi}{3} \le \frac{2\pi}{3}$.

$$\therefore \ g_{\max} = \sqrt{3} \cos \frac{\pi}{3} = \frac{\sqrt{3}}{2} \ \text{in} \ \left[0, \frac{4}{3} \right].$$

11. (1)

$$\because \ f(x) = \frac{1}{2} \left[1 + \cos \left(2x + \frac{\pi}{6} \right) \right].$$

$\because \ x = x_0$ is axis of symmetry of $y = f(x)$

$$\therefore \ 2x_0 + \frac{\pi}{6} = k\pi, \ \text{namely} \ 2x_0 = k\pi - \frac{\pi}{6}.$$

$$\therefore \ g(x_0) = 1 + \frac{1}{2} \sin 2x_0 = 1 + \frac{1}{2} \sin \left(k\pi - \frac{\pi}{6} \right).$$

When k is an even number,

$$g(x_0) = 1 + \frac{1}{2} \sin \left(-\frac{\pi}{6} \right) = 1 - \frac{1}{4} = \frac{3}{4}.$$

When k is an odd number, $g(x_0) = 1 + \frac{1}{2} \sin \frac{\pi}{6} = 1 + \frac{1}{4} = \frac{5}{4}.$

(2)

$$h(x) = f(x) + g(x) = \frac{1}{2} \left[1 + \cos \left(2x + \frac{\pi}{6} \right) \right] + 1 + \frac{1}{2} \sin 2x$$

$$= \frac{1}{2} \left(\frac{\sqrt{3}}{2} \cos 2x + \frac{1}{2} \sin 2x \right) + \frac{3}{2}$$

$$= \frac{1}{2} \sin \left(2x + \frac{\pi}{3} \right) + \frac{3}{2}.$$

When $2k\pi - \frac{\pi}{2} \le 2x + \frac{\pi}{3} \le 2k\pi + \frac{\pi}{2}$, namely

$$k\pi - \frac{5\pi}{12} \le x \le k\pi + \frac{\pi}{12} \ (k \in Z).$$

$\because$ The function $h(x) = \frac{1}{2}\sin\left(2x + \frac{\pi}{3}\right)\frac{3}{2}$

is increasing function.

$\therefore$ Increasing interval of function $h(x)$ is

$$\left[k\pi - \frac{5\pi}{12}, k\pi + \frac{\pi}{12}\right] \quad (k \in Z).$$

12. Suppose $f(x) = \sin(\cos x) - x$, $x \in [0, \frac{\pi}{2}]$.

$\therefore$ $f(x)$ is continuous function in definition domain.

Take arbitrary $x_1, x_2 \in \left[0, \frac{\pi}{2}\right]$ to $0 \le x_1 < x_2 \le \frac{\pi}{2}$

$f(x_1) = \sin(\cos x_1) - x_1$, $f(x_2) = \sin(\cos x_2) - x_2$

$\because$ $y = \cos x$ is a decreasing function in $\left[0, \frac{\pi}{2}\right]$

$\therefore$ $\cos x_1 > \cos x_2$

$\therefore$ $\sin(\cos x_1) > \sin(\cos x_2)$

$\therefore$ $x_1 < x_2$

$\therefore$ $\sin(\cos x_1) - x_1 > \sin(\cos x_2) - x_2$

$\therefore$ $f(x_1) > f(x_2)$

$\therefore$ $f(x)$ is a decreasing function in $\left[0, \frac{\pi}{2}\right]$

$\because$ $f(0) = \sin 1 > 1, f\left(\frac{\pi}{2}\right) = -\frac{\pi}{2} < 0.$

$\therefore$ There is only one intersection of x-axis and image of

function $f(x)$ in $\left(0, \frac{\pi}{2}\right)$.

$\therefore$ There exists only one real number $c \in \left(0, \dfrac{\pi}{2}\right)$

to make equation $\sin(\cos c) = c$ hold.

Similarly, there exists only one real number $d \in \left(0, \dfrac{\pi}{2}\right)$

to make equation $\cos(\sin d) = d$ hold.

$\because \cos(\sin d) = d$,

$\therefore \sin(\cos(\sin d)) = \sin d$

$\because \sin d \in \left(0, \dfrac{\pi}{2}\right)$.

$\therefore$ There is a unique solution c of equation

$\qquad \sin(\cos c) = c$ in $\left(0, \dfrac{\pi}{2}\right)$.

$\therefore c = \sin d$

When $x > 0, \sin x < x$,

$\therefore \sin d < d$

$\therefore c < d$.

13. Take arbitrary $0 < x_1 < x_2 < \frac{\pi}{2}$.

Inequality $f(x_1) > f(x_2)$ always holds equal to

$\dfrac{m - 2\sin x_1}{\cos x_1} > \dfrac{m - 2\sin x_2}{\cos x_2}$ always holds

$\therefore m\cos x_2 - 2\sin x_1 \cos x_2 > m\cos x_1 - 2\sin x_2 \cos x_1$

$\therefore m(\cos x_2 - \cos x_1) > 2\sin(x_1 - x_2)$

$\because 0 < x_1 < x_2 < \dfrac{\pi}{2}$

$\therefore \cos x_2 - \cos x_1 < 0$

$\therefore m < \dfrac{2\sin(x_1 - x_2)}{\cos x_2 - \cos x_1}$ always holds.

Let

$$t = \frac{2\sin(x_1 - x_2)}{\cos x_2 - \cos x_1} = \frac{4\sin\frac{x_1-x_2}{2}\cos\frac{x_1-x_2}{2}}{2\sin\frac{x_1+x_2}{2}\sin\frac{x_1-x_2}{2}} = \frac{2\cos\frac{x_1-x_2}{2}}{\sin\frac{x_1+x_2}{2}}$$

$$= \frac{2\left(\cos\frac{x_1}{2}\cos\frac{x_2}{2} + \sin\frac{x_1}{2}\sin\frac{x_2}{2}\right)}{\sin\frac{x_1}{2}\cos\frac{x_2}{2} + \cos\frac{x_1}{2}\sin\frac{x_2}{2}} = \frac{2\left(1 + \tan\frac{x_1}{2}\tan\frac{x_2}{2}\right)}{\tan\frac{x_1}{2} + \tan\frac{x_2}{2}}.$$

When $0 < x_1 < x_2 < \dfrac{\pi}{2}$, $0 < \dfrac{x_1}{2} < \dfrac{x_2}{2} < \dfrac{\pi}{4}$

$\therefore\ 0 < \tan\dfrac{x_1}{2} < \tan\dfrac{x_2}{2} < 1$

$\therefore\ 1 + \tan\dfrac{x_1}{2}\tan\dfrac{x_2}{2} - \left(\tan\dfrac{x_1}{2} + \tan\dfrac{x_2}{2}\right)$

$$= \left(1 - \tan\dfrac{x_1}{2}\right)\left(1 - \tan\dfrac{x_2}{2}\right) > 0$$

$\therefore\ \dfrac{2\left(1 + \tan\frac{x_1}{2}\tan\frac{x_2}{2}\right)}{\tan\frac{x_1}{2} + \tan\frac{x_2}{2}} > 2$

$\therefore\ m \in (-\infty, 2]$.

14. Suppose $\mu = F(a, \theta)$

$\therefore$ Original type $= 2a\sin\theta - 2\mu a\cos\theta = (\mu - 1)(a^2 + 2)$

$\Rightarrow \sqrt{4a^2 + 4\mu^2 a^2}\,\sin(\theta + \varphi) = (\mu - 1)(a^2 + 2)$

$\Rightarrow \sin(\theta + \varphi) = \dfrac{(\mu - 1)(a^2 + 2)}{\sqrt{4a^2 + 4\mu^2 a^2}}$

$\therefore\ \left|\dfrac{(\mu - 1)(a^2 + 2)}{\sqrt{4a^2 + 4\mu^2 a^2}}\right| \le 1 \Rightarrow \dfrac{|\mu - 1|(a^2 + 2)}{2|a|\sqrt{\mu^2 + 1}} \le 1 \Rightarrow \dfrac{|\mu - 1|}{\sqrt{\mu^2 + 1}}$

$\le \dfrac{2|a|}{a^2 + 2} \le \dfrac{2|a|}{2\sqrt{2}|a|} = \dfrac{1}{\sqrt{2}}$

$$\therefore \frac{|\mu - 1|}{\sqrt{\mu^2 + 1}} \leq \frac{1}{\sqrt{2}} \Rightarrow 2\mu^2 - 4\mu + 2 \leq \mu^2 + 1$$

$$\therefore \mu^2 - 4\mu + 1 \leq 0$$

$$\therefore 2 - \sqrt{3} \leq \mu \leq 2 + \sqrt{3}$$

$$\therefore F(a, \theta) \in [2 - \sqrt{3}, \ 2 + \sqrt{3}].$$

15. $\because \tan^2 \alpha = \frac{1}{\cos^2 \alpha} - 1$

$$\therefore \tan^2 \alpha + \tan^2 \beta + 8 \tan^2 \gamma = \frac{1}{\cos^2 \alpha} + \frac{1}{\cos^2 \beta} + \frac{8}{\cos^2 \gamma} - 10$$

$$\because (a + b + c)\left(\frac{1}{\cos^2 \alpha} + \frac{1}{\cos^2 \beta} + \frac{8}{\cos^2 \gamma}\right)$$

$$\geq \left(\frac{\sqrt{a}}{\cos \alpha} + \frac{\sqrt{b}}{\cos \beta} + \frac{\sqrt{8c}}{\cos \gamma}\right)^2$$

$$= \left(\left(\frac{\sqrt{a}}{\cos \alpha} + \frac{\sqrt{b}}{\cos \beta} + \frac{\sqrt{8c}}{\cos \gamma}\right)(\cos \alpha + \cos \beta + \cos \gamma)\right)^2$$

$$\geq \left(\sqrt[4]{a} + \sqrt[4]{b} + \sqrt[4]{8c}\right)^2$$

Equally holds if and only if $a = b = \dfrac{c}{2}$ and $\cos \alpha = \cos \beta$

$$= \frac{\cos \gamma}{2} = \frac{1}{4}.$$

$$\therefore \frac{1}{\cos^2 \alpha} + \frac{1}{\cos^2 \beta} + \frac{8}{\cos^2 \gamma} \geq 64$$

$$\therefore \tan^2 \alpha + \tan^2 \beta + 8 \tan^2 \gamma \geq 54.$$

16. When $x_1 = x_2 = x_3 = x_4 = \frac{\pi}{4}$, $f(x_1, x_2, x_3, x_4) = 81$.

If x_1, x_2, x_3, x_4 are not all equal, suppose $x_1 > \dfrac{\pi}{4} > x_2$.

Take $x'_1 = \dfrac{\pi}{4}$, $\quad x'_2 = x_1 + x_2 - \dfrac{\pi}{4}$, $\quad x'_3 = x_3$, $\quad x'_4 = x_4$

$\therefore x'_1 + x'_2 + x'_3 + x'_4 = x_1 + x_2 + x_3 + x_4 = \pi$

and $\quad x'_1 = \dfrac{\pi}{4} < x_1, x'_2 = x_1 + x_2 - \dfrac{\pi}{4} > x_2$

$\therefore \sin x_1 \sin x_2 - \sin x'_1 \sin x'_2$

$\qquad = \dfrac{1}{2}(\cos(x_1 - x_2) - \cos(x'_1 - x'_2)) < 0$

and $\sin^2 x_1 > \sin^2 x'_1, \sin^2 x_2 < \sin^2 x'_2$

$\therefore \dfrac{\sin^2 x_2}{\sin^2 x_1} < \dfrac{\sin^2 x'_2}{\sin^2 x'_1}$

$\because \left(2\sin^2 x_1 + \dfrac{1}{\sin^2 x_1} \right) \left(2\sin^2 x_2 + \dfrac{1}{\sin^2 x_2} \right)$

$= 2 \left(2\sin^2 x_1 \sin^2 x_2 + \dfrac{1}{2\sin^2 x_1 \sin^2 x_2} \right) + 2 \left(\dfrac{\sin^2 x_1}{\sin^2 x_2} + \dfrac{\sin^2 x_2}{\sin^2 x_1} \right)$

$\because y = x + \dfrac{1}{x}$ is decreasing function in $(0,1)$

$\therefore \left(2\sin^2 x_1 + \dfrac{1}{\sin^2 x_1} \right) \left(2\sin^2 x_2 + \dfrac{1}{\sin^2 x_2} \right)$

$\qquad > \left(2\sin^2 x'_1 + \dfrac{1}{\sin^2 x'_1} \right) \left(2\sin^2 x'_2 + \dfrac{1}{\sin^2 x'_2} \right)$

$\because f(x_1, x_2, x_3, x_4)_{\min} = 81$.

17. Note $f(x) = x^3 \sin\theta - (\sin\theta - 2)x^2 + 6x - 4$, $\therefore f(1) = 0$

$\quad \therefore f(x) = (x - 1)(\sin\theta \cdot x^2 - 2x - 4)$

$\quad \therefore f(x) = 0$ has three real roots

$\quad \therefore \Delta \geq 0 \quad$ and $\quad \dfrac{2}{\sin\theta} > 0$, $\quad \dfrac{4}{\sin\theta} > 0$,

$$\therefore \ 0 < \sin \theta \leq \frac{1}{4}$$

$$\therefore \ 9 \sin^2 \theta - 4 \sin \theta + 3 = 9 \left(\sin \theta - \frac{2}{9} \right)^2 + \frac{23}{9} \geq \frac{23}{9}$$

$$(1 - \cos \theta)(2 \cos \theta - 6 \sin \theta - 3 \sin 2\theta + 2)$$

$$= 2(1 - \cos \theta)(1 + \cos \theta)(1 - 3 \sin \theta) = 2 \sin^2 \theta (1 - 3 \sin \theta)$$

$$= \frac{8}{9} \cdot \frac{3}{2} \sin \theta \cdot \frac{3}{2} \sin \theta (1 - 3 \sin \theta) \leq \frac{8}{9} \cdot \left(\frac{1}{3} \right)^3 = \frac{8}{243}$$

$$\therefore \ \mu \geq \frac{\frac{23}{9}}{\frac{8}{243}} = \frac{621}{8}; \text{ the equally holds if and only if } \sin \theta = \frac{2}{9}.$$

18. $f(0) = a$

When $x = y = 0$, $f(a) = a$.

When $y = 0$, $f(x + a) = f(x)$ $(x \in R)$.

Let $g(x) = f(x) - a$, so $g(0) = 0$.

$$g(x + g(y)) = f(x + g(y)) - a = f(x + f(y) - a) - a$$

$$= f(x + f(y)) - a = f(x) + \sin y - a$$

$$= g(x) + \sin y \quad (x, y \in R)$$

When $x = 0$, $g(g(y)) = \sin y$. \hfill (1)

When $y = g(x)$, $g(\sin x) = \sin(g(x))$

When $x = \sin \mu$, $y = g(v)$, $g(\sin u + \sin v)$

$$= \sin(g(u)) + \sin(g(v)) \quad (u, v \in R) \hfill (2)$$

$\sin(\sin u + \sin v) = g(g(\sin u + \sin v))$ \quad (according to type (1))

$$= g(\sin(g(u)) + \sin(g(v)))$$

\hfill (according to type (2))

$$= \sin(g(g(u))) + \sin(g(g(v)))$$

$$\text{(according to type (2))}$$

$$= \sin(\sin u) + \sin(\sin v)$$

$$\text{(according to type (1))}.$$

When $u = v = \dfrac{\pi}{2}$, $\sin 2 = 2\sin 1$, $\cos 1 = 1$ (contradiction).

$\therefore$ There is no function which meets the condition.

19. $-\frac{\pi}{2} < f(x) < \frac{\pi}{2}, -\frac{\pi}{2} < g(x) < \frac{\pi}{2}$

(i)   If $0 \le f(x) < \dfrac{\pi}{2}, -\dfrac{\pi}{2} < g(x) < \dfrac{\pi}{2} - f(x) \le \dfrac{\pi}{2}$

According to monotonousness of $y = \sin x$ in $\left[-\dfrac{\pi}{2}, \dfrac{\pi}{2}\right]$:

$$\sin g(x) < \sin\left(\dfrac{\pi}{2} - f(x)\right) = \cos f(x).$$

$\therefore$ For all $x \in R$, the proposition is true.

When $x \in R, |\sin x \pm \cos x| \le \sqrt{2} < \dfrac{\pi}{2}.$

$\therefore \cos(\cos x) > \sin(\sin x)$

20. Note $g(x) = \frac{f(x)+f(-x)}{2}$, $h(x) = \frac{f(x)-f(-x)}{2}$.

$\therefore$ $f(x) = g(x) + h(x)$, $g(x)$ is even function and $h(x)$ is odd function.

For arbitrary $x \in R$, $g(x + 2\pi) = g(x)$, $h(x + 2\pi) = h(x)$.

Let

$$f_1(x) = \dfrac{g(x) + g(x + \pi)}{2},$$

$$f_2(x) = \begin{cases} \dfrac{g(x) - g(x + \pi)}{2\cos x}, & x \ne k\pi + \dfrac{\pi}{2}, \\[2mm] 0, & x = k\pi + \dfrac{\pi}{2}, \end{cases}$$

$$f_3(x) = \begin{cases} \dfrac{h(x) - h(x + \pi)}{2\sin x}, & x \neq k\pi, \\ \\ 0, & x = k\pi, \end{cases}$$

$$f_4(x) = \begin{cases} \dfrac{h(x) + h(x + \pi)}{2\sin 2x}, & x \neq \dfrac{k\pi}{2}, \\ \\ 0, & x = \dfrac{k\pi}{2}, \end{cases} \qquad (k \in Z).$$

It is easy to prove that $f_i(x)$ $(i = 1, 2, 3, 4)$ is even function and $f_i(x + \pi) = f_i(x)$ $(i = 1, 2, 3, 4)$ to arbitrary $x \in R$.

Then prove: for arbitrary $x \in R$, $f_1(x) + f_2(x)\cos x = g(x)$.

When $x \neq k\pi + \dfrac{\pi}{2}$, clearly established.

When $x = k\pi + \dfrac{\pi}{2}$, $f_1(x) + f_2(x)\cos x = f_1(x) = \dfrac{g(x) + g(x + \pi)}{2}$.

$$\because g(x + \pi) = g\left(k\pi + \frac{3\pi}{2}\right) = g\left(-k\pi - \frac{\pi}{2}\right)$$

$$= g\left(k\pi + \frac{\pi}{2}\right) = g(x)$$

$\therefore$ For arbitrary $x \in R$, $f_1(x) + f_2(x)\cos x = g(x)$.

Prove: For arbitrary $x \in R$, $f_3(x)\sin x + f_4(x)\sin 2x = h(x)$.

When $x \neq \dfrac{k\pi}{2}$, clearly established.

When $x = k\pi$, $h(x) = h(k\pi) = h(-k\pi) = -h(k\pi) = 0$

$$= f_3(x)\sin k\pi + f_4(x)\sin 2k\pi$$

$$= f_3(x)\sin x + f_4(x)\sin 2x.$$

When $x = k\pi + \dfrac{\pi}{2}$, $h(x + \pi) = h\left(k\pi + \dfrac{3\pi}{2}\right) = h\left(-k\pi - \dfrac{\pi}{2}\right)$

$$= -h\left(k\pi + \frac{\pi}{2}\right) = -h(x).$$

$\therefore$ For arbitrary $x \in R$, $f_3(x)\sin x + f_4(x)\sin 2x = h(x)$.

$\therefore$ Conclusion is evident.

21. Firstly, prove that $f(x)$ is not identical to 0.

$\because$ For all real number $x, \cos(a_i + x) \geq -1$

$\therefore f(-a_1) = 1 + \dfrac{1}{2}\cos(a_2 - a_1) + \dfrac{1}{4}\cos(a_3 - a_1) + \cdots$

$$+ \dfrac{1}{2^{n-1}}\cos(a_n - a_1)$$

$\geq 1 - \dfrac{1}{2} - \dfrac{1}{4} - \cdots - \dfrac{1}{2^{n-1}} = \dfrac{1}{2^{n-1}} > 0.$

$\therefore$ There is at least one real number $x = -a_1$ to $f(x) \neq 0$.

Then, according to addition theorem:

$$f(x) = \sum_{k=1}^{n} \dfrac{1}{2^{k-1}}(\cos a_k \cdot \cos x - \sin a_k \cdot \sin x)$$

$$= \left(\sum_{k=1}^{n} \dfrac{1}{2^{k-1}}\cos a_k\right)\cos x - \left(\sum_{k=1}^{n} \dfrac{1}{2^{k-1}}\sin a_k\right)\sin x$$

$$= A \cdot \cos x$$

$$-B \cdot \sin x \left(A = \sum_{k=1}^{n} \dfrac{1}{2^{k-1}}\cos a_k, B = \sum_{k=1}^{n} \dfrac{1}{2^{k-1}}\sin a_k\right).$$

If A and B are equal to 0 at the same time, $f(x)$ is identical to 0, which is contradictory with above conclusion.

If $A \neq 0$, $f_1(x) = A \cdot \cos x_1 - B\sin x_1 = 0$ and

$$f_2(x) = A \cdot \cos x_2 - B\sin x_2 = 0$$

$\therefore \cot x_1 = \cot x_2 = \dfrac{B}{A}.$

If $A = 0, B \neq 0$.

$\because f_1(x) = f_2(x) = 0$

$\therefore \sin x_1 = \sin x_2 = 0$

$\therefore x_2 - x_1 = m\pi, m \in Z.$

22. Suppose $\sin\theta + \cos\theta = x, \theta \in \left[0, \dfrac{\pi}{2}\right]$,

$\therefore x \in [1, \sqrt{2}], \sin 2\theta = x^2 - 1$

$\therefore x^2 - 1 - (2 + a)x - \dfrac{4}{x} + 3 + 2a > 0$

$\therefore (x - 2)\left(x + \dfrac{2}{x} - a\right) > 0$

$\therefore x \in [1, \sqrt{2}]$

$\therefore x + \dfrac{2}{x} - a < 0, \text{ namely, } a > x + \dfrac{2}{x}.$

Note $f(x) = x + \dfrac{2}{x}$

$\because f(x)$ is decreasing function in $[1, \sqrt{2}]$

$\therefore f(x)_{\max} = f(1) = 1 + \dfrac{2}{1} = 3$

$\therefore a > 3.$

23. When $n = 1, \cos\theta_1 = \dfrac{\sqrt{3}}{3}, \lambda_{\min} = \dfrac{\sqrt{3}}{3}.$

When $n = 2, \cos\theta_1 + \cos\theta_2 \leq \dfrac{2\sqrt{3}}{3}.$

The Equality holds if and only if $\theta_1 = \theta_2$.

$\therefore \lambda_{\min} = \dfrac{2\sqrt{3}}{3}.$

When $n \geq 3, \cos\theta_1 + \cos\theta_2 + \cdots + \cos\theta_n \leq n - 2.$

In fact, suppose $\theta_1 \geq \theta_2 \geq \theta_3 \geq \cdots \geq \theta_n,$

so $\cos\theta_1 \leq \cos\theta_2 \leq \cdots \leq \cos\theta_n;$

only need to prove: $\cos\theta_1 + \cos\theta_2 + \cos\theta_3 \leq 2;$

$\because \tan\theta_1 \cdot \tan\theta_2 \cdot \tan\theta_3 \geq 2\sqrt{2},$

$\therefore \tan^2\theta_1 \geq \dfrac{8}{\tan^2\theta_2\tan^2\theta_3} = \dfrac{8\cos^2\theta_2\cos^2\theta_3}{\sin^2\theta_2\sin^2\theta_3},$

$$\therefore \ \cos\theta_1 = \frac{1}{\sqrt{1+\tan^2\theta_1}} \leq \frac{\sin\theta_2\sin\theta_3}{\sqrt{8\cos^2\theta_2\cos^2\theta_3 + \sin^2\theta_2\sin^2\theta_3}},$$

$$\because \ \cos\theta_2 = \sqrt{1-\sin^2\theta_2} \leq 1 - \frac{1}{2}\sin^2\theta_2,$$

$$\cos\theta_3 = \sqrt{1-\sin^2\theta_3} \leq 1 - \frac{1}{2}\sin^2\theta_3,$$

$$\therefore \ \cos\theta_2 + \cos\theta_3 \leq 2 - \frac{1}{2}(\sin^2\theta_2 + \sin^2\theta_3) < 2 - \sin\theta_2\sin\theta_3.$$

(i) If $8\cos^2\theta_2\cos^2\theta_3 + \sin^2\theta_2\sin^2\theta_3 \geq 1$, then $\cos\theta_1 \leq \sin\theta_2\sin\theta_3$.

$$\therefore \ \cos\theta_1 + \cos\theta_2 + \cos\theta_3 \leq 2.$$

(ii) If $8\cos^2\theta_2\cos^2\theta_3 + \sin^2\theta_2\sin^2\theta_3 < 1$, then
$9\cos^2\theta_2\cos^2\theta_3 - \cos^2\theta_2 - \cos^2\theta_3 < 0$.

$$\therefore \ \tan^2\theta_2 + \tan^2\theta_3 > 7,$$

$$\therefore \ \tan^2\theta_2 > \frac{7}{2}, \cos\theta_1 \leq \cos\theta_2 = \frac{1}{\sqrt{1+\tan^2\theta_2}} \leq \frac{1}{\sqrt{1+\frac{7}{2}}} = \frac{\sqrt{2}}{3}$$

$$\therefore \ \cos\theta_1 + \cos\theta_2 + \cos\theta_3 < \frac{2\sqrt{2}}{3} + 1 < 2.$$

In addition, when $\theta_1 = \theta_2 = \cdots = \theta_{n-1} \to 0, \theta_n \to \frac{\pi}{2}$,

$$\cos\theta_1 + \cos\theta_2 + \cdots + \cos\theta_n \to n - 1.$$

$$\therefore \ \lambda_{min} = n - 1.$$

Exercise 4

1. We have

$$\cos 5x + \cos x > 2\cos 2x \Rightarrow 2\cos 2x(1 - \cos 3x) < 0$$

$$\Rightarrow \cos 3x < 1 \quad \text{and} \quad \cos 2x < 0$$

$$\Rightarrow n\pi + \frac{\pi}{4} < x < n\pi + \frac{3\pi}{4} \quad \text{and } x \ne \frac{2m\pi}{3}, \quad m, n \in Z$$

$$\frac{x^2 + x - 2}{x^2 + x - 6} < 0 \Rightarrow -3 < x < -2 \quad \text{or} \quad 1 < x < 2$$

$$\therefore x \in \left(-\frac{3\pi}{4}, -\frac{2\pi}{3}\right) \cup \left(-\frac{2\pi}{3}, -2\right) \cup (1, 2).$$

2. $\sin 3x = \cos x - \cos 2x \Rightarrow \sin \dfrac{3x}{2} \left(\cos \dfrac{3x}{2} - \sin \dfrac{x}{2}\right) = 0$

$$\Rightarrow \sin \frac{3x}{2} = 0$$

or $\cos \dfrac{3x}{2} = \cos \left(\dfrac{\pi}{2} - \dfrac{x}{2}\right) \Rightarrow x = \dfrac{2k\pi}{3}$ or $x = k\pi + \dfrac{\pi}{4}$ or

$$x = 2k\pi - \frac{\pi}{2}, \ k \in Z.$$

After inspection, $2k\pi + \dfrac{\pi}{4}, k \in Z$ is the root of origin equation.

3. Substitute $C = 180° - (A + B)$ into $\cos 3A + \cos 3B + \cos 3C = 1$.

$\therefore \ \cos 3A + \cos 3B - \cos 3(A + B) = 1$, namely,

$$\sin 3A \sin 3B = (1 - \cos 3A)(1 - \cos 3B).$$

Squaring both sides, we get

$$\sin^2 3A \sin^2 3B = (1 - \cos 3A)^2 (1 - \cos 3B)^2$$

$$\Rightarrow (1 - \cos 3A)(1 - \cos 3B)(\cos 3A + \cos 3B) = 0$$

$\therefore \ \cos 3A = -\cos 3B = \cos(180° - 3B)$

$$\Rightarrow 3A = 180° - 3B \Rightarrow C = 120°.$$

4. Suppose the height of BC is $AD = h$, so

$$\tan(\angle BAC) = \tan(\angle BAD + \angle CAD)$$

$$\therefore \ \frac{22}{7} = \frac{\frac{17}{h} + \frac{3}{h}}{1 - \frac{17}{h} \times \frac{3}{h}} \Rightarrow h = 11 \Rightarrow S_{\triangle ABC} = \frac{1}{2} \times 20 \times 11 = 110.$$

5. $\arcsin(\sin\alpha + \sin\beta) = \frac{\pi}{2} - \arcsin(\sin\alpha - \sin\beta)$,

$$\therefore \ \sin\alpha + \sin\beta = \cos[\arcsin(\sin\alpha - \sin\beta)].$$

$$(\sin\alpha + \sin\beta)^2 = 1 - \sin^2[\arcsin(\sin\alpha - \sin\beta)]$$

$$= 1 - (\sin\alpha - \sin\beta)^2$$

$$\therefore \ \sin^2\alpha + \sin^2\beta = \frac{1}{2}.$$

6. Definition domain: $x \neq -1$.

    ①   When $x < -1$, $\dfrac{1-x}{1+x} = -1 + \dfrac{2}{1+x} < -1$

$$\Rightarrow -\frac{\pi}{2} < \arctan x < -\frac{\pi}{4},$$

$$-\frac{\pi}{2} < \arctan\frac{1-x}{1+x} < -\frac{\pi}{4}$$

$$\Rightarrow -\pi < \arctan x + \arctan\frac{1-x}{1+x} < -\frac{\pi}{2}$$

$$\therefore \ \tan\left(\arctan x + \arctan\frac{1-x}{1+x}\right) = \frac{x + \frac{1-x}{1+x}}{1 - x\cdot\frac{1-x}{1+x}}$$

$$= -1 \Rightarrow \arctan x + \arctan\frac{1-x}{1+x} = -\frac{3\pi}{4}.$$

    ②   When $x > -1$,

$$\frac{1-x}{1+x} = -1 + \frac{2}{1+x} > -1 \Rightarrow -\frac{\pi}{4} < \arctan x < \frac{\pi}{2},$$

$$-\frac{\pi}{4} < \arctan\frac{1-x}{1+x} < \frac{\pi}{2}$$

$$\Rightarrow -\frac{\pi}{2} < \arctan x + \arctan\frac{1-x}{1+x} < \pi$$

$$\therefore \ \tan\left(\arctan x + \ \ \arctan\frac{1-x}{1+x}\right) = \frac{x + \frac{1-x}{1+x}}{1 - x\cdot\frac{1-x}{1+x}} = 1$$

$$\Rightarrow \arctan x + \arctan\frac{1-x}{1+x} = \frac{\pi}{4}$$

$$\therefore \ y \in \left\{\frac{\pi}{4}, -\frac{3\pi}{4}\right\}.$$

7. (1) Suppose the side of inscribed square $PQRS$ of the right-angled triangle ABC is x,

$\because PS//BC$, namely $\angle APS = \theta$.

In the right-angled triangles APS and the

right-angled triangles $PQB, AP = x\cos\theta$,

$PB\sin\theta = x$,

$\therefore x = \dfrac{a\sin\theta}{1+\sin\theta\cos\theta}$,

$\because AC = AB\tan\theta = a\tan\theta$,

$\therefore S_1 = \dfrac{a^2}{2}\tan\theta, S_2 = \dfrac{a^2\sin^2\theta}{(1+\sin\theta\cos\theta)^2}$.

(2) Suppose $y = f(\theta) = \dfrac{S_1}{S_2} = \dfrac{(2+\sin 2\theta)^2}{4\sin 2\theta}$ and $\sin 2\theta = t$

$\because 0 < \theta < \dfrac{\pi}{2} \Rightarrow 0 < t \le 1,$ (*)

$\because y = \dfrac{1}{4}\left(t + 4 + \dfrac{4}{t}\right).$

The above type is equal to $t^2 - 4t(y-1) + 4 = 0 \Rightarrow t = 2[(y-1) \pm \sqrt{y^2 - 2y}]$.

By (*), we have $\begin{cases} y > 0, & \text{(i)} \\[2mm] y^2 - 2y \ge 0, & \text{(ii)} \\[2mm] 0 < (y-1) \pm \sqrt{y^2 - 2y} \le \dfrac{1}{2}. & \text{(iii)} \end{cases}$

According to (ii) and (iii): $y \ge 2, \therefore y - 1 + \sqrt{y^2 - 2y} > \dfrac{1}{2}$, which is impossible.

$\therefore$ Front of radical can only be negative

$\therefore -2\sqrt{y^2 - 2y} \le 3 - 2y \Rightarrow y \ge \dfrac{9}{4}.$

The equality holds, if and only if $\sin 2\theta = t = 2\left(1.25 \pm \frac{3}{4}\right) = 1$, namely, $\theta = \frac{\pi}{4}$.

8. Suppose $f(n) = \frac{n}{n+2}$, so $\frac{f(n)-f(n-1)}{1+f(n)f(n-1)} = \frac{\frac{n}{n+2} - \frac{n-1}{n+1}}{1 + \frac{n}{n+2} \cdot \frac{n-1}{n+1}} = \frac{1}{n^2+n+1}$.

$\therefore$ $\tan\left(\arctan\dfrac{1}{n^2+n+1}\right)$

$$= \frac{\tan\left(\arctan f(n)\right) - \tan\left(\arctan f(n-1)\right)}{1 + \tan\left(\arctan f(n)\right) \cdot \tan\left(\arctan f(n-1)\right)}$$

$$= \tan\left(\arctan f(n) - \arctan f(n-1)\right)$$

$\therefore$ $\arctan \dfrac{1}{n^2+n+1} = \arctan f(n) - \arctan f(n-1)$

$$= \arctan \frac{n}{n+2} - \arctan \frac{n-1}{n+1}$$

$\therefore$ $\arctan \dfrac{1}{3} + \arctan \dfrac{1}{7} + \arctan \dfrac{1}{13} + \cdots$
$+ \arctan \dfrac{1}{n^2+n+1}$

$$= \arctan \frac{1}{3} + \sum_{k=2}^{n}\left(\arctan \frac{k}{k+2} - \arctan \frac{k-1}{k+1}\right)$$

$$= \arctan \frac{n}{n+2}.$$

9. Suppose $f(x)$ is odd function in $\left(-\frac{1}{4}, \frac{1}{4}\right)$, so $f(0) = 0$.

$\therefore$ $\arctan 2 + c = 0 \Rightarrow c = -\arctan 2$.

If there is a c which meets the condition, $c = -\arctan 2$.

Following we prove: when $c = -\arctan 2$, $f(x)$ is odd function in $\left(-\frac{1}{4}, \frac{1}{4}\right)$.

Note when $x \in \left(-\frac{1}{4}, \frac{1}{4}\right)$, then function $u(x) = \frac{2-2x}{1+4x} = \frac{1}{2}\left(\frac{5}{1+4x} - 1\right)$.

$$\therefore \ u\left(x\right) \in \left(\frac{3}{4} + \infty\right)$$

$$\therefore \ f\left(x\right) = \arctan u\left(x\right) - \arctan 2$$

$$\in \left(\arctan \frac{3}{4} - \arctan 2, \frac{\pi}{2} - \arctan 2\right) \subset \left(-\frac{\pi}{2}, \frac{\pi}{2}\right).$$

The above discussion shows, when $x \in \left(-\frac{1}{4}, \frac{1}{4}\right)$.

To prove: when $x \in \left(-\frac{1}{4}, \frac{1}{4}\right)$, $f\left(x\right) \in \left(-\frac{\pi}{2}, \frac{\pi}{2}\right)$,

$$-f\left(-x\right) \in \left(-\frac{\pi}{2}, \frac{\pi}{2}\right).$$

Only to prove: $\tan(-f(-x)) = \tan f(x)$:

$$\because \ \tan f\left(x\right) = \tan(\arctan u\left(x\right) - \arctan 2) = \frac{u(x) - 2}{1 + 2u(x)}$$

$$= -2x$$

$$\tan(-f(-x)) = \tan\left(\arctan 2 - \arctan u(-x)\right)$$

$$= \frac{2 - u(-x)}{1 + 2u(-x)} = -2x$$

$$\therefore \ \text{if and only if } c = -\arctan 2, \ f\left(x\right) \text{ is odd function}$$

$$\text{in } \left(-\frac{1}{4}, \frac{1}{4}\right).$$

10. we have

$$\because \ \begin{cases} \sin^3 x \le \sin^2 x, \\ \cos^3 x \le \cos^2 x, \end{cases} \quad \therefore \ \sin^3 x + \cos^5 x \le 1$$

$$\therefore \ \begin{cases} \sin x = 1 \\ \cos x = 0 \end{cases} \text{ or } \begin{cases} \sin x = 0, \\ \cos x = 1. \end{cases}$$

$\therefore$ The solution of origin equation is

$$\left\{ x \mid x = 2k\pi \ or \ x = 2k\pi + \frac{\pi}{2}, k \in Z \right\}.$$

11. Suppose $\arctan x_1 = \alpha$, $\arctan x_2 = \beta$, so $\alpha, \beta \in (-\frac{\pi}{2}, \frac{\pi}{2})$

and $\tan \alpha = x_1$, $\tan \beta = x_2$, $\alpha + \beta \in (-\pi, \pi)$.

$$\because \begin{cases} x_1 + x_2 = 6 \\ x_1 \cdot x_2 = 7 \end{cases}$$

$$x_1, x_2 > 0, \alpha + \beta \in (0, \pi),$$

$$\therefore \ \tan(\alpha + \beta) = \frac{x_1 + x_2}{1 - x_1 \cdot x_2} = -1,$$

namely, $\arctan x_1 + \arctan x_2 = \alpha + \beta = \dfrac{3\pi}{4}.$

12. Let $n = \csc \alpha_n$, $\alpha_n \in (0, \frac{\pi}{2}]$, $\sin \alpha_n = \frac{1}{n}$,

so $\arcsin \dfrac{\sqrt{n^2-1}-\sqrt{(n-1)^2-1}}{n(n-1)}$ $=$ $\arcsin \dfrac{\cot \alpha_n - \cot \alpha_{n-1}}{\csc \alpha_n \csc \alpha_{n-1}}$ $=$

$\arcsin(\sin(\alpha_{n-1} - \alpha_n)) = \alpha_{n-1} - \alpha_n$

$$\therefore \ \ origin \ type = (\alpha_1 - \alpha_2) + (\alpha_2 - \alpha_3) + \cdots$$

$$+ (\alpha_n - \alpha_{n+1}) = \alpha_1 - \alpha_{n+1}$$

$$= \frac{\pi}{2} - \arcsin \frac{1}{n+1} = \arccos \frac{1}{n+1}.$$

13. The origin equation equals to $\sin^2 x - \sin x = a - 1$, namely, $(\sin x - \frac{1}{2})^2 = a - \frac{3}{4}.$

Suppose $y_1 = (\sin x - \frac{1}{2})^2$, $y_2 = a - \frac{3}{4}.$

Make the image of y_1:

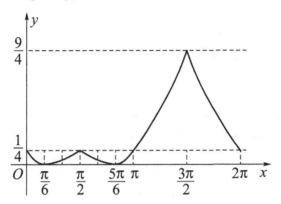

(1) When $a - \frac{3}{4} < 0$ or $a - \frac{3}{4} > \frac{9}{4}$, namely, $a < \frac{3}{4}$ or $a > 3$, no solution.

(2) When $a - \frac{3}{4} = \frac{9}{4}$, namely, $a = 3$, the equation has one unique solution $x = \frac{3\pi}{2}$.

(3) When $\frac{1}{4} < a - \frac{3}{4} < \frac{9}{4}$ or $a - \frac{3}{4} = 0$, namely, $1 < a < 3$ or $a = \frac{3}{4}$, the equation has two solutions.

(4) When $a - \frac{3}{4} = \frac{1}{4}$, namely, $a = 1$, the equation has four solutions: $x = 0, \frac{\pi}{2}, \pi, 2\pi$.

(5) When $0 < a - \frac{3}{4} < \frac{1}{4}$, namely, $\frac{3}{4} < a < 1$, the equation has four solutions.

14. (i) When n is even number, $\cos^n x = 1 + \sin^n x \geq 1$.
    If $\cos^n x \leq 1$, then $\cos^n x = 1$. $\therefore$ $\cos x = \pm 1, \sin x = 0$.
    $\therefore$ $x = k\pi$ $(k \in Z)$.

   (ii) When n is odd number, $\cos^n x = 1 + \sin^n x \geq 0$ and $\sin^n x = \cos^n x - 1 \leq 0$
    $\therefore$ The original equation is equal to $|\cos x|^n + |\sin x|^n = 1$.

    Compare this equation with $\sin^2 x + \cos^2 x = 1$:

    If $x \neq k \cdot \frac{\pi}{2}$, then $0 < |\cos x| < 1, 0 < |\sin x| < 1$.

When $n > 2, |\cos x|^n + |\sin x|^n < \cos^2 x + \sin^2 x = 1$.

When $n = 1, |\cos x|^n + |\sin x|^n = |\cos x| + |\sin x|$

$$= \sqrt{(|\cos x| + |\sin x|)^2} > \sqrt{|\cos x|^2 + |\sin x|^2} = 1.$$

$\therefore$ When $x \neq k \cdot \frac{\pi}{2}$, no solution.

When $x = k \cdot \frac{\pi}{2}, x = 2k\pi$ and $x = 2k\pi - \frac{\pi}{2}$ ($k \in Z$) are solutions of equation after inspection.

$\therefore$ When n is even number, $x = k\pi$ ($k \in Z$).

When n is odd number, $x = 2k\pi$ or $x = 2k\pi - \frac{\pi}{2}$ ($k \in Z$).

15. **Method 1:** $\cos^3 3x = \frac{1}{2}(1 + \cos 6x), \cos^2 x = \frac{1}{2}(1 + \cos 2x)$.
Substitute it into original equation:

$$2\cos^2 2x + \cos 6x + \cos 2x = 0 \qquad\qquad (*)$$

$\because \cos 6x + \cos 2x = 2\cos 4x \cdot \cos 2x$.

$\therefore$ Type (*) is equal to $\cos 2x(\cos 2x + \cos 4x) = 0$.

(i) If $\cos 2x = 0$, then $x = \frac{1}{4}(2k + 1)\pi, k \in Z$.

(ii) If $\cos 2x + \cos 4x = 0$, then $\cos 4x = -\cos 2x = \cos(\pi - 2x)$

$$4x = 2k\pi \pm (\pi - 2x), k \in Z$$

$\therefore x = \dfrac{1}{2}(2k + 1)\pi$ or $\dfrac{1}{6}(2k + 1)\pi, k \in Z$.

Method 2: $2\cos^2 x = 1 + \cos 2x, 2\cos^2 2x = 1 + \cos 4x$.

Through substitution and simplification, we get

$\cos 2x + \cos 4x + 2\cos^2 3x = 0$.

$\therefore 2\cos 3x \cdot \cos x + 2\cos^2 3x = 0,$

$\therefore 2\cos 3x(\cos x + \cos 3x) = 0.$

According to Sum-to-Product Formula:

$$4\cos x \cdot \cos 2x \cdot \cos 3x = 0. \tag{†}$$

$\therefore$ Solution is $x_1 = \pm 90° + k \cdot 360°$, $x_2 = \pm 45° + k \cdot 180°$,

$x_3 = \pm 30° + k \cdot 120°$, $k \in Z$.

After inspection, these solutions all meet the origin equation.
Method 3:

$$\cos^2 x + \cos^2 2x - \sin^2 3x = 0$$

$$\Rightarrow \cos^2 2x + (\cos x + \sin 3x)(\cos x - \sin 3x) = 0$$

$$\Rightarrow \cos^2 2x + \left(\sin\left(\frac{\pi}{2} + x\right) + \sin 3x\right)\left(\sin\left(\frac{\pi}{2} + x\right) - \sin 3x\right)$$

$$= 0$$

$$\Rightarrow \cos^2 2x + 2\sin\left(\frac{\pi}{4} + 2x\right) \cdot \cos\left(\frac{\pi}{4} - x\right) \cdot 2\cos\left(\frac{\pi}{4} + 2x\right)$$

$$\cdot \sin\left(\frac{\pi}{4} - x\right) = 0$$

$$\Rightarrow \cos^2 2x + \sin\left(\frac{\pi}{2} + 4x\right) \cdot \sin\left(\frac{\pi}{2} - 2x\right) = 0$$

$$\Rightarrow \cos 2x(\cos 2x + \cos 4x) = 0.$$

Type (†) is equal to $2\cos 2x \cdot \cos x \cdot \cos 3x = 0$.

The result follows as Method 2.

Exercise 5

1. **Method 1:** $B = 2A$, $C = \pi - 3A$.
   According to sine law:

$$\frac{a}{b} = \frac{\sin A}{\sin 2A} = \frac{1}{2\cos A},$$

$$\frac{a+b}{a+b+c} = \frac{\sin A + \sin 2A}{\sin A + \sin 2A + \sin(\pi - 3A)}$$

$$= \frac{\sin A + \sin 2A}{\sin 2A + 2\sin 2A \cos A} = \frac{1}{2\cos A}.$$

$$\therefore \frac{a}{b} = \frac{a+b}{a+b+c}.$$

Method 2:

As above:

$$\frac{a}{b} = \frac{1}{2\cos A},$$

$$\cos A = \frac{b^2 + c^2 - a^2}{2bc}; \text{ substitute it into above type, we get}$$

$$\frac{a}{b} = \frac{bc}{b^2 + c^2 - a^2} \Rightarrow ab^2 + ac^2 - a^3 = b^2 c$$

$$\Rightarrow b^2(c - a) = a(c - a)(c + a)$$

$$\Rightarrow b^2 = a^2 + ac.$$

$$\therefore \frac{a}{b} = \frac{b}{a+c} = \frac{a+b}{a+b+c}.$$

2.

$$\because \cot C = 2004(\cot A + \cot B),$$

$$\therefore \frac{\cos C}{\sin C} = 2004 \left(\frac{\cos A}{\sin A} + \frac{\cos B}{\sin B} \right)$$

$$= 2004 \cdot \frac{\sin(A + B)}{\sin A \sin B} = 2004 \cdot \frac{\sin C}{\sin A \sin B}$$

$$\therefore \cos C = \frac{2004 \sin^2 C}{\sin A \sin B} = \frac{2004 c^2}{ab}, \text{ namely,}$$

$$\frac{a^2 + b^2 - c^2}{2ab} = \frac{2004 c^2}{ab}.$$

$$\therefore a^2 + b^2 - c^2 = 4008 c^2, \text{ namely } a^2 + b^2 = 4009 c^2.$$

3.

$$c^2 = AD^2 + BD^2 - 2AD \cdot BD \cdot \cos \angle ADB$$

$$= AD^2 + p^2 - 2AD \cdot p \cdot \cos \angle ADB. \tag{1}$$

Similarly, $b^2 = AD^2 + q^2 - 2AD \cdot q \cdot \cos \angle ADC. \tag{2}$

$\because \angle ADB + \angle ADC = \pi \Rightarrow \cos\angle ADB + \cos \angle ADC = 0.$

Substitute types (1) and (2) into above type, we get

$$qc^2 + pb^2 = (p+q)AD^2 + pq(p+q).$$

$$\therefore \ AD^2 = \frac{b^2p + c^2q}{p+q} - pq.$$

4.

(1) $\because a + c = 2b, \ \therefore \ 2\sin B = \sin A + \sin C,$

namely, $4\sin\dfrac{A+C}{2} \cos \dfrac{A+C}{2} = 2\sin \dfrac{A+C}{2} \cos \dfrac{A-C}{2}.$

$\because 0 < \dfrac{A+C}{2} < \pi, \ \sin \dfrac{A+C}{2} \neq 0$

$\therefore \ \cos \dfrac{A-C}{2} = 2\cos \dfrac{A+C}{2}.$

$\therefore \ 3\sin \dfrac{A}{2} \sin \dfrac{C}{2} = \cos \dfrac{A}{2} \cos \dfrac{C}{2}, \ \therefore \ \tan \dfrac{A}{2} \tan \dfrac{C}{2} = \dfrac{1}{3}.$

(2) $5\cos A - 4\cos A \cos C + 5 \cos C$

$$= 5(\cos A + \cos C) - 2(\cos(A + C) + \cos(A - C))$$

$$= 10 \cos \frac{A+C}{2} \cos \frac{A-C}{2}$$

$$-2\left(2\cos^2\frac{A+C}{2} - 1 + 2\cos^2\frac{A-C}{2} - 1\right).$$

According to solution of (1):

$$\cos\frac{A-C}{2} = 2\cos\frac{A+C}{2}.$$

$$\therefore \text{ Above type} = 20\cos^2\frac{A+C}{2} - 4\cos^2\frac{A+C}{2}$$

$$-16\cos^2\frac{A+C}{2} + 4 = 4.$$

(3) $\because A-C = \dfrac{\pi}{2}$ and $\cos\dfrac{A-C}{2} = 2\cos\dfrac{A+C}{2}$

$$\therefore \sin\frac{B}{2} = \frac{\sqrt{2}}{4} \Rightarrow \cos\frac{B}{2} = \frac{\sqrt{14}}{4}$$

$$\therefore \sin B = 2\sin\frac{B}{2}\cos\frac{B}{2} = \frac{\sqrt{7}}{4}, \text{ namely, } \sin A + \cos C = \frac{\sqrt{7}}{2}.$$

5. $$A = \frac{\pi}{7},\ B = \frac{2\pi}{7},\ C = \frac{4\pi}{7}.$$

$$\therefore \cos A\cos B\cos C = \cos\frac{\pi}{7}\cos\frac{2\pi}{7}\cos\frac{4\pi}{7}$$

$$= \frac{4}{8\sin\frac{\pi}{7}}\left(2\sin\frac{\pi}{7}\cos\frac{\pi}{7}\right)\cos\frac{2\pi}{7}\cos\frac{4\pi}{7}$$

$$= \frac{1}{8\sin\frac{\pi}{7}}\sin\frac{8\pi}{7} = -\frac{1}{8}.$$

6. Suppose $AP = PQ = QB = BC = x$.

In $\triangle APQ$, $AQ = 2x\cos A$, so $AB = x + 2x\cos A$.

In $\triangle ABC$, $AB\cos B = \dfrac{x}{2}$, namely, $AB = \dfrac{x}{2\sin\frac{A}{2}}$.

$$\therefore x + 2x\cos A = \frac{x}{2\sin\frac{A}{2}}$$

$$\therefore\ 1 + 2\cos A = \frac{1}{2\sin\frac{A}{2}} \Rightarrow 3 - 4\sin^2\frac{A}{2} = \frac{1}{2\sin\frac{A}{2}}$$

$$\Rightarrow 3\sin\frac{A}{2} - 4\sin^3\frac{A}{2} = \frac{1}{2} \Rightarrow \sin\frac{3A}{2} = \frac{1}{2}.$$

$$\therefore\ A = 20°.$$

7. $$\because\ S = a^2 - b^2 - c^2 + 2bc = 2bc(1 - \cos A) = \frac{1}{2}bc\sin A.$$

$$\therefore\ 4 - 4\cos A = \sin A \Rightarrow (4 - \sin A)^2 = 16\cos^2 A$$

$$\therefore\ \sin A = \frac{8}{17}$$

$$\therefore\ S = \frac{4}{17}bc \le \frac{4}{17}\cdot\left(\frac{b+c}{2}\right)^2 = \frac{64}{17}.$$

8. The radius of the inscribed circle of $\triangle ABC$ is $r = 2\sqrt{3}$. Suppose two sides are $8x$ and $5x$, so the third side is

$$\sqrt{(8x)^2 + (5x)^2 - 2\cdot 8x\cdot 5x\cos 60°} = 7x.$$

$$S = S_{\triangle ABC} = p\cdot r = 10x\cdot 2\sqrt{3} = \frac{1}{2}\cdot 8x\cdot 5x\sin 60°.$$

$$\therefore\ 20\sqrt{3}x = 10\sqrt{3}x^2 \Rightarrow x = 2.\ \therefore\ S = 4\sqrt{3}.$$

9. $$S = S_{\triangle ABD} + S_{\triangle CDB} = \frac{1}{2}AB\cdot AD\cdot\sin A + \frac{1}{2}BC\cdot CD\cdot\sin C$$

$$= \frac{1}{2}(2\cdot 4 + 6\cdot 4)\sin A = 16\sin A.$$

In $\triangle ABD$, according to cosine law,

$$BD^2 = AB^2 + AD^2 - 2AB\cdot AD\cdot\cos A = 20 - 16\cos A.$$
In $\triangle BCD$, $BD^2 = CB^2 + CD^2 - 2CB\cdot CD\cdot\cos C$

$$= 52 - 48\cos C$$

$$\therefore\ 20 - 16\cos A = 52 - 48\cos C$$

$$\therefore\ \cos A = -\frac{1}{2},\ A = 120°,\ \therefore\ S = 16\sin 120° = 8\sqrt{3}.$$

10. Suppose that car bottom is square $ABCD$, line BC and corridor wall intersect at H,

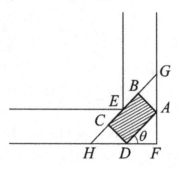

$\angle ADF = \theta.$

$$\therefore\ GH = EH + EG = \frac{3}{2}\left(\frac{1}{\sin\theta} + \frac{1}{\cos\theta}\right)$$

$$CH = CD \cdot \cot\theta = \cot\theta,\ BG = AB \cdot \tan\theta = \tan\theta.$$

$$\therefore\ AD = GH - CH - BG = \frac{3(\sin\theta + \cos\theta) - 2}{2\sin\theta\cos\theta}.$$

Let $\sin\theta + \cos\theta = t \in (1, \sqrt{2}]$, so $AD = f(t) = \dfrac{3t - 2}{t^2 - 1}.$

$\because\ f(t)$ is decreasing function in $(1, \sqrt{2}]$

$\therefore\ f(t)_{\min} = f(\sqrt{2}) = 3\sqrt{2} - 2.$

11.

$\qquad\qquad \because\ \log_{\sqrt{5}} x = \log_b(4x - 4)$

$\qquad\qquad \therefore\ x^2 = 4x - 4 \Rightarrow x = 2.$

$\qquad\qquad \therefore\ \dfrac{C}{A} = \dfrac{\sin B}{\sin A} = 2$

$\qquad\qquad \therefore\ \begin{cases} C = 2A \\ \sin B = 2\sin A \end{cases} \Rightarrow \begin{cases} \sin C = 2\sin A\cos A \\ b = 2a \end{cases}$

$$\Rightarrow \begin{cases} c = 2a \cdot \dfrac{b^2 + c^2 - a^2}{2bc} \\[2mm] b = 2a \end{cases}$$

$$\therefore \begin{cases} c = \sqrt{3}a \\ b = 2a \end{cases}$$

$\therefore a^2 + c^2 = b^2$, namely, $\triangle ABC$ is right triangle.

12. Suppose $\alpha = \angle BOC$, $\beta = \angle COA$, $\gamma = \angle AOB$ and unit vector $\overrightarrow{e_1} = \overrightarrow{OA}$, $\overrightarrow{e_2} = \overrightarrow{OB}$, $\overrightarrow{e_3} = \overrightarrow{OC}$,

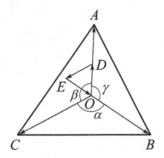

so $S_A \cdot \overrightarrow{OA} = \dfrac{1}{2}|OA| \cdot |OB| \cdot |OC| \sin \alpha \cdot \overrightarrow{e_1}$.

Similarly, $S_B \cdot \overrightarrow{OB} = \dfrac{1}{2}|OB| \cdot |OC| \cdot |OA|$

$\sin \beta \cdot \overrightarrow{e_2}$, $S_C \cdot \overrightarrow{OC} = \dfrac{1}{2}|OA| \cdot |OB| \cdot |OC| \sin \gamma \cdot \overrightarrow{e_3}$

$\therefore \overrightarrow{e_1} \sin \alpha + \overrightarrow{e_2} \sin \beta + \overrightarrow{e_3} \sin \gamma = \overrightarrow{0}$.

Take a point D on OA, where $\overrightarrow{OD} = \overrightarrow{e_1} \sin \alpha$.

Make $DE \parallel OC$, which intersect at E with extension line BO.

In $\triangle EOD$, $|DE| = \sin \gamma$, $|EO| = \sin \beta$.

$$\therefore \overrightarrow{DE} = \overrightarrow{e_3} \sin\gamma, \ \overrightarrow{EO} = \overrightarrow{e_2} \sin\beta$$

$\because OD, OE$ and EO constitute a triangle

$$\therefore \overrightarrow{e_1} \sin\alpha + \overrightarrow{e_2} \sin\beta + \overrightarrow{e_3} \sin\gamma = \overrightarrow{0}.$$

$\therefore$ The original proposition is proved.

13. Suppose O and H are circumcenter and orthocenter of $\triangle ABC$.

Suppose R is the radius of circumcircle of $\triangle ABC$.

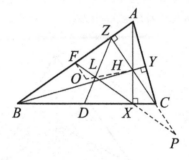

Suppose ZC and FX intersect at P, connect OF, HL and OL.

$\because OF \perp AB, \ PZ \perp AB$

$\therefore OF \parallel PZ, \ \angle OFL = \angle P.$

$\because F$ is the midpoint of hypotenuse of the right-angled

triangle AXB.

$\therefore FX = FB, \ \angle ABC = \angle FXB = \angle CXP,$

$$\angle XPC = \frac{\pi}{2} - \angle ZFP = \frac{\pi}{2} - 2\angle ABC$$

$$PC = \frac{XC \sin \angle CXP}{\sin \angle XPC} = \frac{AC \cos C \sin \angle CXP}{\sin \angle XPC}$$

$$= \frac{2R \sin B \cos C \sin B}{\sin \angle XPC} = \frac{2R \cos C \sin^2 B}{\cos 2B}$$

$$PH = PC + CH = \frac{2R\cos C \sin^2 B}{\cos 2B} + 2R\cos C$$

$$= \frac{2R\cos C \cos^2 B}{\cos 2B}$$

$\because FO = R\cos C,\ \dfrac{PH}{FO} = \dfrac{2\cos^2 B}{\cos 2B}.$

Similarly $\angle B = \angle DZB$.

$\therefore$ In $\triangle PLZ,\ PL = \dfrac{LZ \sin \angle PZL}{\sin \angle LPZ} = \dfrac{LZ \cos B}{\cos 2B}.$

In $\triangle FLZ,\ FL = \dfrac{LZ \sin \angle FZL}{\sin \angle ZFL} = \dfrac{LZ}{2\cos B}.$

$\therefore \dfrac{PL}{FL} = \dfrac{2\cos^2 B}{\cos 2B},$ namely, $\dfrac{PH}{FO} = \dfrac{PL}{FL}.$

$\because \angle OFL = \angle LPH$

$\therefore \triangle OFL \backsim \triangle HPL$

$\therefore \angle LOF = \angle LHP.$

$\therefore$ The three points O, L and H are collinear, namely, point L is on the line OH.

Similarly, M and N are on the line OH.

14.　　(1) Connect OC and OB, and note that CM and AB

intersectat P, BM and AC at Q.

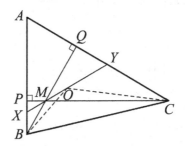

$\therefore \angle CMB = 90° + \angle ABM = 180° - \angle BAM = 120°,$

$\angle COB = 2\angle BAC = 120°.$

$\therefore$ Four points C, O, M and B are on the same circle.

$\because CP \perp AB$

$\therefore \angle AXM = 90° - \angle XMP = 90° - \angle OMC = 90°$

$-\angle OBC = 60°$

$\because \angle ABQ = 90° - 60° = 30°$

$\therefore \angle XMB = 60° - \angle ABQ = 30° = \angle ABQ$

$\therefore MX = XB.$

Similarly, $YM = CY.$

$\therefore$ Perimeter $P = AY + YX + AX$

$= AY + YC + AX + XB = b + c.$

(2) Suppose $AO = R$

$\because$ Four points C, O, M and B are on the same circle.

$\therefore \dfrac{BC}{\sin \angle COB} = \dfrac{OM}{\sin \angle MBO},$ namely

$OM = \dfrac{BC \sin \angle MBO}{\sin \sin \angle COB} = 2BC \sin \sin \angle MBO.$

$\because \angle MBO = \angle ABC - \angle ABQ - \angle COB$

$= \angle ABC - 30° - 30° = \angle ABC - 60°$

$\therefore OM = 4R \sin A \sin(B - 60°) = 2R \sin(B - 60°).$

$\because b - c = 2R(\sin B - \sin C).$

Only need to prove: $\sin(B - 60°)$

$$= \sin B - \sin C$$

$$\Leftrightarrow \sin C - \sin(60° + B) = 0$$

$$\Leftrightarrow 2\cos \frac{C + 60° + B}{2} \sin \frac{C - 60° - B}{2} = 0$$

$$\Leftrightarrow 2\cos \frac{180°}{2} \sin \frac{B + 60° - C}{2} = 0.$$

Above type is clearly established.

$$\therefore \ OM = b - c.$$

15. **Method 1:** In $\triangle ABQ$, according to sine law:

$$\frac{AQ}{\sin \frac{B}{2}} = \frac{AB}{\sin \left(\frac{B}{2} + 60°\right)} = \frac{BQ}{\sin 60°},$$

$$\therefore \ AQ = \frac{AB \cdot \sin \frac{B}{2}}{\sin \left(\frac{B}{2} + 60°\right)}, \ BQ = \frac{AB \cdot \sin 60°}{\sin \left(\frac{B}{2} + 60°\right)}.$$

Similarly, $BP = \dfrac{AB \cdot \sin 30°}{\sin(30° + B)}.$

$\because \ AB + BP = AQ + BQ, \ \angle B + \angle C = 120°, \ 0° < \angle B < 120°$

$$\therefore \ 1 + \frac{\sin 30°}{\sin (B + 30°)} = \frac{\sin \frac{B}{2}}{\sin \left(\frac{B}{2} + 60°\right)} + \frac{\sin 60°}{\sin \left(\frac{B}{2} + 60°\right)}$$

$$\therefore \ 1 + \frac{\sin 30°}{\sin (B + 30°)} = \frac{\cos \left(\frac{B}{4} - 30°\right)}{\cos \left(\frac{B}{4} + 30°\right)}$$

$$\therefore \ \frac{\sin 30°}{\sin (B + 30°)} = \frac{\sin \frac{B}{4}}{\cos \left(\frac{B}{4} + 30°\right)}$$

$$\sin 30° \cdot \cos \left(\frac{B}{4} + 30°\right) = \sin \frac{B}{4} \cdot \sin (B + 30°)$$

$$= \frac{1}{2} \left(\cos \left(\frac{3B}{4} + 30°\right) - \cos \left(\frac{5B}{4} + 30°\right) \right)$$

$$\therefore \cos\left(\frac{B}{4} + 30°\right) + \cos\left(\frac{5B}{4} + 30°\right) = \cos\left(\frac{3B}{4} + 30°\right)$$

$$2\cos\left(\frac{3B}{4} + 30°\right) \cdot \cos\frac{B}{2} = \cos\left(\frac{3B}{4} + 30°\right).$$

When $\cos\left(\frac{3B}{4} + 30°\right) = 0$, $\frac{3B}{4} + 30° = 90°$, $\therefore B = 80°$.

When $\cos\left(\frac{3B}{4} + 30°\right) \neq 0$, $\cos\frac{B}{2} = \frac{1}{2}$, $\therefore B = 120°$

(inconsistent with condition).

$\therefore A = 60°$, $B = 80°$, $C = 40°$.

Method 2: In $\triangle ABQ$,

$$\frac{AQ}{\sin\frac{B}{2}} = \frac{BQ}{\sin 60°} = \frac{AB}{\sin\left(180° - 60° - \frac{B}{2}\right)}.$$

$$\therefore AQ = \frac{AB \cdot \sin\frac{B}{2}}{\sin\left(60° + \frac{B}{2}\right)}, \quad BQ = \frac{\sqrt{3}AB}{2\sin\left(60° + \frac{B}{2}\right)}$$

In $\triangle ABC$, $\dfrac{BP}{\sin 30°} = \dfrac{AB}{\sin(180° - 30° - B)}$

$$\therefore BP = \frac{AB}{2\sin(30° + B)}.$$

Substitute them into $AB + BP = AQ + BQ$:

$$1 + \frac{1}{2\sin(30° + B)} = \frac{\sin\frac{B}{2}}{\sin\left(60° + \frac{B}{2}\right)} + \frac{\sqrt{3}}{2\sin\left(60° + \frac{B}{2}\right)}$$

$$\therefore (2\sin(30° + B) + 1)\sin\left(60° + \frac{B}{2}\right)$$

$$= \left(2\sin\frac{B}{2} + \sqrt{3}\right)\sin(30° + B)$$

$$\therefore \ 2\sin(30° + B)\left(\sin\left(60° + \frac{B}{2}\right) - \sin\frac{B}{2}\right)$$

$$+ \sin\left(60° + \frac{B}{2}\right) - \sqrt{3}\sin(30° + B) = 0$$

$$\text{Right} = 2\sin(30° + B) \cdot \cos\left(30° + \frac{B}{2}\right)$$

$$+ \sin\left(60° + \frac{B}{2}\right) - \sqrt{3}\sin(30° + B)$$

$$= \sin\left(60° + \frac{3B}{2}\right) + \sin\frac{B}{2} + \sin\left(60° + \frac{B}{2}\right)$$

$$- \sqrt{3}\sin(30° + B)$$

$$= \sin\left(30° + \frac{3B}{4}\right) \cdot \cos\left(30° + \frac{3B}{4}\right)$$

$$- 2\sqrt{3}\cos\left(30° + \frac{3B}{4}\right) \cdot \sin\frac{B}{4}$$

$$= 2\cos\left(30° + \frac{3B}{4}\right)\left(\sin\left(30° + \frac{3B}{4}\right) - \sqrt{3}\sin\frac{B}{4}\right)$$

$$\times \sin\left(30° + \frac{3B}{4}\right) - \sqrt{3}\sin\frac{B}{4}$$

$$= \frac{1}{2}\cos\frac{3B}{4} + \frac{\sqrt{3}}{2}\sin\frac{3B}{4} - \sqrt{3}\sin\frac{B}{4}$$

$$= \frac{1}{2}\cos\frac{3B}{2} + \frac{\sqrt{3}}{2}\left(3\sin\frac{B}{4} - \sin^3\frac{B}{4}\right) - \sqrt{3}\sin\frac{B}{4}$$

$$= \frac{1}{2}\cos\frac{3B}{2} + \frac{\sqrt{3}}{2}\sin\frac{B}{4}\left(1 - \sin^2\frac{B}{4}\right)$$

$$\because \ 0° < B < 120°$$

$$\therefore \ 0° < \frac{3B}{4} < 90°, \ \sin\left(30° + \frac{3B}{4}\right) - \sqrt{3}\sin\frac{B}{4} > 0$$

$$\therefore \cos\left(30° + \frac{3B}{4}\right) = 0, \ 30° + \frac{3B}{4} = 90°, \ B = 80°$$

$$\therefore A = 60°, \ B = 80°, \ C = 40°.$$

Method 3: Let $A = 2\theta$, $B = 2\alpha$

In $\triangle ABP$, $\angle BAP = \theta$, $\angle ABP = 2\alpha$,

$\angle APB = 180° - (2\alpha + \theta)$.

According to sine law:

$$\frac{AB}{\sin(2\alpha + \theta)} = \frac{BP}{\sin\theta} = \frac{AP}{\sin 2\alpha}$$

$$\therefore AB + BP = \frac{\sin(2\alpha + \theta) + \sin\theta}{\sin 2\alpha} AP.$$

In $\triangle ABQ$, $\angle BAQ = 2\theta$, $\angle ABQ = \alpha$, $\angle AQB = 180° - (\alpha + 2\theta)$.

According to sine law:

$$\therefore \frac{AQ}{\sin\alpha} = \frac{QB}{\sin 2\theta} = \frac{AB}{\sin(\alpha + 2\theta)}$$

$$\therefore AQ + QB = \frac{\sin\alpha + \sin 2\theta}{\sin(\alpha + 2\theta)} AB.$$

$$\because AB + BP = AQ + QB$$

$$\therefore \frac{\sin(2\alpha + \theta) + \sin\theta}{\sin 2\alpha} AP = \frac{\sin\alpha + \sin 2\theta}{\sin(\alpha + 2\theta)} AB.$$

$$\frac{2\sin(\alpha + \theta)\cdot\cos\alpha}{\sin 2\alpha}\sin 2\alpha = \frac{2\sin\frac{\alpha+2\theta}{2}\cdot\cos\frac{\alpha-2\theta}{2}}{2\sin\frac{\alpha+2\theta}{2}\cdot\cos\frac{\alpha+2\theta}{2}}\sin(2\alpha + \theta)$$

$$2\sin(\alpha + \theta)\cdot\cos\alpha\cdot\cos\frac{\alpha + 2\theta}{2} = \cos\frac{\alpha - 2\theta}{2}\cdot\sin(2\alpha + \theta).$$

$$\therefore \sin(\alpha + \theta)\cdot\cos\frac{3\alpha + 2\theta}{2} + \cos\frac{\alpha - 2\theta}{2}$$

$$\cdot(\sin(\alpha + \theta) - \sin(2\alpha + \theta)) = 0$$

$$\therefore\ 0 = \sin(\alpha + \theta) \cdot \cos\frac{3\alpha + 2\theta}{2} + \cos\frac{\alpha - 2\theta}{2}$$

$$\cdot\, 2\cos\frac{3\alpha + 2\theta}{2} \cdot \sin\left(-\frac{\alpha}{2}\right)$$

$$= \cos\frac{3\alpha + 2\theta}{2}\left(\sin(\alpha + \theta) - 2\cos\frac{\alpha - 2\theta}{2} \cdot \sin\frac{\alpha}{2}\right)$$

$$= \cos\frac{3\alpha + 2\theta}{2}\left(\sin(\alpha + \theta) - \sin(\alpha - \theta) - \sin\theta\right)$$

$$= \cos\frac{3\alpha + 2\theta}{2}\left(2\cos\alpha \cdot \sin\theta - \sin\theta\right).$$

$\because\ \theta$ is an acute angle.

$\therefore\ \sin\theta \neq 0$

$$\cos\frac{3\alpha + 2\theta}{2}(2\cos\alpha - 1) = 0 \qquad\qquad (*)$$

$\because\ A = 2\theta = 60°$

$\therefore\ B = 2\alpha < 120°,\ \alpha < 60°$

$\therefore\ \cos\alpha > \dfrac{1}{2},\ 2\cos\alpha - 1 > 0.$

According to type$(*)$: $\cos\dfrac{3\alpha + 60°}{2} = 0$

$\therefore\ 0° < 3\alpha + 60° < 240°$

$\therefore\ 3\alpha + 60° = 180°$

$\therefore\ B = 2\alpha = 80°,\ C = 40°,\ A = 60°.$

Exercise 6

1. $\because\ \dfrac{1}{4} \leq x \leq 2,$

$\therefore\ 0 \leq x - \dfrac{1}{4} \leq \dfrac{7}{4},\ 2 - x = \dfrac{7}{4} - \left(x - \dfrac{1}{4}\right)$

Let $x - \dfrac{1}{4} = \dfrac{7}{4}\sin^2\alpha,\ \alpha \in \left[0, \dfrac{\pi}{2}\right],\ 2 - x = \dfrac{7}{4}\cos^2\alpha,$

$$\therefore \; y = \sqrt{7}\sin\alpha + \frac{\sqrt{7}}{2}\cos\alpha = \frac{\sqrt{35}}{2}\sin(\alpha+\phi)\left(\phi = \arcsin\frac{\sqrt{5}}{5}\right)$$

$$\therefore \; \frac{\sqrt{5}}{5} \le \sin(\alpha+\phi) \le 1, \; \Rightarrow y \in \left[\frac{\sqrt{7}}{2}, \frac{\sqrt{35}}{2}\right].$$

2. $\quad \because \; -3x^2 + 6x + 12 \ge 0 \Rightarrow 3(5 - (x-1)^2) \ge 0 \Rightarrow |x - 1| \le \sqrt{5}.$

Let $x - 1 = \sqrt{5}\sin\alpha, \; \therefore \; x = 1 + \sqrt{5}\sin\alpha, \; \alpha \in \left[-\frac{\pi}{2}, \frac{\pi}{2}\right]$

$\therefore y = x + 3 + \sqrt{-3x^2 + 6x + 12}$

$\quad = \sqrt{5}\sin\alpha + \sqrt{15}\cos\alpha + 4 = 2\sqrt{5}\sin\left(\alpha + \frac{\pi}{3}\right) + 4,$

$\because \; \alpha + \frac{\pi}{3} \in \left[-\frac{\pi}{6}, \frac{5\pi}{6}\right]$

$\therefore \; y \in [4 - \sqrt{5}, 4 + 2\sqrt{5}].$

3. $\quad \therefore \; \tan C = -\tan(A+B) = -\dfrac{\frac{1}{4} + \frac{3}{5}}{1 - \frac{1}{4} \times \frac{3}{5}} = -1.$

$\because \; 0 < C < \pi, \; \therefore \; C = \frac{3}{4}\pi, \; \therefore \; AB$ is maximum,

namely, $AB = \sqrt{17}$

$\because \; \tan A < \tan B, A, B \in \left(0, \frac{\pi}{2}\right), \; \therefore \;$ angle A is minimum, side

BC is minimum.

$$\because \; \begin{cases} \tan A = \dfrac{\sin A}{\cos A} = \dfrac{1}{4}, \\ \sin^2 A + \cos^2 A = 1, \end{cases} \quad \text{and} \quad A \in \left(0, \frac{\pi}{2}\right)$$

$\therefore \; \sin A = \dfrac{\sqrt{17}}{17}.$

$\because \; \dfrac{AB}{\sin C} = \dfrac{BC}{\sin A}$

$\therefore \; BC = \sqrt{2}$

$\therefore \;$ the minimum of side is given by $BC = \sqrt{2}.$

4. Suppose $a_n = \tan\theta_n$.

$$\because \ 2 - \sqrt{3} = \frac{1}{2+\sqrt{3}} = \frac{\frac{1}{2}}{1+\frac{\sqrt{3}}{2}} = \frac{\sin\frac{\pi}{6}}{1+\cos\frac{\pi}{6}} = \tan\frac{\pi}{12},$$

$$\therefore \ a_{n+1} = \tan\left(\theta_n + \frac{\pi}{12}\right), \ \therefore \ \theta_{n+1} = \theta_n + \frac{\pi}{12},$$

$$\therefore \ a_{n+12} = \tan\left(\theta_{n+11} + \frac{\pi}{12}\right) = \tan\left(\theta_n + \frac{\pi}{12}\times 12\right)$$

$$= \tan(\theta_n + \pi) = \tan\theta_n = a_n$$

$$\therefore \ a_{2005} = a_{12\times 167+1} = a_1 = 1.$$

5. Let $a = \cos\theta$, $b = \sin\theta$.

$$\left(a^2 + \frac{1}{a^2}\right)\left(b^2 + \frac{1}{b^2}\right) = \frac{\sin^4\theta\cos^4\theta + \sin^4\theta + \cos^4\theta + 1}{\sin^2\theta\cos^2\theta}$$

$$= \frac{(\frac{1}{4}\sin^2 2\theta - 1)^2 + 1}{\frac{1}{4}\sin^2 2\theta} \geq \frac{(\frac{1}{4}-1)^2 + 1}{\frac{1}{4}} = \frac{25}{4}.$$

6. Let $x = \sqrt{5}\sin\theta, -\frac{\pi}{2}\leq\theta\leq\frac{\pi}{2}$.

$$\therefore \ \sqrt{5-x^2} + \sqrt{3}x = \sqrt{5}\cos\theta + \sqrt{15}\sin\theta = 2\sqrt{5}\sin\left(\theta + \frac{\pi}{6}\right).$$

$\therefore$ The maximum is $2\sqrt{5}$; The minimum is $-\sqrt{15}$.

7. Let $x = \tan\theta, \theta \in \left(-\frac{\pi}{2}, \frac{\pi}{2}\right)$.

$$-2\sin^2\theta + \sin\theta + 1 > 0 \Rightarrow \left(\sin\theta - \frac{1}{4}\right)^2 < \frac{9}{16}$$

$$\Rightarrow -\frac{\pi}{6} < \theta < \frac{\pi}{2} \Rightarrow x > -\frac{\sqrt{3}}{3}.$$

8. We have

$$y = \frac{1 + x - 2x^2 + x^3 + x^4}{1 + 2x^2 + x^4} = \frac{1 - 2x^2 + x^4}{1 + 2x^2 + x^4}$$

$$+ \frac{x + x^3}{1 + 2x^2 + x^4} = \left(\frac{1 - x^2}{1 + x^2}\right)^2 + \frac{x}{1 + x^2}.$$

Let $x = \tan\theta, \theta \in \left(-\dfrac{\pi}{2}, \dfrac{\pi}{2}\right)$.

$$y = \cos^2 2\theta + \frac{1}{2}\sin 2\theta = -\left(\sin 2\theta - \frac{1}{4}\right)^2 + \frac{17}{16}.$$

When $\sin 2\theta = \dfrac{1}{4}$, $y_{\max} = \dfrac{17}{16}$; when $\sin 2\theta = -1$, $y_{\min} = -\dfrac{1}{2}$.

9. Let $x^2 + y^2 + z^2 = r^2$ $(r > 0)$;

suppose $x = r\cos\alpha\cos\beta$, $y = r\cos\alpha\sin\beta$, $z = r\sin\alpha$

$$\therefore \ u = \cos^2\alpha\cos\beta\sin\beta + 2\cos\alpha\sin\alpha\sin\beta + 2\sin\alpha\cos\alpha\cos\beta$$

$$= \frac{1}{2}\cos^2\alpha\sin 2\beta + (\sin\beta + \cos\beta)\sin 2\alpha$$

$$= \frac{1}{4}(1 + \cos 2\alpha)\sin 2\beta + (\sin\beta + \cos\beta)\sin 2\alpha$$

$$= \frac{1}{4}\sin 2\beta + (\sin\beta + \cos\beta)\sin 2\alpha + \frac{1}{4}\sin 2\beta\cos 2\alpha$$

$$\leq \frac{1}{4}\sin 2\beta + \sqrt{(\sin\beta + \cos\beta)^2 + \frac{1}{16}\sin^2 2\beta}$$

$$= \frac{1}{4}\sin 2\beta + \sqrt{\frac{1}{16}(\sin 2\beta + 8)^2 - 3}$$

$$\leq \frac{1}{4} + \sqrt{\frac{1}{16}(1 + 8)^2 - 3} = \frac{\sqrt{33} + 1}{4}$$

$u_{\max} = \dfrac{\sqrt{33} + 1}{4}$, if and only if

$$\sin 2\beta = 1, \ \sin 2\alpha = \frac{4\sqrt{66}}{33}, \ \cos 2\alpha = \frac{\sqrt{33}}{33}.$$

10.   Suppose $z < \dfrac{1}{2}$, $x = \sin^2\alpha\cos^2\beta$, $y = \cos^2\alpha\cos^2\beta$ z

$$= \sin^2\beta\left(0 \leq \beta \leq \frac{\pi}{4}\right).$$

$$\therefore xy + yz + zx - 2xyz = xy(1 - z) + z(x + y - xy)$$

$$= \sin^2 \alpha \cos^2 \alpha \cos^6 \beta + \sin^2 \beta \cos^2 \beta$$

$$(1 - \sin^2 \alpha \cos^2 \alpha \cos^2 \beta) \geq 0$$

$$\because xy + yz + zx - 2xyz = xy(1 - 2z) + z(y + x)$$

$$= \sin^2 \alpha \cos^2 \alpha \cos^4 \beta \cos 2\beta + \frac{1}{4} \sin^2 2\beta$$

$$= \frac{1}{4} \sin^2 2\alpha \cdot \frac{1}{4} (1 + \cos 2\beta)^2 \cos 2\beta + \frac{1}{4} (1 - \cos^2 2\beta)$$

$$\leq \frac{1}{16} (1 + \cos 2\beta)^2 \cos 2\beta + \frac{1}{4} (1 - \cos^2 2\beta)$$

$$= \frac{1}{4} + \frac{1}{32} (2 \cos 2\beta)(1 - \cos 2\beta)(1 - \cos 2\beta)$$

$$\leq \frac{1}{4} + \frac{1}{32} \left(\frac{2 \cos 2\beta + 1 - \cos 2\beta + 1 - \cos 2\beta}{3} \right)^3 = \frac{7}{27}.$$

If and only if $\beta = \arccos \dfrac{\sqrt{6}}{3}$, $\alpha = k\pi + \dfrac{\pi}{4} \ (k \in Z)$,

namely, $x = y = z = \dfrac{1}{3}$, then equality holds.

$$\therefore \ 0 \leq xy + yz + zx - 2xyz \leq \frac{7}{27}.$$

11. Suppose 13 real numbers as $\tan \theta_i, \theta_i \in \left(-\frac{\pi}{2}, \frac{\pi}{2} \right)$, $i = 1, 2, \ldots, 13$. Interval $\left(-\frac{\pi}{2}, \frac{\pi}{2} \right)$ is equally divided into 12 intervals, so θ_i is at least in one interval.

Suppose α, β and $\alpha - \beta \in \left[0, \dfrac{\pi}{12} \right]$

$$\therefore \ 0 \leq \tan(\alpha - \beta) \leq \tan \frac{\pi}{12} = 2 - \sqrt{3}.$$

Let $x = \tan \alpha$ and $y = \tan \beta$, so $\dfrac{x - y}{1 + xy} = \tan(\alpha - \beta)$.

$$\therefore \ 0 \leq \frac{x - y}{1 + xy} \leq 2 - \sqrt{3}.$$

12.   Let $a_0 = \cos \theta = \dfrac{1}{3}$, $\theta \in \left(0, \dfrac{\pi}{2} \right)$.

$$a_1 = \cos \frac{\theta}{2}, \ a_2 = \cos \frac{\theta}{4}, a_3 = \cos \frac{\theta}{8}, \ldots, a_n = \cos \frac{\theta}{2^n}.$$

$$\because \quad \frac{\pi}{2} > \frac{\theta}{2} > \frac{\theta}{4} > \cdots > \frac{\theta}{2^n} > 0$$

$$\therefore \quad \cos\frac{\theta}{2} < \cos\frac{\theta}{4} < \cdots < \cos\frac{\theta}{2^n}.$$

$\therefore$ Sequence $\{a_n\}$ is monotonically increasing.

13. Suppose $x = \sec^2\alpha,\ y = \tan^2\alpha,\ 0 < \alpha < \dfrac{\pi}{2}$.

$$\therefore\ A = \left(\sec\alpha - \frac{1}{\sec\alpha}\right)\left(\tan\alpha + \frac{1}{\tan\alpha}\right)\cdot\frac{1}{\sec^2\alpha} = \sin\alpha \in (0,1).$$

$\therefore\ 0 < A < 1.$

14.   Let $a_n = \sec\alpha_n\left(0 < \alpha_n < \dfrac{\pi}{2}\right)$, so $\alpha_{n+1} = \dfrac{1}{2}\alpha_n$.

namely, $\alpha_n = \dfrac{\pi}{2^{n+1}},\ a_n = \sec\dfrac{\pi}{2^{n+1}}$.

Let $b_n = \csc\beta_n\left(0 < \beta_n < \dfrac{\pi}{2}\right)$, so $\beta_{n+1} = \dfrac{1}{2}\beta_n$,

namely, $\beta_n = \dfrac{\pi}{2^{n+1}},\ b_n = \csc\dfrac{\pi}{2^{n+1}}$.

$$\therefore\ \frac{1}{a_n^2} + \frac{1}{b_n^2} = \cos^2\frac{\pi}{2^{n+1}} + \sin^2\frac{\pi}{2^{n+1}} = 1.$$

15. Suppose area of quadrilateral $ABCD$ is $S = 32\,\text{cm}^2$, $AD = y$, $AC = x$, $BC = z$,

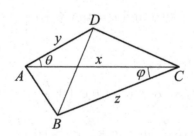

$\therefore\ x+y+z=16.$

$S=\dfrac{1}{2}xy\sin\theta+\dfrac{1}{2}xz\sin\varphi(|\sin\varphi|,|\sin\theta|\le 1)$

$S\le\dfrac{1}{2}xy+\dfrac{1}{2}xz=\dfrac{1}{2}x(y+z)=\dfrac{1}{2}x(16-x)$

$\le\dfrac{1}{2}\left(\dfrac{x+16-x}{2}\right)^2=32$

$\because\ S=32,\ \therefore\ \sin\theta=1,\ \sin\varphi=1,\ \text{namely},\ x-8=0,$

$\theta=\varphi=\dfrac{\pi}{2}$ where $BD=8\sqrt{2}.$

16. $\because\ \dfrac{1}{\sqrt{x}}+\dfrac{3}{\sqrt{y}}=1.$

$\therefore$ Suppose $\dfrac{1}{\sqrt{x}}=\cos^2\varphi,\ \dfrac{3}{\sqrt{y}}=\sin^2\varphi\left(0<\varphi<\dfrac{\pi}{2}\right),$

namely, substitute

$\begin{cases}x=\dfrac{1}{\cos^4\varphi}\\[2mm] y=\dfrac{1}{\sin^4\varphi}\end{cases}$ into $1+\dfrac{12}{3x+y}=\dfrac{2}{\sqrt{x}}$

$\therefore\ \cos^4 2\varphi-4\cos^3 2\varphi-6\cos^2 2\varphi-4\cos 2\varphi+1=0.$

Suppose $\cos 2\varphi=t$

$\therefore\ t^4-4t^3-6t^2-4t+1=0$

$\Leftrightarrow t^2+\dfrac{1}{t^2}-4\left(t+\dfrac{1}{t}\right)-6=0$

$\Leftrightarrow\left(t+\dfrac{1}{t}\right)^2-4\left(t+\dfrac{1}{t}\right)-8=0$

$\Leftrightarrow t+\dfrac{1}{t}=2+2\sqrt{3}$ or $t+\dfrac{1}{t}=2-2\sqrt{3}$ (rejection)

$$\because \ t + \frac{1}{t} = 2 + 2\sqrt{3}$$

$$\therefore \ t = \sqrt{3} + 1 \pm \sqrt{3 + 2\sqrt{3}}$$

$$\because \ |t| \leq 1, \ \therefore \ t = \sqrt{3} + 1 - \sqrt{3 + 2\sqrt{3}}$$

$$\therefore \ \begin{cases} x = \dfrac{1}{\cos^4 \varphi} = \dfrac{4}{t^2 + 2t + 1} = \dfrac{2\left(2 - \sqrt{3}\right)}{\sqrt{3} + 1 - \sqrt{3 + 2\sqrt{3}}}, \\[4mm] y = \dfrac{9}{\sin^4 \varphi} = \dfrac{36}{t^2 - 2t + 1} = \dfrac{6\sqrt{3}}{\sqrt{3} + 1 - \sqrt{3 + 2\sqrt{3}}}. \end{cases}$$

17. $\because \ a + c = (1 - ac)b$ and $1 - ac \neq 0, \ \therefore \ b = \dfrac{a + c}{1 - ac}.$

Let $\alpha = \arctan a, \ \beta = \arctan b, \ \gamma = \arctan c, \ \alpha, \beta, \gamma \in \left(0, \dfrac{\pi}{2}\right).$

$$\therefore \ \tan \beta = \frac{\tan \alpha + \tan \gamma}{1 - \tan \alpha \cdot \tan \gamma} = \tan(\alpha + \gamma)$$

$$\because \ \beta, \alpha + \gamma \in (0, \pi), \ \therefore \ \beta = \alpha + \gamma$$

$$\therefore \ p = \frac{2}{1 + \tan^2 \alpha} - \frac{2}{1 + \tan^2 \beta} + \frac{3}{1 + \tan^2 \gamma}$$

$$= 2\cos^2 \alpha - 2\cos^2(\alpha + \gamma) + 3\cos^2 \gamma$$

$$= (\cos 2\alpha + 1) - [\cos(2\alpha + 2\gamma) + 1] + 3\cos^2 \gamma$$

$$= 2\sin \gamma \cdot \sin(2\alpha + \gamma) + 3\cos^2 \gamma \leq 2\sin \gamma + 3\cos^2 \gamma$$

$$= \frac{10}{3} - 3\left(\sin \gamma - \frac{1}{3}\right)^2 \leq \frac{10}{3}.$$

18.   In $\triangle ABC$,

$$\cot \frac{A}{2} = \tan \frac{B+C}{2} = \frac{\tan \frac{B}{2} + \tan \frac{C}{2}}{1 - \tan \frac{B}{2} \tan \frac{C}{2}},$$

$$\therefore \ \tan \frac{A}{2} \tan \frac{B}{2} + \tan \frac{B}{2} \tan \frac{C}{2} + \tan \frac{C}{2} \tan \frac{A}{2} = 1.$$

According to power mean inequality:

$$S \leq \sqrt{3\left(\left(3\tan \frac{A}{2} \tan \frac{B}{2} + 1\right) + \left(3\tan \frac{B}{2} \tan \frac{C}{2} + 1\right) + \left(3\tan \frac{C}{2} \tan \frac{A}{2} + 1\right)\right)}$$

$$= \sqrt{3 \times 6} = 3\sqrt{2} < 5.$$

When $0 < x < 1$, $x^2 < x$.

$$\therefore \ \sqrt{3\tan \frac{B}{2} \tan \frac{C}{2} + 1} > \tan \frac{B}{2} \tan \frac{C}{2} + 1$$

$$\sqrt{3\tan \frac{C}{2} \tan \frac{A}{2} + 1} > \tan \frac{C}{2} \tan \frac{A}{2} + 1$$

$$\sqrt{3\tan \frac{A}{2} \tan \frac{B}{2} + 1} > \tan \frac{A}{2} \tan \frac{B}{2} + 1$$

$$\therefore \ S > 3 + \tan \frac{B}{2} \tan \frac{C}{2} + \tan \frac{C}{2} \tan \frac{A}{2} + \tan \frac{A}{2} \tan \frac{B}{2} = 4.$$

$\therefore$ Integral part of S is 4.

19. Make the circumcircle O of $\triangle ABC$, and suppose its radius is R. Take one point D on AB, and suppose extension line of CD and circumcircle intersect at E, so $CD \cdot DE = AD \cdot DB$. CD is mean term of proportional of AD and DB which is equal to $CD = DE$.

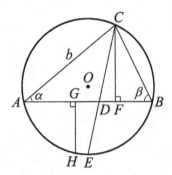

D which meets the nature of proposition is equal to find one point E on the circle where segment CE is divide equally by AB.

Find this point E if and only if $CF \leq HG$ where CF is the height of $\triangle ABC$ and HG is the height of chord AB.

$$CF = b \cdot \sin \alpha = 2R \cdot \sin \alpha \cdot \sin \beta$$

$\because$ radius $ON \perp AB$ at M,

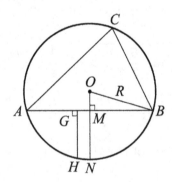

$$\therefore \ GH \leq MN = ON - OM = R - R \cdot \cos \gamma = 2R \cdot \sin^2 \frac{\gamma}{2}$$

$$\therefore \ \sin \alpha \cdot \sin \beta \leq \sin^2 \frac{\gamma}{2}.$$

20. $\sin B \sin C = \dfrac{1}{2}(\cos(B+C) - \cos(B-C))$

$\leq -\dfrac{1}{2}(\cos(B+C) - 1) = \dfrac{1}{2}(1 + \cos A)$

$\therefore\ 1 - \sin B \sin C \geq 1 - \dfrac{1}{2}(1 + \cos A) = \dfrac{1}{2}(1 - \cos A) = \sin^2 \dfrac{A}{2}.$

$\therefore\ \sqrt{1 - \sin B \sin C} \geq \sin \dfrac{A}{2},\ \dfrac{\sin A}{\sqrt{1 - \sin B \sin C}} \leq 2\cos \dfrac{A}{2}.$

Similarly, $\dfrac{\sin B}{\sqrt{1 - \sin C \sin C}} \leq 2\cos \dfrac{B}{2}$ and

$\dfrac{\sin C}{\sqrt{1 - \sin A \sin B}} \leq 2\cos \dfrac{C}{2}.$

Add these three inequalities,

$$f(A, B, C) \leq 2\left(\cos \dfrac{A}{2} + \cos \dfrac{B}{2} + \cos \dfrac{C}{2}\right).$$

$\because\ \cos \dfrac{A}{2} + \cos \dfrac{B}{2} + \cos \dfrac{C}{2} + \cos \dfrac{\pi}{6}$

$= 2\cos \dfrac{A+B}{4} \cos \dfrac{A-B}{4} + 2\cos \dfrac{C+\frac{\pi}{3}}{4} \cos \dfrac{C+\frac{\pi}{3}}{4}$

$\leq 2\left(\cos \dfrac{A+B}{4} + \cos \dfrac{C+\frac{\pi}{3}}{4}\right)$

$= 4\cos \dfrac{A+B+C+\frac{\pi}{3}}{8} \cos \dfrac{A+B-C-\frac{\pi}{3}}{8}$

$\leq 4\cos \dfrac{A+B+C+\frac{\pi}{3}}{8}$

$= 4\cos \dfrac{\pi}{6}.$

$\therefore\ f(A, B, C)_{\max} = 3\sqrt{3}.$

21. $b(az + cx - b) + c(bx + ay - c) - a(cy + bz - a) = 0$, namely, $x = \frac{b^2+c^2-a^2}{2bc}$.

According to symmetry:

$$y = \frac{a^2 + c^2 - b^2}{2ac}, \quad z = \frac{a^2 + b^2 - c^2}{2ab}.$$

$\because \ x, y, z, a, b, c \in R^+,$

$\therefore \ b^2 + c^2 > a^2, \ a^2 + c^2 > b^2, \ a^2 + b^2 > c^2.$

$\therefore \ \triangle ABC$ is acute triangle whose sides are a, b, c.

$\therefore \ x = \cos A, \ y = \cos B, \ z = \cos C.$

$\therefore$ The question equals to find the minimum of

$$f(\cos A, \cos B, \cos C) = \frac{\cos^2 A}{1 + \cos A} + \frac{\cos^2 B}{1 + \cos B} + \frac{\cos^2 C}{1 + \cos C}$$

in acute triangle ABC.

$$\frac{\cos^2 A}{1 + \cos A} = \frac{1 - \sin^2 A}{2\cos^2 \frac{A}{2}} = \frac{1 - (2\sin \frac{A}{2}\cos \frac{A}{2})^2}{2\cos^2 \frac{A}{2}}$$

$$= \frac{1}{2} + \frac{1}{2}\tan^2 \frac{A}{2} - 2\sin^2 \frac{A}{2}.$$

Similarly, $\dfrac{\cos^2 B}{1 + \cos B} = \dfrac{1}{2} + \dfrac{1}{2}\tan^2 \dfrac{B}{2} - 2\sin^2 \dfrac{B}{2};$

$$\frac{\cos^2 C}{1 + \cos C} = \frac{1}{2} + \frac{1}{2}\tan^2 \frac{C}{2} - 2\sin^2 \frac{C}{2}.$$

According to inequality Carfunkel:

$$\tan^2 \frac{A}{2} + \tan^2 \frac{B}{2} + \tan^2 \frac{C}{2} \geq 2 - 8\sin \frac{A}{2}\sin \frac{B}{2}\sin \frac{C}{2}$$

and identity: $\sin^2 \dfrac{A}{2} + \sin^2 \dfrac{B}{2} + \sin^2 \dfrac{C}{2}$

$$= 1 - 2\sin\frac{A}{2}\sin\frac{B}{2}\sin\frac{C}{2}:$$

$$\therefore \ f(\cos A, \cos B, \cos C) = \frac{\cos^2 A}{1 + \cos A} + \frac{\cos^2 B}{1 + \cos B} + \frac{\cos^2 C}{1 + \cos C}$$

$$= \frac{3}{2} + \frac{1}{2}\left(\tan^2\frac{A}{2} + \tan^2\frac{B}{2} + \tan^2\frac{C}{2}\right.$$

$$\left. - 2\left(\sin^2\frac{A}{2} + \sin^2\frac{B}{2} + \sin^2\frac{C}{2}\right)\right)$$

$$\geq \frac{3}{2} + \frac{1}{2}\left(2 - 8\sin\frac{A}{2}\sin\frac{B}{2}\sin\frac{C}{2}\right)$$

$$-2\left(1 - 2\sin\frac{A}{2}\sin\frac{B}{2}\sin\frac{C}{2}\right) = \frac{1}{2}.$$

22. $\because \ x \in (-1, 1), \ \therefore \ $ let $1 + x = 2\sin^2\theta, \ \theta \in \left(0, \dfrac{\pi}{2}\right);$

$$\therefore \ 1 - x = 2\cos^2\theta,$$

$$y = \frac{1}{(\sqrt{2})^n}\left(\frac{a}{\sin^n\theta} + \frac{b}{\cos^n\theta}\right), \ \theta \in \left(0, \frac{\pi}{2}\right)$$

Suppose A and B are undetermined positive constants.

$$y = \frac{1}{(\sqrt{2})^n}\left[\frac{a}{\sin^n\theta} + A\underbrace{\sin\theta + \sin\theta + \cdots + \sin\theta}_{n\text{ terms}} + \frac{b}{\cos^n\theta}\right.$$

$$\left. + B\underbrace{(\cos\theta + \cos\theta + \cdots + \cos\theta)}_{n\text{ terms}} - n\sqrt{A^2 + B^2}\sin\left(\theta + \arctan\frac{B}{A}\right)\right]$$

$$\geq \frac{1}{(\sqrt{2})^n}[(n + 1)\sqrt[n+1]{aA^n} + (n + 1)\sqrt[n+1]{bB^n} - n\sqrt{A^2 + B^2}].$$

The equality holds if and only if

$$\begin{cases} \dfrac{a}{\sin^n \theta} = A \sin \theta, & \text{(i)} \\[2mm] \dfrac{b}{\cos^n \theta} = B \sin \theta, & \text{(ii)} \\[2mm] \sin\left(\theta + \arctan \dfrac{B}{A}\right) = 1. & \text{(iii)} \end{cases}$$

According to type (i) and (ii): $\tan^{n+1}\theta = \dfrac{aB}{bA}$.

According to type (iii): $\theta = \dfrac{\pi}{2} - \arctan\dfrac{B}{A}$, namely, $\tan\theta = \dfrac{A}{B}$.

By the above equation, we have: $\tan^{n+2}\theta = \dfrac{a}{b}$ namely,

$$\tan\theta = \left(\frac{a}{b}\right)^{\frac{1}{n+2}}, \quad x = 2\sin^2\theta = 1 = \frac{2\tan^2\theta}{1+\tan^2\theta} - 1$$

$$= \frac{2a^{\frac{2}{n+2}}}{a^{\frac{2}{n+2}} + b^{\frac{2}{n+2}}} - 1.$$

$\therefore$ When $x = \dfrac{2a^{\frac{2}{n+2}}}{a^{\frac{2}{n+2}} + b^{\frac{2}{n+2}}} - 1,$

$$y_{\min} = \frac{1}{(\sqrt{2})^n}\left[\frac{a(a^{\frac{2}{n+2}} + b^{\frac{2}{n+2}})^{\frac{n}{2}}}{a^{\frac{n}{n+2}}} + \frac{b(a^{\frac{2}{n+2}} + b^{\frac{2}{n+2}})^{\frac{n}{2}}}{b^{\frac{n}{n+2}}}\right]$$

$$= \frac{1}{(\sqrt{2})^n}(a^{\frac{2}{n+2}} + b^{\frac{2}{n+2}})^{\frac{n+2}{2}}.$$

23. Let $g(t) = \frac{\sin t}{3 + 2\sin t}$. Suppose $-\frac{\pi}{2} < t_1 < t_2 < \frac{\pi}{2}$, so

$$g(t_1) - g(t_2) = \frac{\sin t_1}{3 + 2\sin t_1} - \frac{\sin t_2}{3 + 2\sin t_2}$$

$$= \frac{3(\sin t_1 - \sin t_2)}{(3 + 2\sin t_1)(3 + 2\sin t_2)} < 0.$$

$\therefore g(t_1) < g(t_2)$, $g(t)$ is increasing function in $\left(-\frac{\pi}{2}, \frac{\pi}{2}\right)$.

Let $x_1 \in \left(-\frac{\pi}{2}, \frac{\pi}{2}\right)$,

so

$$\left(x_1 - \frac{\pi}{6}\right)\left(\frac{\sin x_1}{3 + 2\sin x_1} - \frac{\sin \frac{\pi}{6}}{3 + 2\sin \frac{\pi}{6}}\right) \geq 0,$$

$$\frac{6x_1 - \pi}{6} \times \frac{\sin x_1}{3 + 2\sin x_1} \geq \frac{1}{8} \times \frac{6x_1 - \pi}{6},$$

namely, $\dfrac{6x_1 \sin x_1 - \pi \sin x_1}{3 + 2\sin x_1} \geq \dfrac{6x_1 - \pi}{8}.$

Similarly,

$$\frac{6x_2 \sin x_2 - \pi \sin x_2}{3 + 2\sin x_2} \geq \frac{6x_2 - \pi}{8}, \ldots, \frac{6x_n \sin x_n - \pi \sin x_n}{3 + 2\sin x_n}$$

$$\geq \frac{6x_n - \pi}{8}.$$

Add those above types:

$$\sum_{i=1}^{n} \frac{6x_i \sin x_i - \pi \sin x_i}{3 + 2\sin x_i} \geq \frac{6(x_1 + x_2 + \cdots + x_n) - n\pi}{8} = 0.$$

$\therefore$ The minimum of

$$y = \sum_{i=1}^{n} \frac{6x_i \sin x_i - \pi \sin x_i}{3 + 2\sin x_i} \text{ is } 0.$$

24. As in figure, connect FD,

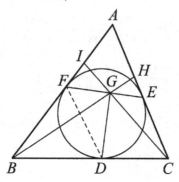

$$\because FG = FD\cos\angle DFG$$

$$= 2FB\cos\angle BFD\cos\angle DFG$$

$$= 2FB\cos\frac{\pi - B}{2}\cos\frac{\pi - C}{2}$$

$$= FB\cdot 2\sin\frac{B}{2}\sin\frac{C}{2},$$

$$\therefore \frac{FG}{FB} = 2\sin\frac{B}{2}\sin\frac{C}{2}.$$

Similarly,

$$\frac{EG}{EC} = 2\sin\frac{B}{2}\sin\frac{C}{2}. \quad\therefore\ \frac{FG}{FB} = \frac{EG}{EC}.$$

$\because \angle GFB = \angle GEC, \therefore \triangle GFB \backsim \triangle GEC.$

$\therefore \triangle IFG \backsim \triangle HEG$

$$\therefore \frac{FB}{EC} = \frac{FG}{EG} = \frac{IF}{HE} = \frac{4}{3}.$$

$\because BC = FB + EC = 21, \therefore\ FB = 12,\ EC = 9.$

Suppose $AF = AE = x$.

$\therefore$ In $\triangle AEF$, according to Menelaus law:

$$\frac{AB}{BF} \cdot \frac{FG}{GE} \cdot \frac{EH}{HA} = \frac{x+12}{12} \cdot \frac{4}{3} \cdot \frac{3}{x-3} = 1,$$

$$\therefore x = \frac{21}{2}.$$

$\therefore$ Three sides of $\triangle ABC$ are $AC = \dfrac{39}{2}$, $BC = 21$, $AB = \dfrac{45}{2}$.

According to Heron's formula: $S_{\triangle ABC} = 189$.

25. As in figure, suppose the radius of inscribed circle of $\triangle ABC$ is 1, and O is incenter of the triangle. Connect OA, OB, OC, so OA, OB and OC are three angular bisectors in $\triangle ABC$.

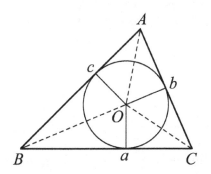

$$a = AB = \cot\frac{A}{2} + \cot\frac{B}{2}, \quad b = AC = \cot\frac{A}{2} + \cot\frac{C}{2},$$

$$S_{\triangle ABC} = \frac{1}{2}\left(\cot\frac{A}{2} + \cot\frac{B}{2}\right)\left(\cot\frac{A}{2} + \cot\frac{C}{2}\right) \cdot \sin A$$

$$= \frac{\cos\frac{A}{2}\cos\frac{B}{2}\cos\frac{C}{2}}{\sin\frac{A}{2}\sin\frac{B}{2}\sin\frac{C}{2}} = \frac{1}{\tan\frac{A}{2}\tan\frac{B}{2}\tan\frac{C}{2}}.$$

$$A + B + C = \pi$$

$$\Rightarrow \frac{A+B}{2} = \frac{\pi}{2} - \frac{C}{2}$$

$$\Rightarrow \tan\left(\frac{A}{2}+\frac{B}{2}\right)=\tan\left(\frac{\pi}{2}-\frac{C}{2}\right)=\frac{1}{\tan\frac{C}{2}}$$

$$\Leftrightarrow \frac{\tan\frac{A}{2}+\tan\frac{B}{2}}{1-\tan\frac{A}{2}\tan\frac{B}{2}}=\frac{1}{\tan\frac{C}{2}}$$

$$\Leftrightarrow \tan\frac{A}{2}\tan\frac{B}{2}+\tan\frac{B}{2}\tan\frac{C}{2}+\tan\frac{C}{2}\tan\frac{A}{2}=1.$$

$\because \tan\dfrac{A}{2},\ \tan\dfrac{B}{2}$ and $\tan\dfrac{C}{2}$ are positive.

According to mean value inequality:

$$\frac{1}{3}=\frac{\tan\frac{A}{2}\tan\frac{B}{2}+\tan\frac{B}{2}\tan\frac{C}{2}+\tan\frac{C}{2}\tan\frac{A}{2}}{3}$$

$$\geq \sqrt[3]{\left(\tan\frac{A}{2}\tan\frac{B}{2}\tan\frac{C}{2}\right)^2}$$

$$\Leftrightarrow \tan\frac{A}{2}\tan\frac{B}{2}\tan\frac{C}{2}\leq \frac{\sqrt{3}}{9}$$

$$\Rightarrow \frac{1}{\tan\frac{A}{2}\tan\frac{B}{2}\tan\frac{C}{2}}\geq 3\sqrt{3}.$$

The minimum of $S_{\triangle ABC}$ is $3\sqrt{3}$ if and only if

$$\tan\frac{A}{2}\tan\frac{B}{2}=\tan\frac{B}{2}\tan\frac{C}{2}=\tan\frac{C}{2}\tan\frac{A}{2}$$

$$\Leftrightarrow \tan\frac{A}{2}=\tan\frac{B}{2}=\tan\frac{C}{2}$$

$$\Leftrightarrow A=B=C=\frac{\pi}{3}.$$

$\therefore$ When triangle is equilateral triangle, the area of triangle is minimum which is $3\sqrt{3}$.

26. Firstly, prove the following lemma (Moscow mathematics competition in 1963): $a,b,c\in R^+$, prove: $\frac{a}{b+c}+\frac{b}{c+a}+\frac{c}{a+b}\geq\frac{3}{2}$.

Proof: According to symmetry, suppose:

$a \geq b \geq c > 0$, so $\dfrac{1}{b+c} \geq \dfrac{1}{c+a} \geq \dfrac{1}{a+b}$.

$\therefore$ (ordered sum) $\dfrac{a}{b+c} + \dfrac{b}{c+a} + \dfrac{c}{a+b}$

$\geq \dfrac{b}{b+c} + \dfrac{c}{c+a} + \dfrac{a}{a+b}$ (disordered sum)

(ordered sum) $\dfrac{a}{b+c} + \dfrac{b}{c+a} + \dfrac{c}{a+b}$

$\geq \dfrac{c}{b+c} + \dfrac{a}{c+a} + \dfrac{b}{a+b}$ (disordered sum)

Add above two inequalities: $\dfrac{a}{b+c} + \dfrac{b}{c+a} + \dfrac{c}{a+b} \geq \dfrac{3}{2}$.

Back to this question: in acute $\triangle ABC$,

$$\frac{\cos(B-C)}{\cos A} = -\frac{\cos(B-C)}{\cos(B+C)} = \frac{\sin B \sin C + \cos B \cos C}{\sin B \sin C - \cos B \cos C}$$

$$= \frac{\tan B \tan C + 1}{\tan B \tan C - 1}$$

$$= \frac{\tan A \tan B \tan C + \tan A}{\tan A \tan B \tan C - \tan A}$$

$$= \frac{2\tan A + \tan B + \tan C}{\tan B + \tan C}$$

$$= \frac{2\tan A}{\tan B + \tan C} + 1.$$

Similarly,

$$\frac{\cos(C-A)}{\cos B} = \frac{2\tan B}{\tan C + \tan A} + 1,$$

$$\frac{\cos(A-B)}{\cos C} = \frac{2\tan C}{\tan A + \tan B} + 1.$$

$\therefore$ Inequality (1) is equal to:

$$3 + 2\left(\frac{\tan A}{\tan B + \tan C} + \frac{\tan B}{\tan C + \tan A} + \frac{\tan C}{\tan A + \tan B}\right) \geq 6,$$

only to prove:

$$\frac{\tan A}{\tan B + \tan C} + \frac{\tan B}{\tan C + \tan A} + \frac{\tan C}{\tan A + \tan B} \geq \frac{3}{2}$$

$\therefore$ According to lemma, the above inequality holds.

27. Construct two vectors:

$$\vec{a} = \left(\frac{\sin^2 \alpha}{\sqrt{\sin \alpha \sin \beta}}, \frac{\sin^2 \beta}{\sqrt{\sin \beta \sin \gamma}}, \frac{\sin^2 \gamma}{\sqrt{\sin \gamma \sin \alpha}}\right),$$

$$\vec{b} = \left(\sqrt{\sin \alpha \sin \beta}, \sqrt{\sin \beta \sin \gamma}, \sqrt{\sin \gamma \sin \alpha}\right)$$

According to scalar product inequality:

$$|\vec{a}|^2 \times |\vec{b}|^2 \geq (\vec{a} \cdot \vec{b})^2, \text{namely,}$$

$$\left(\frac{\sin^4 \alpha}{\sin \alpha \sin \beta} + \frac{\sin^4 \beta}{\sin \beta \sin \gamma} + \frac{\sin^4 \gamma}{\sin \gamma \sin \alpha}\right)$$

$$(\sin \alpha \sin \beta + \sin \beta \sin \gamma + \sin \gamma \sin \alpha)$$

$$\geq (\sin^2 \alpha + \sin^2 \beta + \sin^2 \gamma)^2$$

$$\therefore \frac{\sin^3 \alpha}{\sin \beta} + \frac{\sin^3 \beta}{\sin \gamma} + \frac{\sin^3 \gamma}{\sin \alpha}$$

$$\geq \frac{(\sin^2 \alpha + \sin^2 \beta + \sin^2 \gamma)^2}{\sin \alpha \sin \beta + \sin \beta \sin \gamma + \sin \gamma \sin \alpha}$$

$$\geq \frac{(\sin^2 \alpha + \sin^2 \beta + \sin^2 \gamma)^2}{\sin^2 \alpha + \sin^2 \beta + \sin^2 \gamma} = \sin^2 \alpha + \sin^2 \beta + \sin^2 \gamma = 1.$$

28. Prove: connect PD and QE. Suppose $\angle ACD = \alpha$, $\angle BCE = \beta$, $\angle DCE = \gamma$, R is the radius of $\odot CDE$.

$\because \angle ACB = 90°$, $\therefore PQ = 2R$, $PD = 2R\sin\alpha$, $QE = 2R\sin\beta$.

$\because \angle ADP = \angle ACE$, $\angle A$ is common angle, $\therefore \triangle ADP \backsim \triangle ACE$,

$$AP = PD. \frac{AE}{CE} = 2R\sin\alpha \cdot \frac{\sin(\alpha + \gamma)}{\sin A}.$$

Similarly,

$$BQ = 2R\sin\beta \cdot \frac{\beta + \gamma}{\sin B}.$$

$\because \angle A = \angle B = 45°$, $\alpha + \beta + \gamma = 90°$,

$$\sin(\alpha + \gamma) = \cos\beta, \ \sin(\beta + \gamma) = \cos\alpha,$$

$$\therefore AP + BQ$$

$$= 2\sqrt{2}\sin\alpha\cos\beta + 2\sqrt{2}\sin\beta\cos\alpha$$

$$= 2\sqrt{2}\sin(\alpha + \beta).$$

(1) When $\angle DCE = 45°$, $\alpha + \beta = 45°$.

$$\therefore AP + BQ = 2\sqrt{2}R \cdot \frac{1}{\sqrt{2}} = 2R = PQ.$$

(2) When $AP + BQ = PQ$, $2\sqrt{2}R \ \sin(\alpha + \beta) = 2R$,

$$\sin(\alpha + \beta) = \frac{1}{\sqrt{2}}.$$

$\because 0° < \alpha + \beta < 90°$, $\therefore \alpha + \beta = 45°$

$\therefore \angle DCE = \gamma = 45°$.

29. Prove: (1) In $\triangle ABC$:

$$h_a = b\sin C = 2R\sin B\sin C, \ h_b = 2R\sin A\sin C,$$

$$h_c = 2R\sin A\sin B,$$

$$\therefore \sum \frac{\cos A}{h_a} = \sum \frac{\cos A}{2R \sin B \sin C}$$

$$= \frac{1}{4R} \sum \frac{2 \sin A \cos A}{\sin A \sin B \sin C}$$

$$= \frac{1}{4R} \sum \frac{\sin 2A}{\sin A \sin B \sin C}$$

$$= \frac{1}{4R} \cdot \frac{4 \sin A \sin B \sin C}{\sin A \sin B \sin C} = \frac{1}{R}.$$

(2) In $\triangle ABC$:

$$\sin A \sin B \sin C = \frac{\frac{1}{2} ab \sin C}{2R^2} = \frac{\triangle}{2R^2} = \frac{Pr}{2R^2}$$

($\triangle$ express the area of $\triangle ABC$)

$$R = \frac{abc}{4\triangle}, \quad r = \frac{\triangle}{p}, \quad Rr = \frac{abc}{4p},$$

$$\therefore p^2 - 4Rr - r^2 = p^2 - \frac{abc}{p} - \frac{\triangle^2}{p^2}$$

$$= p^2 - \frac{abc}{p} - \frac{(p-a)(p-b)(p-c)}{p}$$

$$= \frac{1}{p}\{p^3 - abc - [p^3 - (a+b+c)p^2$$

$$+ (ab + bc + ca)p - abc]\}$$

$$= (a + b + c)p - (ab + bc + ca)$$

$$= \frac{1}{2}[(a+b+c)^2 - 2(ab+bc+ca)] = \frac{1}{2} \sum a^2.$$

$$\therefore \sum \sin^3 A = \sum \frac{a^2}{4R^2} = \frac{p^2 - 4Rr - r^2}{2R^2}.$$

$$\sum \frac{\sin A}{h_a} = \sum \frac{\sin A}{2R \sin B \sin C} = \frac{\sum \sin^2 A}{2R \sin A \sin B \sin C}$$

$$= \frac{\frac{p^2 - 4Rr - r^2}{2R^2}}{\frac{pr}{R}} = \frac{p^2 - 4Rr - r^2}{2pRr}.$$

30. Lemma: Suppose O is excenter of $\triangle ABC$. P is a point on circumcircle of $\triangle ABC$ and $PO = d$. Make $PD \perp BC$, $PE \perp CA$ and $PF \perp AB$, foot points are D, E and F. R is the radius of circumcircle of $\triangle ABC$, so $S_{\triangle DEF} = \frac{1}{2}(R^2 - d^2) \sin A \sin B \sin C$. Connect and expand AP and intersect $\odot O$ at A'.

$\because PE \perp AC$, $PF \perp AB$,

$\therefore EF = AP \sin A$ \hfill (i)

Similarly:

$FD = BP \sin B$ \hfill (ii)

$ED = CP \sin C$ \hfill (iii)

In $\triangle PBA'$, according to sine law:

$$\frac{PB}{PA'} = \frac{\sin \angle PA'B}{\sin \angle PBA'} - = \frac{\sin C}{\sin \angle PBA'} \qquad \text{(iv)}$$

According to type (iii) and (iv): $PB = PA' \dfrac{\sin C}{\sin \angle EFD}$ \hfill (v)

According to type (ii) and (v): $FD = PA' \dfrac{\sin B \sin C}{\sin \angle EFD}$ \hfill (vi)

According to type (i) and (vi):

$$S_{\triangle DEF} = \frac{1}{2} EF \times FD \sin \angle EFD = \frac{1}{2} PA \times PA' \sin A \sin B \sin C$$

$$= \frac{1}{2}(R^2 - d^2) \sin A \sin B \sin C.$$

Then prove original proposition:

$\because \triangle ABC$ is equilateral triangle, $\therefore d_1 + d_2 + d_3 = 3r$.

According to lemma:

$$S_{\triangle DEF} = \frac{\sqrt{3}}{4}(d_1 d_2 + d_2 d_3 + d_3 d_1) = \frac{1}{2}(R^2 - d^2)\sin A \sin B \sin C$$

$$= \frac{1}{12}\left[(2r)^2 - \left(\frac{1}{2}r\right)^2\right]\sin A \sin B \sin C = \frac{45\sqrt{3}}{64}r$$

$$d_1 d_2 + d_2 d_3 + d_3 d_1 = \frac{45}{16}r^2.$$

According to Heron's formula:

$$(\sqrt{d_1} + \sqrt{d_2} + \sqrt{d_3})(\sqrt{d_1} + \sqrt{d_2} - \sqrt{d_3})$$

$$(\sqrt{d_1} - \sqrt{d_2} + \sqrt{d_3})(-\sqrt{d_1} + \sqrt{d_2} + \sqrt{d_3})$$

$$= 2(d_1 d_2 + d_2 d_3 + d_3 d_1) - (d_1^2 + d_2^2 + d_3^2)$$

$$= -(d_1 + d_2 + d_3)^2 + 4(d_1 d_2 + d_2 d_3 + d_3 d_1) = \frac{9}{4}r^2 > 0 \quad (*)$$

$$(\sqrt{d_1} + \sqrt{d_2} - \sqrt{d_3})(\sqrt{d_1} - \sqrt{d_2} + \sqrt{d_3})$$

$$(-\sqrt{d_1} + \sqrt{d_2} + \sqrt{d_3}) > 0.$$

$\therefore$ Any sum of two is more than the third among

$$\sqrt{d_1}, \sqrt{d_2}, \sqrt{d_3}.$$

$\therefore \sqrt{d_1}, \sqrt{d_2}, \sqrt{d_3}$ construct three sides of triangle.

31. Prove: suppose $\theta_1 \geq \theta_2 \geq \theta_3$,

$$\because \tan\theta_1 \tan\theta_2 \tan\theta_3 = 2\sqrt{2}$$

$$\therefore \tan^2\theta_1 = \frac{8}{\tan^2\theta_2 \tan^2\theta_3},$$

$$\cos \theta_1 = \frac{1}{\sqrt{1 + \tan^2 \theta_1}} = \frac{1}{\sqrt{1 + \frac{8}{\tan^2 \theta_2 \tan^2 \theta_3}}}$$

$$= \frac{\tan^2 \theta_2 \tan^2 \theta_3}{\sqrt{8 + \tan^2 \theta_2 \tan^2 \theta_3}}$$

$$= \frac{\sin \theta_2 \sin \vartheta_3}{\sqrt{8 \cos^2 \theta_2 \cos^2 \theta_3 + \sin^2 \theta_2 \sin^2 \theta_3}}$$

$$\cos \theta_i = \sqrt{1 - \sin^2 \theta_i} < 1 - \frac{1}{2} \sin^2 \theta_i \quad (i = 1, 2),$$

$$\therefore \quad \cos \theta_2 + \cos \theta_3 < 2 - \frac{1}{2} \sin^2 \theta_2 - \frac{1}{2} \sin^2 \theta_3 \le 2 - \sin \theta_2 \sin \theta_3$$

$$\cos \theta_1 + \cos \theta_2 + \cos \theta_3 < 2 - \sin \theta_2 \sin \theta_3$$

$$\left[1 - \frac{1}{\sqrt{8 \cos^2 \theta_2 \cos^2 \theta_3 + \sin^2 \theta_2 \sin^2 \theta_3}} \right] \tag{i}$$

$$\because \quad 8 \cos^2 \theta_2 \cos^2 \theta_3 + \sin^2 \theta_2 \sin^2 \theta_3 \ge 1$$

$$\Leftrightarrow 8 + \tan^2 \theta_2 \tan^2 \theta_3 \ge \frac{1}{\cos^2 \theta_2 \cos^2 \theta_3}$$

$$= (1 + \tan^2 \theta_2)(1 + \tan^2 \theta_3)$$

$$\Leftrightarrow \tan^2 \theta_2 + \tan^2 \theta_3 \le 7 \tag{ii}$$

If inequality (ii) holds, according to inequality (i): $\cos \theta_1 + \cos \theta_2 + \cos \theta_3 < 2$.

If inequality (ii) does not hold,

$$\tan^2 \theta_2 + \tan^2 \theta_3 > 7 \Rightarrow \tan^2 \theta_1 \ge \tan^2 \theta_2 > \frac{7}{2}$$

$$\Rightarrow \cos \theta_1 \le \cos \theta_2 < \frac{\sqrt{2}}{3}$$

$$\therefore \quad \cos \theta_1 + \cos \theta_2 + \cos \theta_3 < \frac{2\sqrt{2}}{3} + 1 < 2.$$

32. Prove:

$$8\sin\frac{A}{2}\sin\frac{B}{2}\sin\frac{C}{2} = \frac{\sin A \sin B \sin C}{\cos\frac{A}{2}\cos\frac{B}{2}\cos\frac{C}{2}}$$

$$= \frac{\sqrt{\sin A \sin B}\cdot\sqrt{\sin B \sin C}\cdot\sqrt{\sin C \sin A}}{\cos\frac{A}{2}\cos\frac{B}{2}\cos\frac{C}{2}}$$

$$\leq \frac{\frac{\sin A+\sin B}{2}\cdot\frac{\sin B+\sin C}{2}\cdot\frac{\sin C+\sin A}{2}}{\cos\frac{A}{2}\cos\frac{B}{2}\cos\frac{C}{2}}$$

$$= \frac{\sin\frac{A+B}{2}\cos\frac{A-B}{2}\cdot\sin\frac{B+C}{2}\cos\frac{B-C}{2}\cdot\sin\frac{C+A}{2}\cos\frac{C-A}{2}}{\cos\frac{A}{2}\cos\frac{B}{2}\cos\frac{C}{2}}$$

$$= \cos\frac{A-B}{2}\cos\frac{B-C}{2}\cos\frac{C-A}{2}.$$

The equality holds if and only if $A = B = C$.

(2) (a) When $\triangle ABC$ is obtuse triangle or right triangle,

$$8\cos A \cos B \cos C \leq \cos^2\frac{A-B}{2}\cos^2\frac{B-C}{2}\cos^2\frac{C-A}{2}.$$

(b) When $\triangle ABC$ is acute triangle, $\cos A, \cos B, \cos C > 0$.
According to mean value inequality and sum-to-product law:

$$8\cos A \cos B \cos C$$

$$= 2\sqrt{\cos A \cos B}\cdot 2\sqrt{\cos B \cos C}\cdot 2\sqrt{\cos C \cos A}$$

$$\leq (\cos A + \cos B)(\cos B + \cos C)(\cos C + \cos A)$$

$$= 2\cos\frac{A+B}{2}\cos\frac{A-B}{2}\cdot 2\cos\frac{B+C}{2}\cos\frac{B-C}{2}$$

$$\cdot 2\cos\frac{C+A}{2}\cos\frac{C-A}{2}$$

$$= \left(8\sin\frac{C}{2}\sin\frac{A}{2}\sin\frac{B}{2}\right)\cdot\left(\cos\frac{A-B}{2}\cos\frac{B-C}{2}\cos\frac{C-A}{2}\right)$$

$$\le \cos^2\frac{A-B}{2}\cos^2\frac{B-C}{2}\cos^2\frac{C-A}{2}.$$

The equality holds if and only if $A = B = C$.

33. Prove:

$$\because \frac{a}{h_a} = \cot B + \cot C, \quad \frac{b}{h_b} = \cot C + \cot A,$$

$$\frac{c}{h_c} = \cot A + \cot B.$$

$\therefore$ Above type is equal to $\dfrac{1}{\cot B + \cot C} + \dfrac{1}{\cot C + \cot A}$

$$+\frac{1}{\cot A + \cot B} \ge \frac{5}{2}.$$

Let $x = \cot A$, $y = \cot B$, $z = \cot C$.

$\because \triangle ABC$ is not obtuse triangle

$\therefore$ The above type is equal to that $\frac{1}{y+z} + \frac{1}{z+x} + \frac{1}{x+y} \ge \frac{5}{2}$ when $x \ge 0$, $y \ge 0$, $z \ge 0$ and $xy + yz + zx = 1$.

According to symmetry: suppose $x \ge y \ge z \ge 0$, so

$0 \le yz \le \frac{1}{3}$, $x = \frac{1-yz}{y+z}$.

$\therefore$ The above type is equal to

$$2[(z+x)(x+y) + (x+y)(y+z) + (y+z)(z+x)]$$

$$\ge 5(y+z)(z+x)(x+y)$$

$$\Longleftrightarrow 2[x^2 + y^2 + z^2 + 3(yz + zx + xy)]$$

$$\ge 5[(x+y+z)(xy+yz+zx) - xyz]$$

$$\Longleftrightarrow 2[(x+y+z)^2 + 1] \geq 5(x+y+z-xyz)$$

$$\Longleftrightarrow 2(x+y+z)^2 - 5(x+y+z) + 2 + 5xyz \geq 0$$

$$\Longleftrightarrow 2[(x+y+z) - 2]^2 + 3(z+y+z) + 5xyz - 6 \geq 0$$

$$\Longleftrightarrow 2[(x+y+z) - 2]^2 + \frac{3(1-yz)}{y+z} + 3(y+z)$$

$$+ 5 \cdot \frac{1-yz}{y+z} \cdot yz - 6 \geq 0$$

$$\Longleftrightarrow 2[(x+y+z) - 2]^2 + \frac{1}{y+z}[3(y+z)^2 - 6(y+z)$$

$$+ 3 + 2yz - 5(yz)^2] \geq 0$$

$$\Longleftrightarrow 2[(x+y+z) - 2]^2 + \frac{1}{y+z}[3(y+z-1)^2$$

$$+ yz(2-5yz)] \geq 0.$$

$$\because \ yz \leq \frac{1}{3} < \frac{2}{5}, \ 2 - 5yz > 0, \ yz \geq 0, \ \frac{1}{y+z} > 0.$$

$\therefore$ The original inequality holds.

34. According to symmetry: suppose $\alpha \leq \beta \leq \gamma$, α, β and γ correspond to sides a, b, c.

According to sine law:

$$\frac{a}{\sin\alpha} = \frac{b}{\sin\beta} = \frac{c}{\sin\beta}, \ \therefore \ \sin\alpha \leq \sin\beta \leq \sin\gamma,$$

$$\therefore \ \frac{1}{\beta} \leq \frac{1}{\alpha}, \ \sin\alpha - \sin\beta \leq 0, \ \therefore \ \left(\frac{1}{\beta} - \frac{1}{\alpha}\right)(\sin\alpha - \sin\beta) \geq 0.$$

After arrangement:

$$\frac{\sin\alpha}{\beta} + \frac{\sin\beta}{\alpha} \geq \frac{\sin\alpha}{\alpha} + \frac{\sin\beta}{\beta}. \tag{i}$$

Similarly,

$$\frac{\sin\alpha}{\gamma} + \frac{\sin\gamma}{\alpha} \geq \frac{\sin\alpha}{\alpha} + \frac{\sin\gamma}{\gamma}, \tag{ii}$$

$$\frac{\sin\beta}{\gamma} + \frac{\sin\gamma}{\beta} \geq \frac{\sin\beta}{\beta} + \frac{\sin\gamma}{\gamma}. \tag{iii}$$

Equations (i) + (ii) + (iii) imply

$$\left(\frac{1}{\beta} + \frac{1}{\gamma}\right)\sin\alpha + \left(\frac{1}{\alpha} + \frac{1}{\gamma}\right)\sin\beta + \left(\frac{1}{\alpha} + \frac{1}{\beta}\right)\sin\gamma$$

$$\geq 2\left(\frac{\sin\alpha}{\alpha} + \frac{\sin\beta}{\beta} + \frac{\sin\gamma}{\gamma}\right).$$

$\because \alpha, \beta, \gamma > 0, \sin\alpha, \sin\beta, \sin\gamma > 0,$

$$u = \frac{\left(\frac{1}{\beta} + \frac{1}{\gamma}\right)\sin\alpha + \left(\frac{1}{\gamma} + \frac{1}{\alpha}\right)\sin\beta + \left(\frac{1}{\alpha} + \frac{1}{\beta}\right)\sin\gamma}{\frac{\sin\alpha}{\alpha} + \frac{\sin\beta}{\beta} + \frac{\sin\gamma}{\gamma}}$$

$\geq 2, \therefore u_{\min} = 2.$

35. Suppose $x = \sin^3\alpha, y = \sin^3\beta, z = \sin^3\gamma$.
$\therefore$ The original question is equal to that $x, y, z > 0, x+y+z = 1$.
Prove:

$$\frac{x^{\frac{2}{3}}}{1 - x^{\frac{2}{3}}} + \frac{y^{\frac{2}{3}}}{1 - y^{\frac{2}{3}}} + \frac{z^{\frac{2}{3}}}{1 - z^{\frac{2}{3}}} \geq \frac{3}{\sqrt[3]{9} - 1}. \tag{1}$$

Note $f(x) = \dfrac{x^{\frac{2}{3}}}{1 - x^{\frac{2}{3}}}$, so $f'(x) = \dfrac{\frac{2}{3}x^{-\frac{1}{3}}}{(1 - x^{\frac{2}{3}})^2}$ where $0 < x < 1$.

$\therefore$ Tangential equation of $f(x)$ at $x = \dfrac{1}{3}$, where $0 < x < 1$,

$$y = \frac{\frac{2}{3}(\frac{1}{3})^{-\frac{1}{3}}}{[1 - (\frac{1}{3})^{\frac{2}{3}}]^2}\left(x - \frac{1}{3}\right) + \frac{(\frac{1}{3})^{\frac{2}{3}}}{1 - (\frac{1}{3})^{\frac{2}{3}}}.$$

Consider using tangent values to estimate the value of the curve, so that we transform the "curve" into "straight". So guess: when

$0 < x < 1,$

$$\frac{x^{\frac{2}{3}}}{1 - x^{\frac{2}{3}}} \geq \frac{\frac{2}{3}(\frac{1}{3})^{-\frac{1}{3}}}{[1 - (\frac{1}{3})^{\frac{2}{3}}]^2}\left(x - \frac{1}{3}\right) + \frac{(\frac{1}{3})^{\frac{2}{3}}}{1 - (\frac{1}{3})^{\frac{2}{3}}}. \tag{2}$$

To prove (2), let $p = x^{\frac{1}{3}}$, $q = (\frac{1}{3})^{\frac{1}{3}}$, so $0 < p, q < 1$.

$$\therefore (2) \Leftrightarrow \frac{p^2}{1 - p^2} - \frac{q^2}{1 - q^2} \geq \frac{2(p^3 - q^3)}{3q(1 - p^2)^2}$$

$$\Leftrightarrow \frac{p^2 - q^2}{(1 - p^2)(1 - q^2)} \geq \frac{2(p^3 - q^3)}{3q(1 - p^2)^2}$$

$$\Leftrightarrow 3q(1 - q^2)^2(p^2 - q^2)$$

$$\geq 2(p^3 - q^3)(1 - p^2)(1 - q^2)$$

$$\Leftrightarrow (p - q)^2(1 - q^2)[(2p^3 + 4p^2q) + (3q^2 - 1)(2p + q)] \geq 0.$$

$$1 - q^2 > 0, \ (p - q)^2 \geq 0,$$

$$(2p^3 + 4p^2q) + (3q^2 - 1)(2p + q) \geq 0.$$

$\therefore$ Type (2) holds.

The equality holds if and only if $p = q$, namely, $x = \frac{1}{3}$.

$$\frac{x^{\frac{2}{3}}}{1 - x^{\frac{2}{3}}} + \frac{y^{\frac{2}{3}}}{1 - y^{\frac{2}{3}}} + \frac{z^{\frac{2}{3}}}{1 - z^{\frac{2}{3}}}$$

$$\geq \frac{\frac{2}{3}(\frac{1}{3})^{-\frac{1}{3}}}{[1 - (\frac{1}{3})^{\frac{2}{3}}]^2}(x + y + z - 1) + \frac{3(\frac{1}{3})^{\frac{2}{3}}}{1 - (\frac{1}{3})^{\frac{2}{3}}}$$

$$= \frac{3}{\sqrt[3]{9} - 1}.$$

The equality holds if and only if $x = y = z = \frac{1}{3}$.

$\therefore$ The original proposition holds.

36. According to

(i) :
$$\sqrt{a} > \frac{\sin\theta\cos\theta}{\sin\theta + \cos\theta} \qquad \text{(a)}$$

Suppose $\dfrac{a}{\sin^2\theta} + \dfrac{a}{\cos^2\theta} \le 1$.

(ii)   There exists $x \in \left[1 - \dfrac{\sqrt{a}}{\sin\theta}, \dfrac{\sqrt{a}}{\cos\theta}\right]$ which meets

$$2(1-x)\sin\theta\sqrt{a - x^2\cos^2\theta} + 2x\cos\theta\sqrt{a - (1-x)^2\sin^2\theta} \ge a, \qquad \text{(b)}$$

$$2\sin\theta\cos\theta\left[(1-x)\sqrt{\frac{a}{\cos^2\theta} - x^2} + x\sqrt{\frac{a}{\sin^2\theta} - (1-x)^2}\right] \ge a.$$

Firstly, prove one lemma: suppose

$0 < p < 1,\ 0 < q < 1,\ p + q > 1,\ p^2 + q^2 \le 1,$

$f(x) = (1-x)\sqrt{p^2 - x^2} + x\sqrt{q^2 - (1-x)^2}\,(1 - q \le x \le p),$

so $f(x)_{\max}$ when $\sqrt{p^2 - x^2} = \sqrt{q^2 - (1-x)^2}$,

namely, $x = \dfrac{p^2 - q^2 + 1}{2} \in [1 - p, q].$

$\because 1 - q \le x \le p.$

$\therefore$ Let $x = p\sin\alpha,\ 1 - x = q\sin\beta,\ 0 < \alpha < \dfrac{\pi}{2},$

$0 < \beta < \dfrac{\pi}{2},\ \alpha + \beta < \pi.$

$\therefore\ f(x) = pq(\sin\beta\cos\alpha + \sin\alpha\cos\beta)$

$\quad = pq\sin(\alpha + \beta)\cos(\alpha + \beta)$

$\quad = \cos\alpha\cos\beta - \sin\alpha\sin\beta$

$$= \frac{\sqrt{p^2 - x^2}\sqrt{q^2 - (1-x)^2} - x(1-x)}{pq}$$

$$2[\sqrt{p^2 - x^2}\sqrt{q^2 - (1-x)^2} - x(1-x)]$$

$$= -[\sqrt{p^2 - x^2} - \sqrt{q^2 - (1-x)^2}]^2$$

$$+ p^2 + q^2 - x^2 - (1-x)^2 - 2x(1-x)$$

$$= p^2 + q^2 - 1 - [\sqrt{p^2 - x^2} - \sqrt{q^2 - (1-x)^2}]^2 \le 0,$$

$$\therefore \frac{\pi}{2} \le \alpha + \beta < \pi,$$

$$\cos(\alpha + \beta)_{\max} = \frac{p^2 + q^2 - 1}{2pq} \le 0 \text{ if and only if}$$

$$\sqrt{p^2 - x^2} = \sqrt{q^2 - (1-x)^2},$$

namely, $x = \dfrac{p^2 - q^2 + 1}{2} \in [1 - p, q];$

$\because$ sine function is decreasing function in $\left[\dfrac{\pi}{2}, \pi\right]$.

$\therefore f(x) = pq \sin(\alpha + \beta)$ takes the maximum if and only if

$$x = \frac{p^2 - q^2 + 1}{2}.$$

According to lemma: the maximum of left of type (b) is

$$2\sin\theta\cos\theta\sqrt{\frac{a}{\cos^2\theta} - \frac{1}{4}\left(\frac{a}{\cos^2\theta} - \frac{a}{\sin^2\theta} + 1\right)^2}, \text{ namely,}$$

$$\sin\theta\cos\theta\sqrt{\frac{4a}{\cos^2\theta} - \left(\frac{a}{\cos^2\theta} - \frac{a}{\sin^2\theta} + 1\right)^2},$$

if and only if $\sqrt{\dfrac{a}{\cos^2\theta} - x^2} = \sqrt{\dfrac{a}{\sin^2\theta} - (1-x)^2}$, namely,

$$x = \frac{1}{2}\left(\frac{a}{\cos^2\theta} - \frac{a}{\sin^2\theta} + 1\right) \in \left[1 - \frac{\sqrt{a}}{\sin\theta}, \frac{\sqrt{a}}{\cos\theta}\right].$$

According to (b): the minimum of a meets type (a):

$$\sqrt{\frac{4a}{\cos^2\theta} - \left(\frac{a}{\cos^2\theta} - \frac{a}{\sin^2\theta} + 1\right)^2} \geq \frac{a}{\cos\theta\sin\theta},$$

namely $a^2\left(\dfrac{1}{\cos^4\theta} + \dfrac{1}{\sin^4\theta} - \dfrac{1}{\cos^2\theta\sin^2\theta}\right)$

$$-2\left(\frac{1}{\cos^2\theta} + \frac{1}{\sin^2\theta}\right)a + 1 \leq 0. \tag{c}$$

$$\because \frac{1}{\cos^4\theta} + \frac{1}{\sin^4\theta} - \frac{1}{\sin^2\theta\cos^2\theta} = \frac{1 - 3\sin^2\theta\cos^2\theta}{\sin^4\theta\cos^4\theta} > 0.$$

$\therefore$ The root of type (c) is

$$\frac{\sin^4\theta\cos^4\theta}{1 - 3\sin^2\theta\cos^2\theta}\left[\frac{1}{\cos^2\theta} + \frac{1}{\sin^2\theta}\right.$$

$$\left. \pm \sqrt{\left(\frac{1}{\cos^2\theta} + \frac{1}{\sin^2\theta}\right)^2 - \frac{1}{\cos^4\theta} - \frac{1}{\sin^4\theta} + \frac{1}{\sin^2\theta\cos^2\theta}}\right]$$

$$= \frac{\sin^2\theta\cos^2\theta}{1 \mp \sqrt{3}\sin\theta\cos\theta}.$$

$\therefore$ According to (c):

$$\frac{\sin^2\theta\cos^2\theta}{1 + \sqrt{3}\sin\theta\cos\theta} \leq a \leq \frac{\sin^2\theta\cos^2\theta}{1 - \sqrt{3}\sin\theta\cos\theta},$$

$$\because \frac{\sin^2\theta\cos^2\theta}{(\sin\theta + \cos\theta)^2} < \frac{\sin^2\theta\cos^2\theta}{1 + \sqrt{3}\sin\theta\cos\theta},$$

$\therefore$ When $a = \dfrac{\sin^2\theta\cos^2\theta}{1 + \sqrt{3}\sin\theta\cos\theta}$, meet type (a).

$\therefore a = \dfrac{\sin^2\theta\cos^2\theta}{1 + \sqrt{3}\sin\theta\cos\theta}.$

37. Note $f_n = \frac{\sin nx}{\sin x}$ $(n = 1, 2, \ldots)$, so

$$f_k^2 - f_{k-1}^2 = \frac{\sin^2 kx}{\sin^2 x} - \frac{\sin^2(k-1)x}{\sin^2 x}$$

$$= \frac{[\sin kx - \sin(k-1)x] \cdot [\sin kx + \sin(k-1)x]}{\sin^2 x}$$

$$= \frac{2 \sin \frac{x}{2} \cos \frac{2k-1}{2}x \cdot 2 \sin \frac{2k-1}{2}x \cos \frac{x}{2}}{\sin^2 x}$$

$$= \frac{\sin(2k-1)x}{\sin x} = f_{2k-1} \qquad \text{(i)}$$

$$2x \le 2kx \le 2nx < \frac{\pi}{2} < \pi \Rightarrow x \le (2k-1)x < \pi - x,$$

$$\therefore \ \sin(k-1)x \ge \sin x, \ \therefore \ f_k^2 - f_{k-1}^2 \ge 1(k = 2, 3, \ldots)$$

$$f_n^2 = (f_n^2 - f_{n-1}^2) + (f_{n-1}^2 - f_{n-2}^2) + \cdots + (f_2^2 - f_1^2) + f_1^2$$

$$\ge 1 + 1 + \cdots + 1 = n$$

$$f_n \ge \sqrt{n} \left(0 < nx < \frac{\pi}{2} \right). \qquad \text{(ii)}$$

Using (i) and (ii),

$$f_k^2 - f_{k-1}^2 \ge \sqrt{2k-1} \ (k = 2, 3, \ldots)$$

$$f_n^2 = (f_n^2 - f_{n-1}^2) + (f_{n-1}^2 - f_{n-2}^2) + \cdots + (f_2^2 - f_1^2) + f_1^2$$

$$\ge \sqrt{2n-1} + \sqrt{2n-3} + \cdots + \sqrt{3} + 1 = \sum_{k=1}^{n} \sqrt{2k-1}.$$

Prove with mathematical inductive method

$$\sum_{k=1}^{n} \sqrt{2k-1} \ge \frac{2n-1}{3}\sqrt{2n-1}, \quad n \in N \qquad \text{(iii)}$$

When $n = 1$, type (iii) holds.

To prove: when $n = k + 1$, type (iii) holds. Only to prove:

$$\frac{2k-1}{3}\sqrt{2k-1} + \sqrt{2k+1} \geq \frac{2(k+1)-1}{3}\sqrt{2(k+1)-1}$$

$$\Leftrightarrow \frac{2k-1}{3}\sqrt{2k-1} \geq \frac{2k-2}{3}\sqrt{2k+1}$$

$$\Leftrightarrow (2k-1)\sqrt{2k-1} \geq (2k-2)(2k+1)$$

$$\Leftrightarrow 8k^3 - 12k^2 + 6k - 1 \geq 8k^3 - 12k^2 + 4. \text{ It is clearly holds.}$$

Namely,

$$\frac{\sin nx}{\sin x} \geq \frac{\sqrt{3}}{3}(2n-1)^{\frac{3}{4}}.$$

38. According to sine law:

$$(a+b)\left(\frac{1}{a} + \frac{1}{b} + \frac{1}{c}\right) - 4 - \frac{1}{\sin \frac{C}{2}}$$

$$= (\sin A + \sin B)\left(\frac{\sin A + \sin B}{\sin A \sin B} + \frac{1}{\sin C}\right) - 4 - \frac{1}{\sin \frac{C}{2}}$$

$$= \frac{(\sin A + \sin B)^2}{\sin A \sin B} - 4 + \frac{\sin A + \sin B}{\sin C} - \frac{1}{\sin \frac{C}{2}}$$

$$= \frac{4\sin^2 \frac{A+B}{2}\cos^2 \frac{A-B}{2}}{\frac{1}{2}[\cos(A-B) - \cos(A+B)]} + \frac{\cos \frac{A-B}{2} - 1}{\sin \frac{C}{2}}$$

$$= \frac{4\cos^2 \frac{A+B}{2}(1 - \cos^2 \frac{A-B}{2})}{[\cos^2 \frac{(A-B)}{2} - \cos^2 \frac{(A+B)}{2}]} - \frac{1 - \cos \frac{A-B}{2}}{\sin \frac{C}{2}}$$

$$\geq \frac{4\cos^2 \frac{A+B}{2}(1 - \cos \frac{A-B}{2})}{\cos \frac{(A-B)}{2} - \cos \frac{(A+B)}{2}} - \frac{1 - \cos \frac{A-B}{2}}{\sin \frac{C}{2}}$$

$$\left(\because 1 + \cos \frac{A-B}{2} \geq \cos \frac{A+B}{2} + \cos \frac{A-B}{2} \right)$$

$$= \frac{1 - \cos \frac{A-B}{2}}{\left(\cos \frac{A-B}{2} - \cos \frac{A+B}{2} \right) \sin \frac{C}{2}}$$

$$\left(4 \sin^3 \frac{C}{2} - \cos \frac{A-B}{2} + \sin \frac{C}{2} \right)$$

$$\geq \frac{1 - \cos \frac{A-B}{2}}{2 \sin \frac{A}{2} \sin \frac{B}{2} \sin \frac{C}{2}} \left(4 \left(\frac{1}{2} \right)^3 - \cos \frac{A-B}{2} + \frac{1}{2} \right)$$

$$\left(\because \sin \frac{C}{2} \geq \frac{1}{2} \right) = \frac{\left(1 - \cos \frac{A-B}{2} \right)^2}{2 \sin \frac{A}{2} \sin \frac{B}{2} \sin \frac{C}{2}} \geq 0.$$

$\therefore$ When $a = b$, $\angle C = 60°$, namely $a = b = c$,

the equality holds.

39. Suppose O is the center of circumcircle of $\triangle ABD$.

$\because \triangle ABD$ is acute triangle, $\therefore O$ is inside $\triangle ABD$.

$\because \angle C = \angle A < 90°$, $\therefore C$ is outside $\odot O$.

Suppose the radius of $\odot O$ is R. If $R \leq 1$, make three vertical line from O, divide $\triangle ABD$ into six right triangles.

$\forall P \in ABD$, so P must fall into one of these six right triangles (including the boundary).

Suppose P is inside the right-angled triangle $\triangle OAH$, so $AP \leq AO = R \leq 1$. $\therefore P$ is covered by $\odot K_A$.

$\therefore \triangle ABD$ is covered by $\odot K_A$, $\odot K_B$ and $\odot K_D$.

Namely, parallelogram $ABCD$ is covered by $\odot K_A$, $\odot K_B$, $\odot K_C$ and $\odot K_D$.

If $R > 1$, $OA = OB = OD = R > 1$, so O is outside $\odot K_A$, $\odot K_B$ and $\odot K_D$.

$\therefore O$ cannot be covered by $\odot K_A$, $\odot K_B$, $\odot K_C$ and $\odot K_D$.

Namely, if parallelogram is covered by $\odot K_A$, $\odot K_B$, $\odot K_C$ and $\odot K_D$, $R \le 1$.

We prove: $R \le 1 \Leftrightarrow a \le \cos\alpha + \sqrt{3}\sin\alpha$.

In $\triangle ABD$, $BD^2 = 1 + a^2 - 2a\cos\alpha = (2R\sin\alpha)^2$, $R \le 1 \Leftrightarrow a^2 - (2\cos\alpha)a + 1 - 4\sin^2\alpha \le 0$,

$$\left(\triangle = 4\cos^2\alpha - 4 + 16\sin^2\alpha = 12\sin^2\alpha > 0, \right.$$

$$\left. a = \frac{2\cos\alpha \pm \sqrt{12\sin^2\alpha}}{2} = \cos\alpha \pm \sqrt{3}\sin\alpha \right).$$

$R \le 1 \Leftrightarrow \cos\alpha - \sqrt{3}\sin\alpha \le a \le \cos\alpha + \sqrt{3}\sin\alpha$.

$R \le 1 \Leftrightarrow -\sqrt{3}\sin\alpha \le a - \cos\alpha \le \sqrt{3}\sin\alpha$.

Make $DE \perp AB$ at E, and E is on AB, $a = AB > AE = AD\cos\alpha = \cos\alpha$, so $a - \cos\alpha > 0$.

$\therefore\ R \le 1 \Leftrightarrow 0 < a \le \cos\alpha + \sqrt{3}\sin\alpha$.

Exercise 7

1. Suppose $z = a + bi$ $(a, b \in R)$, so $\sqrt{a^2 + b^2} = 1$, $\frac{1}{z} = \bar{z} = a - bi$.

$\because\ z^2 + 2z + \frac{1}{z} < 0$

$$\therefore \begin{cases} a^2 - b^2 + 3a < 0 & (1) \\ 2ab + b = 0 & (2) \end{cases}$$

According to (2), $b = 0$ or $a = -\frac{1}{2}$.

When $b = 0$ and $a^2 + b^2 = 1$, $a^2 = 1$.

According to (1), $a^2 + 3a < 0$, $\therefore\ a < -\frac{1}{3}$, $\therefore\ a = -1$.

When $a = -\frac{1}{2}$ and $a^2 + b^2 = 1$, $b^2 = \frac{3}{4}$.

According to (1), $b^2 > -\frac{5}{4}$, so $b = \pm\frac{\sqrt{3}}{2}$.

$\therefore\ z = -1$ or $z = -\frac{1}{2} \pm \frac{\sqrt{3}}{2}i$.

2. Suppose $z = x + yi$ $(x, y \in R)$.

$$z + \frac{14 - z}{z - 1} \in R \Rightarrow z + \frac{14 - z}{z - 1} = \overline{z + \frac{14 - z}{z - 1}}$$

$$\Rightarrow (z - \overline{z})\left[1 - \frac{13}{(z - 1)(\overline{z} - 1)}\right] = 0 \Rightarrow z$$

$$= \overline{z} \text{ or } 1 - \frac{13}{(z - 1)(\overline{z} - 1)} = 0$$

$$\Rightarrow z \in R \text{ or } |z - 1|^2 = 13.$$

Substitute $z = x + yi$ $(x, y \in R)$ into $|z - 4| = |z - 4i|$, so $x = y$.

If $z \in R$, $x = 0$.

If $|z - 1|^2 = 13$, $x^2 - x - 6 = 0 \Rightarrow x = 3$ or -2.

$\therefore z = 0$ or $3 + 3i$ or $-2 - 2i$.

3. (1) When z is real number, imaginary part:

$$1g \frac{x}{10} + 1g (x - 3) = 0 \Rightarrow \begin{cases} x > 0 \\ x - 3 > 0 \\ \dfrac{x}{10}(x - 3) = 1 \end{cases}$$

$$\Rightarrow x = 5 \text{ or } x = -2 \text{ (rejection)}.$$

When $x = 5$, z is real number.

(2) When z is imaginary number, imaginary part:

$$1g \frac{x}{10} + 1g (x - 3) \neq 0 \Rightarrow \begin{cases} x > 0 \\ x - 3 > 0 \\ \dfrac{x}{10}(x - 3) \neq 1 \end{cases}$$

$$\Rightarrow x > 3 \text{ and } x \neq 5.$$

(3) When z is pure imaginary number,

$$\begin{cases} 1g\ (x^2 - 9x + 21) = 0 \\ x \neq 5 \\ x^2 - 9x + 21 > 0 \end{cases} \Rightarrow \begin{cases} x = 4 \text{ or } x = 5 \\ x \neq 5 \\ x^2 - 9x + 21 > 0 \end{cases} \Rightarrow x = 4.$$

$\therefore$ When $x = 4$, z is pure imaginary number.

4. $\because \overline{z_1} + \overline{z_2} = \frac{1}{2} - \frac{1}{4}i$, $\therefore \cos\alpha + \cos\beta - (\sin\alpha + \sin\beta)i = \frac{1}{2} - \frac{1}{4}i$

$$\cos\alpha + \cos\beta = \frac{1}{2}, \tag{3}$$

$$\sin\alpha + \sin\beta = \frac{1}{4}. \tag{4}$$

Therefore $(3)^2 + (4)^2$:

$$\cos^2\alpha + \cos^2\beta + 2\cos\alpha\cos\beta + \sin^2\alpha + \sin^2\beta$$

$$+2\sin\alpha\sin\beta = \frac{1}{4} + \frac{1}{16},$$

$$\therefore 2 + 2\cos(\alpha - \beta) = \frac{5}{16}, \ \therefore \cos(\alpha - \beta) = -\frac{27}{32}.$$

5. Suppose Z_1, Z_2 and Z correspond to z_1, z_1 and z, where $z_1 = r_1(\cos\theta + i\sin\theta)$, $z_1 = r_2(\cos\theta - i\sin\theta)$.

$\because$ Z is center of gravity of $\triangle OZ_1Z_2$, and according to the geometric meaning of addition:

$$3z = z_1 + z_2 = (r_1 + r_2)\cos\theta + i(r_1 - r_2)\sin\theta$$

$$\Rightarrow |3z|^2 = \cdots = (r_1 - r_2)^2 + 4r_1r_2\cos^2\theta.$$

$\because$ the area of $\triangle OZ_1Z_2$ is S and $\sin 2\theta > 0 \left(0 < \theta < \frac{\pi}{2}\right)$

$$\therefore \frac{1}{2}r_1r_2\sin 2\theta = S \Rightarrow r_1r_2 = \frac{2S}{\sin 2\theta},$$

$$\therefore 3^2|z|^2 = (r_1 - r_2)^2 + \frac{8\cos^2\theta}{\sin 2\theta} \cdot S = (r_1 - r_2)^2 + 4S\cot\theta.$$

If and only if $r_1 = r_2$, $|z|_{\min} = \frac{2}{3}\sqrt{S\cot\theta}$.

6. Make full use of complex number $z = x + yi \leftrightarrow$ point $(x, y) \leftrightarrow$ vector $\overrightarrow{OZ} = (x, y)$.

(1) Suppose $P(x, y)$ and $B(x', y')$

$$\because |z| = 1 \quad \therefore x'^2 + y'^2 = 1$$
$$\because \overrightarrow{BP} = 2\overrightarrow{PA}$$
$$\begin{cases} x = \dfrac{x' + 2 \times (-3)}{1 + 2}, \\ y = \dfrac{y' + 2 \times 0}{1 + 2}, \end{cases}$$
$$\therefore \begin{cases} x' = 3x + 6, \\ y' = 3y. \end{cases}$$
$$\therefore (3x + 6)^2 + (3y)^2 = 1$$
$$\therefore \text{The locus of point } P \text{is } (x + 2)^2 + y^2 = \dfrac{1}{9}.$$

(2) Suppose $B(x', y')$, so $x'^2 + y'^2 = 1$.

$\therefore \overrightarrow{OB}$ corresponds to $x' + y'i$.

$\because \overrightarrow{OA} + \overrightarrow{AB} = \overrightarrow{OB} \quad \therefore \overrightarrow{AB} = \overrightarrow{OB} - \overrightarrow{OA}$

$\therefore \overrightarrow{AB}$ corresponds to $x' + y'i - (-3) = x' + y'i + 3$

$\because \overrightarrow{AB} = 3\overrightarrow{AP} \quad \therefore \overrightarrow{AP}$ corresponds to $z' = \dfrac{x' + 3}{3} + \dfrac{y'}{3}i.$

Suppose $M(x, y)$ corresponds to z', so

$$\begin{cases} x = \dfrac{x' + 3}{3}, \\ y = \dfrac{y'}{3}, \end{cases} \qquad \begin{cases} x' = 3x - 3, \\ y' = 3y. \end{cases}$$

$\therefore (3x - 3)^2 + (3y)^2 = 1.$

$\therefore$ The locus of z' is $(x - 1)^2 + y^2 = \frac{1}{9}.$

7. Suppose $z = x + yi$ $(x, y \in R)$; substitute it and according to equality of complex number, we have

$$\begin{cases} x = \sin^2 t, \\ y = a(1-x)^2 + 2b(1-x)x + cx^2 \end{cases} \quad (0 \le x \le 1),$$

namely, $y = (a + c - 2b)x^2 + 2(b-a)x + a$.

$\because$ A, B and C are noncollinear

$\therefore$ $a + c - 2b \neq 0$, which can be seen that the curve is a parabola section.

The midpoints of AB and BC are $D(\frac{1}{4}, \frac{a+b}{2})$, $E(\frac{3}{4}, \frac{b+c}{2})$, so equation of DE is $y = (c-a)x + \frac{1}{4}(3a + 2b - c)$.

Simultaneously $(a + c - 2b)(x - \frac{1}{2})^2 = 0$, so $x = \frac{1}{2}$,

$\because$ $\frac{1}{4} < \frac{1}{2} < \frac{3}{4}$.

$\therefore$ Median line DE is parallel to the AC of the triangle, where it and parabola have only one intersection.

This point is $(\frac{1}{2}, \frac{a+c+2b}{4})$, which corresponds to $z = \frac{1}{2} + \frac{a+c+2b}{4}i$.

8. $\because$ $|z| = 1$, $\therefore$ $\bar{z} = \frac{1}{z}$.

$$\therefore |u| = \left| z^2\left(z^2 - z - 3i - \frac{1}{z} + \frac{1}{z^2}\right) \right| = |(z^2 + \bar{z}^2) - (z + \bar{z}) - 3i|.$$

Suppose $z = x + yi, xy \in R$,

so $x^2 + y^2 = 1$ and $z^2 = x^2 - y^2 + 2xyi = (2x^2 - 1) + 2xyi$

$\therefore$ $|u| = |2(2x^2 - 1) - 2x - 3i| = |(4x^2 - 2x - 2) - 3i|$.

Note $t = 4x^2 - 2x - 2$

$\because$ $x \in [-1, 1]$

$\therefore$ $t \in \left[-\frac{9}{4}, 4\right]$ and $|u| = |t - 3i| = \sqrt{t^2 + 9}$.

$\therefore$ When $t = 0$, namely, $z = 1$ or $-\frac{1}{2} \pm \frac{\sqrt{3}}{2}i$, $|u|_{min} = 3$.

$\therefore$ When $t = 4$, namely, $z = -1$, $|u|_{max} = 5$.

Exercise 8

1. $z_1 = \frac{-1+5i}{1+i} = \frac{(5i-1)(1-i)}{2} = \frac{5i-1+5+i}{2} = \frac{4+6i}{2} = 2+3i.$

$$z_2 = (a-2) - i, \overline{z_2} = (a-2) + i,$$

$$z_1 - \overline{z_2} = (2+3i) - (a-2) - i = (4-a) + 2i,$$

$$\because |z_1 - \overline{z_2}| < |z_1| \therefore 16 - 8a + a^2 + 4 < 9 + 4,$$

$$a^2 - 8a + 7 < 0,$$

$$(a-7)(a-1) < 0, \quad \therefore a \in (1,7).$$

2. Suppose $D(x,y)$, so

$\overrightarrow{AD} = \overrightarrow{OD} - \overrightarrow{OA}$ corresponds to

$(x+yi) - (1+2i) = (x-1) + (y-2)i;$

$\overrightarrow{BC} = \overrightarrow{OC} - \overrightarrow{OB}$ corresponds to $(-1-2i) - (-2+i) = 1-3i;$

$\because \overrightarrow{AD} = \overrightarrow{BC} \therefore (x-1) + (y-2)i = 1-3i$

$$\therefore \begin{cases} x-1=1, \\ y-2=-3, \end{cases} \quad \therefore \begin{cases} x=2, \\ y=-1, \end{cases}$$

$\therefore$ point D corresponds to $2 - i.$

3. $\frac{z_2-z_1}{z_3-z_1} = 1 + \frac{4}{3}i, |\frac{z_2-z}{z_3-z_1}| = |1 + \frac{4}{3}i| = \frac{5}{3}, \therefore \frac{AB}{AC} = \frac{5}{3}.$

$$\because \frac{z_2 - z_1}{z_3 - z_1} - 1 = 1 + \frac{4}{3}i - 1$$

$$\Rightarrow \frac{z_2 - z_3}{z_3 - z_1} = \frac{4}{3}i, \quad \therefore \frac{BC}{AC} = \frac{4}{3},$$

$$\therefore AB{:}BC{:}CA = 5{:}4{:}3.$$

4.

$$\begin{cases} z_1 + z_2 = 6 + 8i, \\ z_2 = z_1 \cdot ai, \end{cases} \quad \therefore \ z_1 = \frac{6+8i}{1+ai},$$

$$\because \ |z_2| = |z_1| \cdot |ai| = a|z_1| \ \because \ a > 0).$$

$$|z_1| = \frac{10}{\sqrt{1+a^2}}, \ \overrightarrow{OZ_1} \perp \overrightarrow{OZ_2},$$

$$S_{\Delta Z_1 O Z_2} = \frac{1}{2}|z_1||z_2| = \frac{a}{2}|z_1|^2 = \frac{50a}{1+a^2}$$

$$S = \frac{50a}{1+a^2}, \ \therefore \ Sa^2 - 50a + S = 0.$$

The sufficient and necessary condition for the existence of positive root a is

$$\Delta = 2500 - 4S^2 \geq 0.$$

The symmetry axis corresponding to the parabola

$$a = \frac{25}{S} > 0, \quad \therefore \ 0 < S \leq 25.$$

When $S = 25$, $a = 1$ where $z_1 = \frac{6+8i}{1+i} = 7 + i$, $z_2 = -1 + 7i$.

5. Method 1: Suppose $z = a + bi, a, b \in R$,

$$\because \ |z + 2 + 3i|^2 + |z - 2 - 3i|^2 = 40$$

$$\therefore \ |a + bi + 2 + 3i|^2 + |a + bi - 2 - 3i|^2 = 40,$$

$$\therefore \ (a + 2)^2 + (b + 3)^2 + (a - 2)^2 + (b - 3)^2 = 40.$$

After arrangement: $a^2 + b^2 = 7$, namely, $|z| = \sqrt{7}$.

Method 2: $\because \ |z + 2 + 3i|^2 + |z - 2 - 3i|^2 = 40$

$$\therefore \ (z + 2 + 3i)\overline{(z + 2 + 3i)} + (z - 2 - 3i)\overline{(z - 2 - 3i)} = 40,$$

$$\therefore \ |z|^2 + (2 - 3i)z + (2 + 3i)\bar{z} + (2 + 3i)(2 - 3i) + |z|^2$$

$$-(2 - 3i)z - (2 + 3i)\bar{z} + (2 + 3i)(2 - 3i)i = 40.$$

$|z|^2 = 7$, namely, $|z| = \sqrt{7}$.

Method 3: $|z + 2 + 3i| = |z - (-2 - 3i)|$ express distance between the point Z corresponding to complex number z and $A(-2, -3)$ corresponding to $-2 - 3i$.

$|z - 2 - 3i| = |z - (2 + 3i)|$ expresses distance between the point Z corresponding to complex number z and $A(2, 3)$ corresponding to $2 + 3i$.

As in the figure, in $\triangle ZOA$ and $\triangle ZPB$:

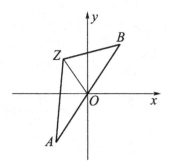

$$|ZA|^2 = |z|^2 + |OA|^2 - 2|z||OA| \cos \angle AOZ$$

$$|ZB|^2 = |z|^2 + |OB|^2 - 2|z||OB| \cos \angle AOZ$$

$$\therefore |ZA|^2 + |ZB|^2 = 2|z|^2 + 2|OA|^2$$

$$\therefore 2|z|^2 + 2 \times 13 = 40, \therefore |z| = \sqrt{7}.$$

6. $\because x^2 - 2x + 2 = 0, \ x = \dfrac{2 \pm \sqrt{-4}}{2} = 1 \pm i,$

$(1 + 2i - 1)(1 + z) = -6, \ 2i(1 + z) = -6, \quad \therefore z = -1 + 3i$

$\therefore A(1, 1), \ B(1, -1), \ C(-1, 3)$

$\therefore \angle A$ is the maximum, $AB = 2, \ AC = 2\sqrt{2}, BC = 2\sqrt{5},$

$\cos \angle A = \dfrac{8 + 4 - 20}{2 \times 2\sqrt{2} \times 2} = -\dfrac{\sqrt{2}}{2}, \ \angle A = 135°,$

so the maximum of interior angle is $135°$.

7. $\because |z_1| = |z_2|, |z_1| \neq 0, \therefore \left|\frac{z_2}{z_1}\right| = 1.$ \hfill (1)

Argument of $\overline{z_1} \cdot z_2$ is $\frac{\pi}{2}$,

$\because \overline{z_1} \cdot z_2 = |z_1|^2 \cdot \frac{z_2}{z_1}, \therefore \arg\left(\frac{z_2}{z_1}\right) = \arg(\overline{z_1} \cdot z_2) = \frac{\pi}{2}.$ \hfill (2)

According to (1) and (2), $\frac{z_2}{z_1} = i$, namely, $z_2 = z_1 i$.

Above type can also be written as follows: $\sqrt{3}b - 1 + (\sqrt{3} - b)i = i(2 - \sqrt{3}a + ai) \Rightarrow \sqrt{3}b - 1 + (\sqrt{3} - b)i = -a + (2 - \sqrt{3}a)i.$

According to the equal of complex number: $a = b = \frac{\sqrt{3}-1}{2}.$

8.

$$\left(\frac{x}{y}\right)^2 + \frac{x}{y} + 1 = 0 \Rightarrow \frac{x}{y} = \frac{-1 \pm \sqrt{3}i}{2}$$

$$\Rightarrow \left(\frac{x}{x+y}\right)^{2010} + \left(\frac{y}{x+y}\right)^{2010}$$

$$= \left(\frac{\frac{x}{y}}{\frac{x}{y}+1}\right)^{2010} + \left(\frac{1}{\frac{x}{y}+1}\right)^{2010}$$

$$= \left(\frac{-\frac{1}{2} + \frac{\sqrt{3}i}{2}i}{\frac{-1+\sqrt{3}i}{2}+1}\right)^{2010} + \left(\frac{1}{\frac{-1+\sqrt{3}i}{2}}\right)^{2010}$$

$$= 1 + 1 = 2.$$

9. Let $z = x + yi(x, y \in R)$, so $|z - \sqrt{3}| + |z + \sqrt{3}| = 4,$

$\because$ point z is on the ellipse whose focus is $(\pm\sqrt{3}, 0)$ and long half axis is 2.

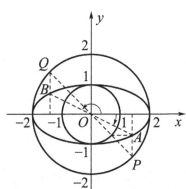

$\therefore$ equation of ellipse is $\frac{x^2}{4}+y^2=1$ and its parameter equation is

$$\begin{cases} x = 2\cos\theta \\ y = \sin\theta \end{cases}$$

$\therefore$ $f(x,y) = |4\cos\theta - 4\sin\theta - 9| = |4\sqrt{2}\cos(\theta+\frac{\pi}{4}) - 9|$.

When $\theta + \frac{\pi}{4} = 0$, namely, $\theta = -\frac{\pi}{4}$, $f(x,y)_{min} = 9 - 4\sqrt{2}$,

where $x = \sqrt{2}$, $y = -\frac{\sqrt{2}}{2}$.

When $\theta + \frac{\pi}{4} = \pi$, namely, $\theta = \frac{3\pi}{4}$, $f(x,y)_{max} = 9 + 4\sqrt{2}$,

where $x = -\sqrt{2}, y = \frac{\sqrt{2}}{2}$.

10. $\because$ The equation of a curve of the known complex number C is ellipse, which is

$$\frac{x^2}{a^2} + \frac{y^2}{b^2} = 1 \qquad\qquad (*)$$

$\because$ $\overrightarrow{Z_1Z_2}i = \overrightarrow{Z_1Z_3} \Rightarrow |Z_1Z_2| = |Z_1Z_3|, \angle Z_2Z_1Z_3 = 90° \Rightarrow$ $\triangle Z_1Z_2Z_3$ is isosceles and right triangle.

Suppose line equation of Z_1Z_2 is

$$y = kx + b \qquad\qquad (**)$$

so line equation of Z_1Z_3 is

$$y = -\frac{1}{k}x + b. \qquad\qquad (***)$$

According to $(*)$ and $(**)$:

$$x_1 = 0, x_2 = -\frac{2a^2bk}{b^2 + a^2k^2} \qquad\qquad (***)$$

$y_2 = kx_2 + b$, change k into $\frac{1}{k}$ in the second solution:

$$x_3 = \frac{2a^2bk}{a^2 + b^2k^2}, y_3 = -\frac{1}{k}x_3 + b,$$

$\because$ $|Z_1Z_2| = |Z_1Z_3| \Rightarrow x_2^2 + (y_2 - b)^2 = x_3^2 + (y_3 - b)^2$

$$\Rightarrow \frac{4a^4b^2k^2}{(b^2 + a^2k^2)^2}(k^2 + 1) = \frac{4a^4b^2k^2}{(a^2 + b^2k^2)^2}(k^2 + 1)\frac{1}{k^2}$$

$$\Rightarrow (k-1)[b^2k^2 + (b^2 - a^2)k + b^2] = 0$$

$$\Rightarrow k = 1 \text{ or } b^2k^2 + (b^2 - a^2)k + b^2 = 0,$$

$$\Delta_k = (b^2 - a^2)^2 - 4b^2 = (a^2 + b^2)(a^2 - 3b^2).$$

- When $a > \sqrt{3}b$, there are three $\Delta Z_1 Z_2 Z_3$.
- When $a = \sqrt{3}b$, there are two $\Delta Z_1 Z_2 Z_3$.
- When $a < \sqrt{3}b$, there is only one $\Delta Z_1 Z_2 Z_3$.

11. (1) $\bar{z}z + 3i(\bar{z} - z) + m = 0 \Rightarrow (z + 3i)(\bar{z} - 3i) = 9 - m$

$$\Rightarrow |z + 3i| = \sqrt{9 - m}$$

$\because$ $m < 9$, $\therefore$ point set of above equation is the circle P_1 where $(0, -3)$ is the center, and $\sqrt{9 - m}$ is the radius.

$\because$ $\varpi = 2iz \Rightarrow z = \frac{\varpi}{2i}$, and substitute it into above equation: $|\varpi - 6| = 2\sqrt{9 - m}$.

P_2 is the circle where $(6, 0)$ is the center, $2\sqrt{9 - m}$ is the radius, so the question equal to find the range of m where $\odot P_1$ and $\odot P_2$ have intersection.

$\therefore$ $\sqrt{9 - m} \le \sqrt{45} \le 3\sqrt{9 - m}$ where $\sqrt{45}$ is distance of centers of two circles.

$\sqrt{9 - m}, 3\sqrt{9 - m}$ express the difference and sum of radius of two circles.

$\therefore$ $-36 \le m \le 4.$

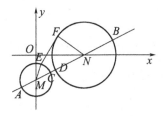

(2) When $m = 5$, P_1 is $\odot M : |Z + 3i| = 2, P_2$ is $\odot N$: $|z - 6| = 4$.

Suppose line MN and circle intersect at A, C, D and B.

E and F are points on two circles, and connect ME, NF.

If one point in E and F is difference with A and B,

$$|AB| = |AM| + |MN| + |NB| = |ME| + |MN| + |NF| > |EF|.$$

At the same time: $|ME| + |EF| + |NF| > |MN|$

$$\therefore \ |ME| + |EF| + |NF| > |MC| + |CD| + |DN|,$$
$$|EF| > |CD|,$$
$$\because \ |MN| = 3\sqrt{5}, |CD| = 3\sqrt{5} - 6,$$
$$\therefore \ |z_1 - z_2|_{\max} = 6 + 3\sqrt{5} \cdot |z_1 - z_2|_{\min} = 3\sqrt{5} - 6.$$

12. Suppose $z = x + yi, (x, y \in R), \because |z| = 5, \therefore x^2 + y^2 = 25,$

$(3 + 4i)z = (3 + 4i)(x + yi) = (3x - 4y) + (4x + 3y)i.$
$\because \ (3 + 4i)z$ is the corresponding points on angular bisector of the second and the fourth quadrant in the complex plane.

$$\therefore \ 3x - 4y + 4x + 3y = 0, \ \therefore \ y = 7x, \ \therefore \ x = \pm\frac{\sqrt{2}}{2},$$
$$y = \pm\frac{7\sqrt{2}}{2}, \text{namely}, \ z = \pm(\frac{\sqrt{2}}{2} + \frac{7\sqrt{2}}{2}i); \ \sqrt{2}z = \pm(1 + 7i).$$

When $\sqrt{2}x = 1 + 7i, |1 + 7i - m| = 5\sqrt{2}$, namely, $(1 - m)^2 + 7^2 = 50$, so $m = 0, m = 2$.

When $\sqrt{2}z = -(1 + 7i), \ m = 0, m = -2.$

13. Let

$$x + \frac{1}{x} = y \Rightarrow x^2 - yx + 1 = 0 \tag{1}$$

$\because \ y \in R, x$ is imaginary number, namely, equation (1)

has imaginary roots.

$\therefore \ \Delta = y^2 - 4 < 0 \Rightarrow -2 < y < 2.$ Conversely, when

$-2 < y < 2,$ equation (1) has no imaginary roots.

$\therefore$ Equation

$$f(y) = y^2 - ay + a + 1 = 0 \tag{2}$$

has at least one real roots in $(-2, 2)$.

(i) If two roots of equation (2) meet $-2 < y < 2$, its necessary and sufficient condition is given by

$$
\begin{cases}
\Delta = a^2 - 4(a+1) \geq 0 \\[2mm]
-2 < \dfrac{a}{2} < 2 \\[2mm]
f(2) = 4 + 2a + a + 1 > 0 \\[2mm]
f(-2) = 4 + 2a + a + 1 > 0
\end{cases}
\Rightarrow -\frac{5}{3} < a \leq 2 - 2\sqrt{2}.
$$

(ii) If one root of equation (2) meets $-2 < y < 2$, its necessary and sufficient condition is given by

$$
f(2)f(-2) \leq 0 \Rightarrow a \leq -\frac{5}{3} \quad \text{or} \quad a \geq 5.
$$

But when $a = 5$, two roots of equation (6) are 2 or 3, out of the question.

$\therefore\ a \leq 2 - 2\sqrt{2}$ or $a > 5$.

14. Suppose $z = x + yi(x, y \in R)$, $\because\ |z| = 1 \Rightarrow x^2 + y^2 = 1$ and $|x| \leq 1$;

$$
u = |(z - \alpha)^2(z + \beta)|
$$

$$
= \sqrt{(-2\alpha x + a^2 + 1)(2\beta x + \beta^2 + 1)}
$$

$$
= \sqrt{4\alpha^2\beta \left(-x + \frac{a^2 + 1}{2a}\right)\left(-x + \frac{a^2 + 1}{2a}\right)\left(2x + \frac{\beta^2 + 1}{\beta}\right)}
$$

$$
\leq \sqrt{4\alpha^2\beta \left(\frac{\frac{a^2+1}{2a} + \frac{a^2+1}{2a} + \frac{\beta^2+1}{\beta}}{3}\right)^3}
$$

$$
= \sqrt{\frac{4[(\alpha\beta + 1)(\alpha + \beta)]^3}{27\alpha\beta^2}}.
$$

(i) If $\left| x = \frac{\beta(\alpha^2+1)-2\alpha(\beta^2+1)}{6\alpha\beta} \right| \leq 1$, $u_{max} = \sqrt{\frac{4((\alpha\beta+1)(\alpha+\beta))^3}{27\alpha\beta^2}}$ if

and only if $x = \frac{\beta(\alpha^2+1)-2\alpha(\beta^2+1)}{6\alpha\beta}$.

(ii) If $\frac{\beta(\alpha^2+1)-2\alpha(\beta^2+1)}{6\alpha\beta} < -1$, $f(x) = (a - x)^2(b + 2x)$ $(a = \frac{\alpha^2+1}{2\alpha} \geq 1, b = \frac{\beta^2+1}{\beta} \geq 2)$ is decreasing function in $[-1, 1]$.

Prove as follows:

Suppose $-1 \leq x_1 < x_2 \leq 1$

$\because f(x) = 2x^3 + (b - 4a)x^2 + 2a(a - b)x + a^2b;$

$f(x_2) - f(x_1)$

$= 2(x_2^3 - x_1^3) + (b - 4a)(x_2^2 - x_1^2) + 2a(a - b)(x_2 - x_1)$

$= (x_2 - x_1)[2(x_2^2 + x_2x_1 + x_1^2) + (b - 4a)(x_2 + x_1)$

$\quad + 2a(a - b)].$

$\because x_1 < x_2, \therefore x_2 - x_1 > 0,$

$\because -1 \leq x_1 < 1, -1 < x_2 \leq 1,$

$\therefore t = 2(x_2^2 + x_2x_1 + x_1^2) + (b - 4a) \cdot (x_2 + x_1) + 2a(a - b)$

$\quad < 6 + 2|b - 4a| + 2a(a - b),$

$\because a - 1 \geq 0, b > a + 3 > 0,$

$\therefore f(x_2) - f(x_1) < 0$

$\therefore f(x)$ is decreasing function in $[-1, 1]$.

$\therefore u_{max} = |(1 + \alpha)^2(1 + \beta)|$ if and only if $x = -1$,

namely, $z = -1$.

(iii) If $\frac{\beta(\alpha^2+1)-2\alpha(\beta^2+1)}{6\alpha\beta} > 1$, $f(x) = (a-x)^2(b+2x)$ $(a = \frac{\alpha^2+1}{2\alpha} \geq 1, b = \frac{\beta^2+1}{\beta} \geq 2)$ is increasing function in $[-1, 1]$.

$\therefore u_{max} = |(1 - \alpha)^2(1 + \beta)|$ if and only if $x = 1$, namely $z = 1$.

15. For given positive numbers a and $c(a > c)$, equation $|z - c| + |z + c| = 2a$ expresses ellipse where equation is $\frac{x^2}{a^2} + \frac{y^2}{b^2} = 1$ ($b = \sqrt{a^2 - c^2}$).

To arbitrary a and b, $\frac{4}{a^2} + \frac{1}{b^2} = 1$, so $b^2 = \frac{a^2}{a^2-4}$.

∵ $a > b$

∴ $a^2 > 5$

∴ $x^2 + (a^2 - 4)y^2 = a^2$, namely,$x^2 - 4y^2 = a^2(1 - y^2)$.

(i) If $1 - y^2 = 0$, $|y| = 1$ and $x^2 - 4y^2 = 0$.

∴ $x = \pm 2, y = \pm 1$.

Namely, $A_1 = \{(2, 1), (2, -1), (-2, 1), (-2, -1)\} \subset A$.

(ii) If $1 - y^2 > 0$, $|y| < 1$ and $x^2 - 4y^2 > 5(1 - y^2)$, namely, $|y| < 1$ and $x^2 + y^2 > 5$.

∴ $A_2 = \{(x, y) | x^2 + y^2 > 5 \text{ and } |y| < 1\} \subset A$.

(iii) If $1 - y^2 < 0$, $|y| > 1$ and $x^2 + y^2 < 5$.

∴ $A_3 = \{(x, y) | x^2 + y^2 < 5 \text{ and } |y| > 1\} \subset A$.

∴ $A = A_1 \cup A_2 \cup A_3$.

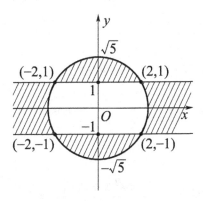

Exercise 9

1. Suppose $z = r(\cos\theta + i\sin\theta)$, $|z^2 + \frac{1}{z^2}| = 1 \Rightarrow 2\cos 4\theta = 1 - (r^4 + \frac{1}{r^4}) \le -1 \Rightarrow \cos 4\theta \le -\frac{1}{2} \Rightarrow \frac{k\pi}{2} + \frac{\pi}{6} \le \theta \le \frac{k\pi}{2} + \frac{\pi}{3}$ ($k \in Z$).

   Let $k = 0, 1, 2, 3$, so the range of $\arg z$ is $[\frac{\pi}{6}, \frac{\pi}{3}] \cup [\frac{2\pi}{3}, \frac{5\pi}{6}] \cup [\frac{7\pi}{6}, \frac{4\pi}{3}] \cup [\frac{5\pi}{3}, \frac{11\pi}{6}]$.

2. $\because z_1 = \sqrt{2}(\cos\frac{\pi}{4} + i\sin\frac{\pi}{4})z_2$, $\therefore$ included angle of $\overrightarrow{OA}$ and $\overrightarrow{OB}$ is $\frac{\pi}{4}$ and $|z_1| = \sqrt{2}|z_2|$, so $S_{\triangle OAB} = \frac{1}{2}|z_1||z_2|\sin\angle AOB = \frac{1}{2}|z_2|^2$.

   $\because |z_2 - 2 - 2i| = 2$, $\therefore 2\sqrt{2} - 2 \le |z_2| \le 2\sqrt{2} + 2$.

   $\therefore (S_{\triangle OAB})_{\max} = 6 + 4\sqrt{2}$, $(S_{\triangle OAB})_{\min} = 6 - 4\sqrt{2}$.

3. **Method 1:**

$$z = \cos x + i\sin x \Rightarrow u = (4 + \cos x)\sin x \Rightarrow u^2 = (4 + \cos x)^2 \sin^2 x$$

$$= \frac{(4 + \cos x)(4 + \cos x)(\sqrt{6} + 1)(1 + \cos x)(\sqrt{6} + 3)(1 - \cos x)}{(\sqrt{6} + 1)(\sqrt{6} + 3)}$$

$$\le \frac{1}{9 + 4\sqrt{6}}\left[\frac{(4 + \cos x) + (4 + \cos x) + (\sqrt{6} + 1)(1 + \cos x) + (\sqrt{6} + 3)(1 - \cos x)}{4}\right]^4$$

$$= \frac{9 + 24\sqrt{6}}{4},$$

if and only if $4 + \cos x = (\sqrt{6} + 1)(1 + \cos x) = (\sqrt{6} + 3)(1 - \cos x)$, namely, $\cos x = \frac{\sqrt{6} - 2}{2}$, and the equality holds.

$$\therefore u_{\max} = \frac{\sqrt{9 + 24\sqrt{6}}}{2}.$$

Method 2: Introduce parameter λ

$$u^2 = \sin^2 x(4 + \cos x)^2 = \frac{\sin^2\theta}{\lambda^2}(\lambda\cos x + 4\lambda)^2$$

$$\le \frac{\sin^2 x}{\lambda^2}(\cos^2 x + \lambda^2)(4 + \lambda^2)$$

$$\le \frac{4 + \lambda^2}{\lambda^2}\left(\frac{\sin^2 x + \cos^2 x + \lambda^2}{2}\right)^2 = \left(\frac{1 + \lambda^2}{2}\right)^2\left(\frac{4 + \lambda^2}{\lambda^2}\right),$$

if and only if

$$\begin{cases} \frac{\cos x}{\lambda} = \frac{\lambda}{4}, \\ \sin^2 x = \cos^2 x + \lambda^2, \end{cases}$$

namely, $\lambda^2 = 2\sqrt{6} - 4$, $\cos x = \frac{\sqrt{6}-2}{2}$, and the equality holds.

$$\therefore u_{max} = \frac{\sqrt{9+24\sqrt{6}}}{2}.$$

4. Suppose

$$A: x_0 + (x_0 + 2)i, C : \cos\theta + i\sin\theta, B: x + yi = x_0$$

$$+(x_0 + 2)i + \cos\theta + i\sin\theta = (x_0 + \cos\theta) + (x_0 + 2 + \sin\theta)i,$$

$$\therefore \begin{cases} x = x_0 + \cos\theta, \\ y = x_0 + 2 + \sin\theta \end{cases}$$

$$\therefore (x - x_0)^2 + (y - x_0 - 2)^2 = 1.$$

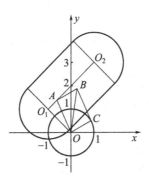

Obviously, locus of B is system of circles which the center is $(x_0, x_0 + 2)$ and the radius is 1. $O_1(-1, 1), O_2(1, 3) O_1 O_2 = 2\sqrt{2}$, so the locus of B is a rectangle and two semicircle.

$$\therefore S = 2 \times 2\sqrt{2} + \pi \times 1^2 = 4\sqrt{2} + \pi.$$

5. $z_n = 48[\frac{\sqrt{2}}{2}(\cos\frac{\pi}{6} + i\sin\frac{\pi}{6})]^{n-1} = 48(\frac{\sqrt{2}}{2})^{n-1}[\cos\frac{(n-1)\pi}{6} + i\sin\frac{(n-1)\pi}{6}], \sin\frac{(n-1)\pi}{6} = 0 \Rightarrow n = 6k + 1 \ (k = 0, 1, 2, \ldots)$, namely, the 1st item, the 7th item and the 13th item of $\{a_n\}$ are

real numbers.

$$a_n = z_{6k+1} \Rightarrow \frac{a_{n+1}}{a_n} = q^6 = \cdots = -\frac{1}{8}, \sum_{k=1}^{\infty} a_k = \frac{48}{1 - (-\frac{1}{8})} = \frac{128}{3}.$$

6. Suppose $z_1 = \cos\alpha + i\sin\alpha$, $z_2 = k(\cos\beta + i\sin\beta)$,

$z_3 = (2-k)(\cos\gamma + i\sin\gamma)$,

$$z_1 + z_2 + z_3 = 0 \Rightarrow \begin{cases} \cos\alpha = (k-2)\cos\gamma - k\cos\beta, \\ \sin\alpha = (k-2)\sin\gamma - k\sin\beta. \end{cases}$$

Add after two square types,

$\because k \neq 0, k \neq 2.$

$\therefore$ After arrangement:

$$\cos(\beta - \gamma) = \frac{(k-2)^2 + k^2 - 1}{2k(k-2)} = 1 + \frac{3}{2(k-1)^2 - 2},$$

$$\because |\cos(\beta - \alpha)| \le 1, \therefore -2 \le \frac{3}{2(k-1)^2 - 2} \le 0 \Rightarrow \frac{1}{2} \le k \le \frac{3}{2}.$$

When $k = 1$, $\cos(\beta - \gamma)_{\max} = -\frac{1}{2}.$

When $k = \frac{1}{2}$ or $\frac{3}{2}$, $\cos(\beta - \gamma)_{\min} = -1.$

7. Suppose $z_1 = r_1(\cos\theta_1 + i\sin\theta_1)$, $z_2 = r_2(\cos\theta_2 + i\sin\theta_2)$ $(r_1, r_2 > 0)$

$$\Rightarrow \begin{cases} r_1\cos\theta_1 + r_2\cos\theta_2 = 0, \\ r_1\sin\theta_1 + r_2\sin\theta_2 = 0, \\ r_1 r_2 = 14. \end{cases}$$

$$\Rightarrow \cos(\theta_1 - \theta_2) = \frac{25 - (r_1^2 + r_2^2)}{2r_1 r_2} = \frac{25 - (r_1^2 + \frac{14^2}{r_1^2})}{28}$$

$$\le \frac{25 - 2 \cdot 14}{28} = -\frac{3}{28}.$$

If and only if $r_1 = r_2 = \sqrt{14}$, the equality holds.

$\because \cos(\theta_1 - \theta_2) \ge -1,$

$\therefore -1 \le \cos(\theta_1 - \theta_2) \le -\frac{3}{28}$.

When $\cos(\theta_1 - \theta_2) = -1$, $r_1^2 + \frac{14^2}{r_1^2} = 53 \Rightarrow r_1 = 7$ or 2,

and $\theta_1 - \theta_2 = (2k+1)\pi$ $(k \in z) \Rightarrow z_1 = 7i, z_2 = -2i$ or $z_1 = -2i, z_2 = 7i$.

8. In the expansion type $\alpha^k + \beta^k$, let $\alpha + \beta = 1, \alpha\beta = 1$.

   The sum of these coefficients is given by $S_k = \alpha^k + \beta^k$, so α, β are two roots of equation $x^2 - x + 1 = 0$.

   $$\therefore \ \alpha = \cos\frac{\pi}{3} + i\sin\frac{\pi}{3}, \beta = \cos\frac{\pi}{3} - i\sin\frac{\pi}{3},$$

   $$\therefore \ \alpha^k + \beta^k = \left(\cos\frac{\pi}{3} + i\sin\frac{\pi}{3}\right)^k + \left(\cos\frac{\pi}{3} - i\sin\frac{\pi}{3}\right)^k$$

   $$= \cos\frac{k\pi}{3} + i\sin\frac{k\pi}{3} + \cos\frac{k\pi}{3} - i\sin\frac{k\pi}{3}$$

   $$= 2\cos\frac{k\pi}{3}.$$

   Take $k = 2012$, so $S_k = -1$.

9. Firstly, prove: $|a_n| = \sqrt{n+1}$

   When $n = 1$, $|a_1| = |1 + i| = \sqrt{2}$.

   Suppose when $n = k$, proposition holds.

   So when $n = k + 1$,

   $$|a_{k+1}| = \left|a_k \cdot \left(1 + \frac{i}{\sqrt{k+1}}\right)\right| = |a_k|\sqrt{1 + \frac{1}{k+1}} = \sqrt{(k+1)+1},$$

   $$\therefore \ |a_n| = \sqrt{n+1}$$

   $$\therefore \ |a_n - a_{n+1}| = \left|a_n - a_n\left(1 + \frac{i}{\sqrt{n+1}}\right)\right|$$

   $$= \left|-a_n \cdot \frac{i}{\sqrt{n+1}}\right| = \sqrt{n+1} \cdot \frac{1}{\sqrt{n+1}} = 1.$$

10.
$$\omega = z^4 - z^3 - 3z^2 i - z + 1 = z^2\left[\left(z^2 + \frac{1}{z^2}\right) - \left(z + \frac{1}{z}\right) - 3i\right]$$

$$= z^2\left[\left(z + \frac{1}{z}\right)^2 - \left(z + \frac{1}{z}\right) - 2 - 3i\right].$$

$\because |x|^2 = z \cdot \overline{z} = 1, \therefore \dfrac{1}{z} = \overline{z}.$

$\therefore \omega = z^2[(z + \overline{z})^2 - (z + \overline{z}) - 2 - 3i].$

Suppose $z = \cos\theta + i\sin\theta$, so $\omega = z^2 \cdot [(4\cos^2\theta - 2\cos\theta - 2) - 3i]$.

$\therefore |\omega| = \sqrt{(4\cos^2\theta - 2\cos\theta - 2)^2 + 9}.$

$\therefore$ Necessary and sufficient condition for the minimum of

$\qquad |\omega|$ is given by

$$4\cos^2\theta - 2\cos\theta - 2 = 2(2\cos\theta + 1)(\cos\theta - 1) = 0.$$

$$\therefore \begin{cases} \cos\theta = 1, \\ \sin\theta = 0, \end{cases} \text{or} \quad \begin{cases} \cos\theta = -\dfrac{1}{2}, \\[2mm] \sin\theta = \pm\dfrac{\sqrt{3}}{2}. \end{cases}$$

Namely, $z = 1$ or $z = -\frac{1}{2} \pm \frac{\sqrt{3}}{2}i$.

11. (1) $\because z_0 - (1+i)z_1 = 0, \quad \therefore z_0 - z_1 = z_1 i,$

$\qquad \therefore |z_0 - z_1| - |z_1| = \sqrt{2}.$

$\qquad \therefore z_1 = \sqrt{2}\left(\cos\frac{5}{12}\pi + i\sin\frac{5}{12}\pi\right),$

$\qquad z_0 = (1+i)z_1 = (1+i)\cdot\sqrt{2}\left(\cos\frac{5}{12}\pi + i\sin\frac{5}{12}\pi\right) = -1 + \sqrt{3}i.$

(2) Suppose Z_1 and Z_0 correspond to complex number z_1 and z_0 in complex plane.

$\qquad \because z_0 - z_1 = z_1 i, \therefore OZ_1 \perp Z_0 Z_1$

$\qquad \because \text{Re } z_0 < \text{Re } z_1, |z_1 - z_0| = \sqrt{2}.$

$\qquad \therefore OZ_1$ is the right tangent to $\odot Z_0$ which passes through the origin.

12. $(b + ci)^2 = [(2\cos\theta + i\sin^2\theta)^2 - 4\cos\theta(1 - \cos 2\theta)i]^2$

$= [(2\cos\theta - i\sin^2\theta)^2 - 8\cos\theta\sin^2\theta i]^2 = (2\cos\theta - i\sin^2\theta)^2.$

$\therefore \begin{cases} b = 2\cos\theta, \\ c = -\sin^2\theta \end{cases}$ or $\begin{cases} b = -2\cos\theta, \\ c = \sin^2\theta. \end{cases}$

The foucs of parabola $y = -x^2 + bx + c = -(x - \frac{b}{2})^2 + \frac{b^2 + 4c}{4}$ is $F(\frac{b}{2}, \frac{-1 + b^2 + 4c}{4})$.

If suppose foucs $F(X, Y)$, so

(i) $\begin{cases} X = \frac{b}{2} = \cos\theta, \\ Y = \frac{-1 + b^2 + 4c}{4} = 2\cos^2\theta - \frac{5}{4}, \end{cases}$

the locus equation of foucs is $Y = 2X^2 - \frac{5}{4}$ $(-1 \le X \le 1)$.

(ii) $\begin{cases} X = \frac{b}{2} = -\cos\theta, \\ Y = \frac{-1 + b^2 + 4c}{4} = \frac{3}{4}, \end{cases}$

the locus equation of foucs is $Y = \frac{3}{4}$ $(-1 \le X \le 1)$.

13. Original function equals to

$$y = \sqrt{\left(x + \frac{1}{2}\right)^2 + \left(\frac{\sqrt{3}}{2}\right)^2} - \sqrt{\left(x - \frac{1}{2}\right)^2 + \left(\frac{\sqrt{3}}{2}\right)^2}$$

$\therefore$ Construct complex $z_1 = \left(x + \frac{1}{2}\right) + \frac{\sqrt{3}}{2}i$,

$z_2 = \left(x - \frac{1}{2}\right) + \frac{\sqrt{3}}{2}i$

$\because \big||z_1| - |z_2|\big| \le |z_1 - z_2| = 1$ $\therefore -1 < y < 1.$

$\therefore$ The range of function is $(-1, 1)$.

14. Construct complex $z_1 = A + Bi, z_2 = x + yi$

$\therefore |z_1| = |(x\cos^2\theta + y\sin^2\theta) + i(x\sin^2\theta + y\cos^2\theta)|$

$= |z_2\cos^2\theta + i z_2\sin^2\theta| \le |z_2\cos^2\theta| + |i\bar{z}_2\sin^2\theta|$

$$= |z_2| \cos^2 \theta + |\overline{z}_z| \sin^2 \theta = |z_2|,$$

$$\therefore \ |z_1{}^2| \leq |z_2|^2, \ \therefore \ x^2 + y^2 \geq A^2 + B^2.$$

15. $\sqrt{(2k-1)^2 + a_k^2} = |(2k-1) + a_k i|, \ k = 1, 2, \ldots, n,$

$$\left(\sqrt{1 + a_1^2} + \sqrt{3 + a_2^2} + \ldots + \sqrt{(2n-1) + a_n^2} \right)_{\min} = S_n$$

$$\sum_{k=1}^{n} \sqrt{(2k-1)^2 + a_k^2}$$

$$= |1 + a_1 i| + |3 + a_2 i| + \cdots + |(2n-1) + a_n i|$$

$$\geq |[(1 + 3 + \cdots + (2n-1)] + (a_1 + a_2 + \cdots + a_n)i]$$

$$= |n^2 + 17i| = \sqrt{n^4 + 17^2}.$$

$\because \ S_n$ is integer number and suppose $S_n = m$;

$\therefore \ n^4 + 17^2 = m^2$, namely, $m^2 - n^4 = 17^2$

$\therefore \ (m + n^2)(m - n^2) = 289$ and $\begin{cases} m - n^2 = 1, \\ m + n^2 = 289, \end{cases} \ \therefore \ n = 12.$

16. In acute $\triangle ABC$, $\tan B = m - 2 \Rightarrow m > 2.$

Suppose $z_1 = 1 + mi$, $z_2 = 1 + (m-2)i$,

so $A = \arg(1 + mi)$, $B = \arg(1 + (m-2)i)$.

$$\because \ z_1 \cdot z_2 = (1 + mi)[1 + (m-2)i]$$

$$= (-m^2 + 2m + 1) + 2(m-1)i,$$

$$\arg(z_1 z_2) = A + B, \ \frac{\pi}{2} < A + B < \pi,$$

$$\therefore \ \tan(A + B) = \frac{2(m-1)}{-m^2 + 2m + 1} < 0,$$

$$\because \ m > 2,$$

$$m > \sqrt{2} + 1.$$

17. According to known inequality and $|z_1 - z_2| + |z_1 + z_2| \leq 2\sqrt{|z_1|^2 + |z_2|^2}$,

$$\therefore \ 4|z_5| = 2|2z_5 - (z_3 + z_4) + (z_3 + z_4)|$$

$$\leq 2(|2z_5 - (z_3 + z_4)| + |z_3 + z_4|)$$

$$\leq 2(|z_3 - z_4| + |z_3 + z_4|)$$

$$= |(2z_3 - (z_1 - z_2)) - (2z_4 - (z_1 + z_2))|$$

$$+ |(2z_3 - (z_1 + z_2)) + (2z_4 - (z_1 + z_2)) + 2(z_1 + z_2)|$$

$$\leq |(2z_3 - (z_1 - z_2)) - (2z_4 - (z_1 + z_2))|$$

$$+ |(2z_3 - (z_1 + z_2)) + (2z_4 - (z_1 + z_2))| + 2|z_1 + z_2|$$

$$\leq 2|z_1 + z_2| + 2\sqrt{|2z_3 - (z_1 + z_2)|^2 + |2z_4 - (z_1 + z_2)|^2}$$

$$\leq 2|z_1 + z_2| + 2\sqrt{2}|z_1 - z_2|$$

$$\leq 2\sqrt{(1^2 + (\sqrt{2})^2) \cdot (|z_1 + z_2|^2 + |z_1 - z_2|^2)}$$

$$= 2\sqrt{3} \cdot \sqrt{2(|z_1|^2 + |z_2|^2)}$$

$$\leq 2\sqrt{3} \cdot \sqrt{4} = 4\sqrt{3}$$

$$\therefore \ |z_5| \leq \sqrt{3}.$$

$\therefore \ |z_5|_{\max} = \sqrt{3}$ if and only if $z_1 = e^{i\theta}, z_2 = e^{-i\theta}, z_3 = \frac{1}{2}(z_1 + z_2) + \frac{\sqrt{6}}{3}e^{i\frac{\pi}{4}}, \ z_4 = \frac{1}{2}(z_1 + z_2) + \frac{\sqrt{6}}{3}e^{-i\frac{\pi}{4}}, z_5 = \sqrt{3} \ (\theta = \arctan\sqrt{2})$.

Exercise 10

1. Suppose $x = a + bi$ $(a, b \in R)$, so $(a + bi)^2 - 3|a + bi| + 2 = 0$.

$$\therefore \ a^2 - b^2 - 3\sqrt{a^2 + b^2} + 2 + 2abi = 0$$

$$\therefore \ \begin{cases} a = 0 \\ b^2 + 3\,|b| - 2 = 0 \end{cases} \quad \text{or} \quad \begin{cases} b = 0, \\ a^2 - 3\,|a| + 2 = 0. \end{cases}$$

$\therefore$ There are six solutions.

2. Suppose $x = x_0 \in R$, so $x_0^2 + zx_0 + 4 + 3i = 0$.

$$\therefore z = \frac{-x_0^2 - 4 - 3i}{x_0} = -\left(x_0 + \frac{4+3i}{x_0}\right) = -\left(x_0 + \frac{4}{x_0} + \frac{3}{x_0}i\right)$$

$$\therefore |z| = \sqrt{\left(x_0 + \frac{4}{x_0}\right)^2 + \left(\frac{3}{x_0}\right)^2} = \sqrt{x_0^2 + \frac{25}{x_0^2} + 8}$$

$$\geq \sqrt{18} = 3\sqrt{2}.$$

$|z|_{\min} = 3\sqrt{2}$ if and only if $x_0^2 = \dfrac{25}{x_0^2}$, namely, $x_0 = \pm\sqrt{5}$.

3. Suppose $x = x_0 \in R$, so $x_0^2 - (2i-1)x_0 + 3m - i = 0$.

$$\therefore (x_0^2 + x_0 + 3m) - (2x_0 + 1)i = 0$$

$$\therefore \begin{cases} x_0^2 + x_0 + 3m = 0, \\ 2x_0 + 1 = 0, \end{cases} \quad \therefore \begin{cases} x_0 = -\frac{1}{2}, \\ m = \frac{1}{12}. \end{cases}$$

4. Suppose $\alpha_1 < \alpha_2 < \cdots < \alpha_6$. $\{\alpha_n\}$ is arithmetic sequence and common difference is $d = \frac{\pi}{6}$.

Add six types: $1 + \tan\alpha_1 \tan\alpha_2 = \frac{\tan\alpha_2 - \tan\alpha_1}{\tan\frac{\pi}{6}}$, $1 + \tan\alpha_2 \tan\alpha_3 = \frac{\tan\alpha_3 - \tan\alpha_2}{\tan\frac{\pi}{6}}, \ldots, 1 + \tan\alpha_6 \tan\alpha_1 = \frac{\tan\alpha_1 - \tan\alpha_6}{\tan\frac{\pi}{6}}$: $\tan\alpha_1 \tan\alpha_2 + \tan\alpha_2 \tan\alpha_3 + \cdots + \tan\alpha_6 \tan\alpha_1 = -6$.

5. $\because x + \frac{1}{x} = 2\cos\theta$

$$\therefore x = \cos\theta \pm i\sin\theta$$

$$x^n + \frac{1}{x^n} = (\cos\theta \pm i\sin\theta)^n + \frac{1}{(\cos\theta \pm i\sin\theta)^n}$$

$$= (\cos n\theta \pm i\sin n\theta) + (\cos n\theta \mp i\sin n\theta) = 2\cos n\theta.$$

6. $\because F_1(-2,0), F_2(2,0), \therefore |F_1 F_2| = 4$.

$\therefore$ The locus of z is a ray $F_1(x)$.

7. Original type $= (x-2)(x+2i)(x-2i)$.

8. Obviously, $z = 0$ is the solution of equation.

When $z \neq 0, z^3 = \bar{z}$.

$\therefore |z^3| = |\bar{z}| = |z|, |z| = 1$

$\therefore z^4 = z\bar{z} = |z|^2 = 1$

$\therefore z = \cos\dfrac{2k\pi}{4} + i\sin\dfrac{2k\pi}{4} = \cos\dfrac{k\pi}{2} + i\sin\dfrac{k\pi}{2} \quad (k = 0, 1, 2, 3).$

9. $\frac{1}{2} + \frac{\sqrt{3}}{2}i$ is one root of equation $x^2 + px + q = 0 (p, q \in R)$, so $\frac{1}{2} - \frac{\sqrt{3}}{2}i$ is another root.

$\therefore \begin{cases} -p = \left(\frac{1}{2} + \frac{\sqrt{3}}{2}i\right) + \left(\frac{1}{2} - \frac{\sqrt{3}}{2}i\right) = 1 \\ q = \left(\frac{1}{2} + \frac{\sqrt{3}}{2}i\right)\left(\frac{1}{2} - \frac{\sqrt{3}}{2}i\right) = 1 \end{cases} \quad \therefore \begin{cases} p = -1 \\ q = 1 \end{cases}$

$\therefore \arg(p + qi) = \arg(-1 + i) = 135°.$

10. Suppose $z_1 = a + bi(a, b \in R)$, so $z_2 = a - bi$.

$\therefore z_1 + z_2 = 2a = -5, a = -\dfrac{5}{2}$

$\therefore |z_1 + z_2| = 2|b| = 3, \therefore b = \pm\dfrac{3}{2}$

$m = z_1 z_2 = a^2 + b^2 = \dfrac{17}{2}.$

11. Suppose $x = x_0 \in R$, so $x_0^2 - (i + \tan\theta)x_0 - (2 + i) = 0$.

$\therefore (x_0^2 - x_0 \tan\theta - 2) - (x_0 + 1)i = 0$

$\therefore \begin{cases} x_0^2 - x_0 \tan\theta - 2 = 0 \\ x_0 + 1 = 0 \end{cases}$

$\therefore \tan\theta = 1, \theta = k\pi + \dfrac{\pi}{4} \ (k \in Z).$

12. Points corresponding to roots of $x^{10} = 1$ are 10 vertexes of regular decagon. Choose one diameter and other 8 points can construct 8 right triangles, so there are $8 \times 5 = 40$ right triangles.

13. $\because$ Equation with real coefficient has imaginary number

$\therefore \Delta = 4(p-q)^2 - 8(p^2+q^2) < 0$

$\therefore -(p+q)^2 < 0, \therefore p+q \neq 0, \ x = -(p-q) \pm (p+q)i$

$x^3 = (-(p-q) \pm (p+q)i)^3$

$\qquad = (p-q)(3(p+q)^2 - (p-q)^2) \pm (p+q)(3(p-q)^2$

$\qquad -(p+q)^2)i$

$\because x^3 \in R, \therefore (p+q)(3(p-q)^2 - (p+q)^2) = 0,$

$3(p-q)^2 - (p+q)^2 = 0, \ p^2 - 4pq + q^2 = 0,$

if $q = 0$, then $p = 0$, it contradicts to $p + q \neq 0$.

$\therefore q \neq 0, \ \left(\dfrac{p}{q}\right)^2 - 4\left(\dfrac{p}{q}\right) + 1 = 0, \ \dfrac{p}{q} = 2 \pm \sqrt{3}.$

14. $\because x^2 + (4+i)x + 4 + ai = 0 (a \in R)$ has a real root b,

$\therefore b^2 + (4+i)b + 4 + ai = 0, \ b^2 + 4b + 4 + (b+a)i = 0$

$\therefore \begin{cases} b^2 + 4b + 4 = 0 \\ b + a = 0 \end{cases} \quad \therefore \begin{cases} a = 2 \\ b = -2 \end{cases}$

$\therefore z = a + bi = 2 - 2i$

$\therefore \bar{z}(1 - ci) = (2 + 2i)(1 - ci) = 2 + 2c + (2 - 2c)i.$

When $0 < c \leq 1$, the complex number $\bar{z}(1 - ci)$, its real part greater than 0 and the imaginary part greater than or equal to 0, the auxiliary angle in $[0, \frac{\pi}{2})$ are as follows:

$\therefore \arg(\bar{z}(1 - ci)) = \arctan \dfrac{2 - 2c}{2 + 2c} = \arctan\left(\dfrac{2}{1+c} - 1\right)$

$\because 0 < c \leq 1, \ \therefore 0 \leq \dfrac{2}{1+c} - 1 < 1$

$\therefore 0 \leq \arctan\left(\dfrac{2}{1+c} - 1\right) < \dfrac{\pi}{4}$

$\therefore 0 \leq \arg(\bar{z}(1 - ci)) < \dfrac{\pi}{4}.$

When $c > 1$, the complex number $\bar{z}(1 - ci)$, its real part greater than 0 and the imaginary part less than 0, the auxiliary angle in $(\frac{3\pi}{2} 2\pi)$ are as follows:

$$\therefore \arg(\bar{z}(1 - ci)) = 2\pi + \arctan \frac{2 - 2c}{2 + 2c}$$

$$= 2\pi + \arctan \left(\frac{2}{1 + c} - 1 \right)$$

$$\because c > 1, \quad \therefore -1 \le \frac{2}{1 + c} - 1 < 0$$

$$\therefore -\frac{\pi}{4} \le \arctan \left(\frac{2}{1 + c} - 1 \right) < 0$$

$$\therefore \frac{7}{4}\pi \le \arg(\bar{z}(1 - ci)) < 2\pi.$$

$\therefore$ The auxiliary angle main value range of $\bar{z}(1 - ci)$ $(c > 0)$ is $[0, \frac{\pi}{4}) \cup (\frac{7}{4}\pi, 2\pi)$.

15. The two sides of the equation are taken from the model:

$$\therefore 2 = |z^n \cos \theta_n + z^{n+1} \cos \theta_{n-1} + \cdots + z \cos \theta_1 + \cos \theta_0|$$

$$\le |z|^n + |z|^{n-1} + |z|^{n-2} + \cdots + |z| + 1$$

if $|z| \le \frac{1}{2}$, then

$$2 \le \left(\frac{1}{2} \right)^n + \left(\frac{1}{2} \right)^{n-1} + \cdots + \frac{1}{2} + 1 = 2 - \left(\frac{1}{2} \right)^n < 2.$$

It is a contradiction, so the point of each complex root is outside of the curve $|z| = \frac{1}{2}$.

16. When $n = 2$, left $= |z_1| + |z_2|$, right $= |z_1 + z_2| + |z_1 - z_2| \ge 2|z_1|$.

$\because$ right $\ge 2|z_2|$,

$\therefore$ right $= |z_1 + z_2| + |z_1 - z_2|$

$\ge \frac{1}{2}(2|z_1| + 2|z_2|) = $ left.

Assume that inequality does not hold when n,

$$\therefore (|z_1|+|z_2|+\cdots+|z_n|) \le |z_1 + z_2 + \cdots + z_n| + \sum_{1\le i<j\le n} |z_i - z_j|$$

When $n + 1$,

$$|z_1 + z_2 + \cdots + z_{n+1}| + |z_{n+1} - z_1| + \cdots + |z_{n+1} - z_n|$$

$$\ge (n + 1)|z_{n+1}|$$

$$\Leftrightarrow \frac{1}{n+1}(|z_1 + z_2 + \cdots + z_{n+1}| + |z_{n+1} - z_1| + \cdots + |z_{n+1}$$

$$-z_n|) \ge |z_{n+1}|. \tag{1}$$

$$\because |z_1 + z_2 + \cdots + z_{n+1}| + |z_{n+1} - z_1| + \cdots + |z_{n+1} - z_n|$$

$$\ge |z_1 + z_2 + \cdots + z_{n+1}| + \frac{1}{n}(|z_{n+1} - z_1| + \cdots + |z_{n+1} - z_n|)$$

$$\ge |z_1 + z_2 + \cdots + z_n + \frac{1}{n}z_1 + \frac{1}{n}z_2 + \cdots + \frac{1}{n}z_n|$$

$$= \frac{n+1}{n}|z_1 + z_2 + \cdots + z_n|$$

$$\therefore \frac{n+1}{n}(|z_1 + z_2 + \cdots + z_{n+1}| + |z_{n+1} - z_1|$$

$$+ \cdots + |z_{n+1} - z_n|).$$

$$\ge |z_1 + z_2 + \cdots + z_n|. \tag{2}$$

Adding (1) and (2), $|z_1 + z_2 + \cdots + z_{n+1}| + |z_{n+1}$

$$-z_1| + \cdots + |z_{n+1} - z_n| \ge (|z_1| + |z_2| + \cdots + |z_n|) + |z_{n+1}|,$$

$$\therefore |z_1 + z_2 + \cdots + z_{n+1}| + |z_{n+1} - z_1| + \cdots + |z_{n+1} - z_n|$$

$$+ \sum_{1\le i<j\le n} |z_i - z_j|$$

$$\le |z_1 + z_2 + \cdots + z_{n+1}| + |z_{n+1} - z_1| + \cdots + |z_{n+1} - z_n|$$

$$++ \sum_{1\le i<j\le n} |z_i - z_j|$$

$\therefore$ The original inequality holds.

17. Suppose $\omega = -\frac{1}{2} + \frac{\sqrt{3}}{2}i$, so $1 + \omega + \omega^2 = 0$,

$$\sqrt{x^2 + y^2} + \sqrt{(x-1)^2 + y^2} + \sqrt{x^2 + (y-1)^2}$$

$$= |x + yi| + |(x-1) + yi| + |x + (y-1)i|$$

$$= |x + yi| + |\omega(x-1) + \omega yi| + |\omega^2 x + \omega^2(y-1)i|$$

$$\geq |x + yi + \omega(x-1) + \omega yi + \omega^2 x + \omega^2(y-1)i|$$

$$= |(1 + \omega + \omega^2)x - \omega + (1 + \omega + \omega^2)yi - \omega^2 i|$$

$$= |-\omega - \omega^2 i| = |1 + \omega i| = \frac{\sqrt{2}}{2}(\sqrt{3} - 1).$$

18. Obviously, $x = 0$ is not the root of original equation. Two coefficients which have the same distance from the first term and the end term are equal, so to every negative root $-x_0$, x_0 must be the positive root of $x^4 - ax^3 + bx^2 - ax + 1 = 0$.

$\because (-a)^2 + b^2 = a^2 + b^2$.

$\therefore$ The problem can be transformed into a case that the original equation at least has one positive root.

Divide x^2 by two sides:

$$x^2 + \frac{1}{x^2} + a\left(x + \frac{1}{x}\right) + b = 0$$

$$\therefore \left(x + \frac{1}{x}\right)^2 + a\left(x + \frac{1}{x}\right) + b - 2 = 0. \tag{3}$$

Let $u = x + \frac{1}{x}$, so equation (3) is changed into

$$u^2 + au + b - 2 = 0. \tag{4}$$

If x_0 is a positive root of original equation,

$$u_0 = x_0 + \frac{1}{x_0} \geq 2.$$

Conversely, if equation (4) has a real root which is not less than 2, so $u_0 = x + \frac{1}{x}$.

$\therefore x^2 - u_0 x + 1 = 0$ must have a real root

(because $\Delta = u_0^2 - 4 \geq 0$).

Namely, through variable substitution and introducing parameter u, the original equation is transformed into research that equation (4) has at least one real root which is not less than 2. If equation (4) has one real root which is not less than 2, then

$$\begin{cases} \Delta = a^2 - 4b + 8 \geq 0, \\ \dfrac{-a + \sqrt{a^2 - 4b + 8}}{2} \geq 2. \end{cases}$$

(i) When $a \geq -4$, $2a + b + 2 \leq 0$.   (5)

In the limited condition (5), find the minimum of $a^2 + b^2$.

Let $a = r\cos\theta, b = r\sin\theta$, where $r = \sqrt{a^2 + b^2}, \theta \in [0, 2\pi]$, so inequality (5) is equal to $2r\cos\theta + r\sin\theta + 2 \leq 0$, namely, $\sqrt{5}r\cos(\theta - \arctan\frac{1}{2}) \leq -2$.

Obviously, the minimum non-negative of r which meets above type is $\frac{2}{\sqrt{5}}$, so the minimum of $a^2 + b^2 = r^2$ is $\left(\frac{2}{\sqrt{5}}\right)^2 = \frac{4}{5}$.

(ii) When $a < -4$, $a^2 + b^2 \geq a^2 > 16$.

$\therefore$ So the minimum of $a^2 + b^2$ is more than $\frac{4}{5}$.

$$\therefore (a^2 + b^2)_{\min} = \frac{4}{5} \text{ if and only if } a = \pm\frac{4}{5}, b = -\frac{2}{5}.$$

19. Suppose $z = x + yi(x, y \in R)$, so

$$\bar{z} = x - yi = \frac{1}{1 + ti} = \frac{1 - ti}{1 + t^2},$$

$$\therefore \begin{cases} x = \dfrac{1}{1 + t^2}, \\ y = \dfrac{t}{1 + t^2}, \end{cases}$$

$$\therefore \left(x - \frac{1}{2}\right)^2 + y^2 = \frac{1}{4}, x \neq 0.$$

∴ Set A is the circle whose center is $\left(\dfrac{1}{2}, 0\right)$

and radius is $\dfrac{1}{2}$ (removing origin)

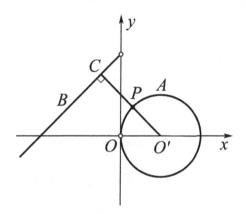

∴ $z' = x' + y'i\,(x', y' \in R) \Rightarrow z' - i = x' + (y' - 1)i$

$$= \sqrt{x'^2 + (y' - 1)^2}\,\frac{x' + (y' - 1)i}{\sqrt{x'^2 + (y' - 1)^2}}$$

∵ $\arg(z' - i) = \dfrac{5\pi}{4}$, and $\cos\dfrac{5\pi}{4} = \sin\dfrac{5\pi}{4}$,

∴ $\dfrac{x'}{\sqrt{x'^2 + (y' - 1)^2}} = \dfrac{y' - 1}{\sqrt{x'^2 + (y' - 1)^2}}$,

∴ $y' = x' + 1\,(x' < 0)$, ∴ Set B is a ray (removing end points).

As in the figure, make $O'C \perp L$ at C, intersect circle at P:

∴ $|CP| = \displaystyle\min_{z \in A,\, z' \in B} |\overrightarrow{ZZ'}|$, $|CP| = \dfrac{|\frac{1}{2} + 1|}{\sqrt{2}} - \dfrac{1}{2} = \dfrac{3\sqrt{2} - 2}{4}$.

20. $\arg z = \arg \dfrac{z_1}{z_2} = \angle Z_1 O Z_2$ or $2\pi - \angle Z_1 O Z_2$.

According to cosine law,

$$\cos(\pi - \arg z) = \frac{|z_1|^2 + |z_2|^2 - |z_1 + z_2|^2}{2|z_1||z_2|} = \frac{|z_1|^2 + |z_2|^2 - 4}{6}$$

$$\geq \frac{|z_1||z_2| - 4}{6} = \frac{1}{3}.$$

$\because \pi - \arg z \in (0, 2\pi)$, $\therefore \cos(\arg z) \leq -\frac{1}{3}$,

$\therefore \pi - \arccos \frac{1}{3} \leq \arg z \leq \pi + \arccos \frac{1}{3}$, $(|z_1| = |z_2| = \sqrt{3})$,

$\therefore (\arg z)_{max} = \pi + \arccos \frac{1}{3}$, $(\arg z)_{min} = \pi - \arccos \frac{1}{3}$.

21. This equation has either real or imaginary roots in pairs.

(i) When n is even number, $f(0) = 0 \Leftrightarrow b = 0$;

$\therefore x^2 + a|x| = 0$.

When $a > 0$, $A = \{0, \pm ai\}$.

$\therefore n = 3$.

When $a = 0$, $A = \{0\}$.

$\therefore n = 1$.

When $a < 0$, $A = \{0, \pm a\}$.

$\therefore n = 3$.

(ii) When n is odd number and $b \neq 0$, to $a = -1$, $b = -2$, $A = \{2, -2\}$. $\therefore n = 2$.

To $a = -2$, $b = 1$, $A = \{\pm 1, \pm(\sqrt{2} - 1)i\}$. $\therefore n = 4$.

To $a = -5$, $b = 6$, $A = \{\pm 2, \pm 3, \pm i\}$. $\therefore n = 6$.

Following prove: when $n = 0$, $n \geq 8$ does not hold.

If $n = 0$ and $a^2 - 4ac < 0 \Rightarrow b > 0 \Rightarrow a^2 + 4b > 0$, $x = \pm \frac{a + \sqrt{a^2 + 4b}}{2} i$ is the root of $f(x) = 0$. ($x = yi(y \neq 0)$, $-|y|^2 + a|y| + b = 0$, $|y|^2 - a|y| - b = 0$) which contradicts with $n = 0$.

If $a^2 - 4ac \geq 0$, $x = \frac{-a \pm \sqrt{a^2 - 4b}}{2}$.

If $n \geq 8$, $x^2 + a|x| + b = 0$ has four roots. $|y|^2 - a|y| - b = 0$ must have four roots.

$\therefore$ $x^2 + ax + b = 0$ has two positive roots. $y^2 - ay - b = 0$ has two positive roots.

This is contradiction between them.

$\therefore$ $n \geq 8$ does not hold. $\therefore$ $B = \{1, 2, 3, 4, 6\}$.

22. Obviously, $x \neq 0$. The original equation equal to $(13 - \frac{1}{x})^{10} = -1$.

Let $y = 13 - \frac{1}{x}$, $\therefore$ $y^{10} = -1$.

Suppose 10 complex roots of $y^{10} = -1$ are $\varepsilon_k, \overline{\varepsilon_k}$ ($k = 1, 2, 3, 4, 5$), and

$\varepsilon_k = \cos\frac{(2k-1)\pi}{10} + i\sin\frac{(2k-1)\pi}{10}$, $k = 1, 2, 3, 4, 5$.

Suppose $13 - \frac{1}{r_k} = \varepsilon_k$, then $\frac{1}{r_k} = 13 - \varepsilon_k$, $k = 1, 2, 3, 4, 5$.

$$\therefore \sum_{k=1}^{5} \frac{1}{r_k \overline{r_k}} = \sum_{k=1}^{5}(13 - \varepsilon_k)(13 - \overline{\varepsilon_k}) = \sum_{k=1}^{5}(170 - 13(\varepsilon_k + \overline{\varepsilon_k}))$$

$$= 850 - 26\left(\cos\frac{\pi}{10} + \cos\frac{3\pi}{10} + \cos\frac{5\pi}{10} + \cos\frac{7\pi}{10} + \cos\frac{9\pi}{10}\right)$$

$$= 850.$$

23. Suppose $\varepsilon_1 = i$ is the unit root of $x^4 = 1$.

According to binomial theorem:

$$(1 + 1)^n = C_n^0 + C_n^1 + C_n^2 + \cdots + C_n^n, \tag{6}$$

$$(1 + \varepsilon_1)^n = C_n^0 + C_n^1\varepsilon_1 + C_n^2\varepsilon_1^2 + \cdots + C_n^n\varepsilon_1^n, \tag{7}$$

$$(1 + \varepsilon_1^2)^n = C_n^0 + C_n^1\varepsilon_1^2 + C_n^2\varepsilon_1^4 + \cdots + C_n^n\varepsilon_1^{2n}, \tag{8}$$

$$(1 + \varepsilon_1^3)^n = C_n^0 + C_n^1\varepsilon_1^3 + C_n^2\varepsilon_1^6 + \cdots + C_n^n\varepsilon_1^{3n}, \tag{9}$$

$(6) + \varepsilon_1^3 \cdot (7) + \varepsilon_1^2 \cdot (8) + \varepsilon_1(9),$

right of sum equation $= 4(C_n^1 + C_n^5 + \cdots + C_n^{4m-3})$

left of sum equation $= 2^n + (1 + \varepsilon_1)^n \cdot \varepsilon_1^3 + (1 + \varepsilon_1^2)^n$

$\cdot \varepsilon_1^2 + (1 + \varepsilon_1^3)^n \cdot \varepsilon_1$

$= 2^n + i((1-i)^n - (1+i)^n) = 2^n + 2^{\frac{n}{2}+1} \sin \dfrac{n\pi}{4}.$

After divide 4 by two sides, it is original equation.

24.

$$x^n - 1 = \prod_{k=0}^{n-1} (x - \varepsilon_k). \tag{10}$$

According to Euler's formula $e^{\theta i} = \cos \theta + i \sin \theta$:

$\therefore\ 2i \sin \theta = e^{\theta i} - e^{-\theta i}$

$\therefore\ 2^n i^n \displaystyle\prod_{k=0}^{n-1} \sin \left(\theta + \dfrac{k\pi}{n} \right) = \prod_{k=0}^{n-1} \left(2i \sin \left(\theta + \dfrac{k\pi}{n} \right) \right)$

$= \displaystyle\prod_{k=0}^{n-1} (e^{\theta + \frac{k\pi}{n} i} - e^{-(\theta + \frac{k\pi}{n})i}) = \prod_{k=0}^{n-1} e^{(\frac{k\pi}{n} - \theta)i} (e^{2\theta i} - \varepsilon_k^{-1})$

$= e^{(\frac{\pi}{n} \sum_{k=0}^{n-1} k - n\theta)i} \displaystyle\prod_{k=0}^{n-1} (e^{2\theta i} - \varepsilon_k^{-1})$

but $\displaystyle\prod_{k=0}^{n-1} k = \dfrac{1}{2} n(n-1),\ e^{\frac{\pi}{2} i} = i$

$\therefore\ e^{(\frac{\pi}{n} \sum\limits_{k=0}^{n-1} k - n\theta)i} = e^{\frac{1}{2}(n-1)\pi i - n\theta i} = i^{n-1} e^{-n\theta i}.$

According to type (10),

$\displaystyle\prod_{k=0}^{n-1} (e^{2\theta i} - \varepsilon_k^{-1}) = e^{2n\theta i} - 1 = 2i e^{n\theta i} \sin n\theta$

$$\therefore \ 2^n i^n \prod_{k=0}^{n-1} \sin\left(\theta + \frac{k\pi}{n}\right) = 2i^n \sin n\theta.$$

$\therefore$ The original proposition holds.

25. (1) Suppose $\omega = e^{\frac{2\pi i}{n}}$

$$\therefore \ \omega^n = 1, \ \omega^{-\frac{n}{2}} = e^{\pi i} = -1, 2\cos\frac{2k\pi}{n} = \omega^k + \omega^{-k}$$

$$\prod_{k=1}^{n}\left(1 + 2\cos\frac{2k\pi}{n}\right) = \prod_{k=1}^{n}(1 + \omega^k + \omega^{-k})$$

$$= \prod_{k=1}^{n} \omega^{-k}(\omega^k + \omega^{2k} + 1)$$

$$= \omega^{-\frac{n(n+1)}{2}} \cdot 3 \prod_{k=1}^{n-1} \frac{1 - \omega^{3k}}{1 - \omega^k} = (-1)^{n+1} \cdot 3 \prod_{k=1}^{n-1} \frac{1 - \omega^{3k}}{1 - \omega^k}$$

$\because \ n$ is a prime number which is more than 3.

$\therefore \ (-1)^{n+1} = 1, \quad \text{and } 3, 3 \times 2, \dots, 3(n-1)$

take all residue class of n.

$$\therefore \ \prod_{k=1}^{n}\left(1 + 2\cos\frac{2k\pi}{n}\right) = 3.$$

(2) $z^{2n} - 1 = 0$ has $2n$ roots which are ± 1 and $z_k = e^{\pm\frac{k\pi i}{n}}$ $(k = 1, 2, \dots, n-1)$

$$\therefore \ z^{2n} - 1 = (z^2 - 1) \prod_{k=1}^{n-1} (z - e^{\frac{k\pi i}{n}})(z - e^{-\frac{k\pi i}{n}})$$

$$= (z^2 - 1) \prod_{k=1}^{n-1}\left(z^2 + 1 - 2z\cos\frac{k\pi}{n}\right).$$

Take $z = e^{\frac{2\pi}{3}i}$, so $z^2 + 1 = -z$.

$$\therefore z^{2n} - 1 = (z^2 - 1)(-z)^{n-1} \prod_{k=1}^{n-1} \left(1 + 2\cos\frac{k\pi}{n}\right)$$

$$\prod_{k=1}^{n-1} \left(1 + 2\cos\frac{k\pi}{n}\right) = \frac{z^{2n} - 1}{(z^2 - 1)(-z)^{n-1}}$$

$$= \begin{cases} 0, & n = 3k, \\[2mm] \dfrac{z^2 - 1}{(z^2 - 1)(-z)^{3k}} = (-1)^{3k} = (-1)^{n-1}, & n = 3k + 1, \\[3mm] \dfrac{z - 1}{(z^2 - 1)(-z)^{3k+1}} = \dfrac{(-1)^{3k+1}}{(z + 1)z}, & n = 3k + 2, \\[3mm] = \dfrac{(-1)^{3k+1}}{-1} = (-1)^n, \end{cases}$$

$$k \in N^*.$$

26. Let $1^2 = a_1$, $3^2 = a_2$, $5^2 = a_3$, $7^2 = a_4$, $2^2 = \lambda_1$, $4^2 = \lambda_2$, $6^2 = \lambda_3$, $8^2 = \lambda_4$.

Construct $f(x) = \prod_{i=1}^{4}(x - a_i) - \prod_{i=1}^{4}(x - \lambda_i)$

$$= \left(\sum_{i=1}^{4}(\lambda_i - a_i)\right) x^3 + \cdots \tag{11}$$

$$\therefore f(\lambda_k) = \prod_{i=1}^{4}(\lambda_i - a_i).$$

According to Lagrange formula:

$$\therefore f(x) = \sum_{j=1}^{4} \prod_{\substack{i \neq j \\ 1 \leq i \leq 4}} \frac{x - a_i}{a_j - a_i} \cdot f(a_j).$$

Let $x = \lambda_k$, divide $f(\lambda_k)$ by both sides.

$$\therefore \frac{A_1}{\lambda_k - a_1} + \frac{A_2}{\lambda_k - a_2} + \frac{A_3}{\lambda_k - a_3} + \frac{A_4}{\lambda_k - a_4} = 1, \tag{12}$$

where

$$A_j = \frac{f(a_j)}{\prod_{\substack{i \neq j \\ 1 \leq i \leq 4}} (a_j - a_i)}, \quad j, k = 1, 2, 3, 4$$

and equation (12) has unique solution.

Compare the coefficient of x^3 in equations (11) and (12):

$$\sum_{i=1}^{4} A_j = \sum_{i=1}^{4} (\lambda_i - a_i)$$

$\because x^2, y^2, z^2, w^2$ are solutions of known equation

$$\therefore x^2 + y^2 + z^2 + w^2 = \sum_{j=1}^{4} A_j = \sum_{j=1}^{4} \lambda_j - \sum_{j=1}^{4} a_j$$

$$= (2^2 + 4^2 + 6^2 + 8^2) - (1^2 + 3^2 + 5^2 + 7^2) = 36.$$

27. Let $\omega = \cos \frac{2\pi}{n} + i \sin \frac{2\pi}{n}$ and η is a complex number which modulus is 1.

If $n \geq 2$ and natural number $k = 1, 2, \ldots, n-1$, so

$$f(\omega^k \eta) = C_0 (\omega^k \eta)^n + C_1 (\omega^k \eta)^{n-1}$$
$$+ C_2 (\omega^k \eta)^{n-2} + \cdots + C_{n-1} (\omega^k \eta) + C_n,$$

$$\sum_{k=1}^{n} f(\omega^k \eta) = C_0 \sum_{k=1}^{n} (\omega^k \eta)^n + C_1 \sum_{k=1}^{n} (\omega^k \eta)^{n-1}$$

$$+ C_2 \sum_{k=1}^{n} (\omega^k \eta)^{n-2} + \cdots + C_{n-1} \sum_{k=1}^{n} (\omega^k \eta) + n C_n \qquad (13)$$

Obviously, $\displaystyle\sum_{k=1}^{n} (\omega^k \eta)^n = n \eta^n$ \qquad (14)

To $j = 1, 2, \ldots, n-1$:

$$\sum_{k=1}^{n} (\omega^k \eta)^j = \eta^j \sum_{k=1}^{n} \omega^{kj} = \eta^j \frac{\omega^j - \omega^{(n+1)j}}{1 - \omega^j} = 0. \qquad (15)$$

Substitute (14) and (15) into (13):

$$\sum_{k=1}^{n} f(\omega^k \eta) = n(C_0 \eta^n + C_n).$$

When $n = 1$, the above type holds.

$\therefore \forall n \in \mathbf{N}^*$, the above type holds.

$$\frac{1}{n} \sum_{k=1}^{n} |f(\omega^k \eta)| \geq \frac{1}{n} \left| \sum_{k=1}^{n} f(\omega^k \eta) \right| = |C_0 \eta^n + C_n|. \tag{16}$$

Note $C_0 = P_0(\cos\theta_0 + i\sin\theta_0)$, where $0 \leq \theta_0 < 2\pi$, $|C_0| = P_0$.

Note $C_n = P_n(\cos\theta_n + i\sin\theta_n)$, where $0 \leq \theta_n < 2\pi$, $|C_n| = P_n$.

Choose θ where $0 \leq \theta < 2\pi$, $\theta_0 + n\theta \equiv \theta_n \pmod{2\pi}$.

Let $\eta = \cos\theta + i\sin\theta$:

$$C_0\eta^n + C_n = \rho_0(\cos\theta_0 + i\sin\theta_0)(\cos\theta_n + i\sin\theta_n)$$

$$+\rho_n(\cos\theta_n + i\sin\theta_n)$$

$$= \rho_0(\cos(\theta_0 + n\theta) + i\sin(\theta_0 + n\theta)) + \rho_n(\cos\theta_n + i\sin\theta_n)$$

$$= (\rho_0 + \rho_n)(\cos\theta_n + i\sin\theta_n) \tag{17}$$

$\therefore |C_0\eta^n + C_n| = \rho_0 + \rho_n = |C_0| + |C_n|.$

Substitute (17) into (16):

$$\therefore \frac{1}{n} \sum_{k=1}^{n} |f(\omega^k \eta)| \geq |C_0| + |C_n|$$

∵ Arithmetic mean of $|f(\omega\eta)|, |f(\omega^2\eta)|, \ldots, |f(\omega^k\eta)|$

is equal or more than $|C_0| + |C_n|$.

∴ There is at least one $|f(\omega^j\eta)| \geq |C_0| + |C_n|$,

where j is among $1, 2, \ldots, n$.

Let $z_0 = \omega^j\eta$, $|z_0| = 1$, so $|f(z_0)| \geq |C_0| + |C_n|$.